Lecture Notes in Physics

The Editorial Policy for Proceedings

The series Lecture Notes in Physics reports new developments in physical research and teaching – quickly, informally, and at a high level. The proceedings to be considered for publication in this series should be limited to only a few areas of research, and these should be closely related to each other. The contributions should be of a high standard and should avoid lengthy redraftings of papers already published or about to be published elsewhere. As a whole, the proceedings should aim for a balanced presentation of the theme of the conference including a description of the techniques used and enough motivation for a broad readership. It should not be assumed that the published proceedings must reflect the conference in its entirety. (A listing or abstracts of papers presented at the meeting but not included in the proceedings could be added as an appendix.)

When applying for publication in the series Lecture Notes in Physics the volume's editor(s) should submit sufficient material to enable the series editors and their referees to make a fairly accurate evaluation (e.g. a complete list of speakers and titles of papers to be presented and abstracts). If, based on this information, the proceedings are (tentatively) accepted, the volume's editor(s), whose name(s) will appear on the title pages, should select the papers suitable for publication and have them refereed (as for a journal) when appropriate. As a rule discussions will not be accepted. The series editors and Springer-Verlag will normally not interfere with the detailed editing except in fairly obvious cases or on technical matters.

Final acceptance is expressed by the series editor in charge, in consultation with Springer-Verlag only after receiving the complete manuscript. It might help to send a copy of the authors' manuscripts in advance to the editor in charge to discuss possible revisions with him. As a general rule, the series editor will confirm his tentative acceptance if the final manuscript corresponds to the original concept discussed, if the quality of the contribution meets the requirements of the series, and if the final size of the manuscript does not greatly exceed the number of pages originally agreed upon.

The manuscript should be forwarded to Springer-Verlag shortly after the meeting. In cases of extreme delay (more than six months after the conference) the series editors will check once more the timeliness of the papers. Therefore, the volume's editor(s) should establish strict deadlines, or collect the articles during the conference and have them revised on the spot. If a delay is unavoidable, one should encourage the authors to update their contributions if appropriate. The editors of proceedings are strongly advised to inform contributors about these points at an early stage.

The final manuscript should contain a table of contents and an informative introduction accessible also to readers not particularly familiar with the topic of the conference. The contributions should be in English. The volume's editor(s) should check the contributions for the correct use of language. At Springer-Verlag only the prefaces will be checked by a copy-editor for language and style. Grave linguistic or technical shortcomings may lead to the rejection of contributions by the series editors.

A conference report should not exceed a total of 500 pages. Keeping the size within this bound should be achieved by a stricter selection of articles and not by imposing an upper limit to the length of the individual papers.

Editors receive jointly 30 complimentary copies of their book. They are entitled to purchase further copies of their book at a reduced rate. As a rule no reprints of individual contributions can be supplied. No royalty is paid on Lecture Notes in Physics volumes. Commitment to publish is made by letter of interest rather than by signing a formal contract. Springer-Verlag secures the copyright for each volume.

The Production Process

The books are hardbound, and quality paper appropriate to the needs of the authors is used. Publication time is about ten weeks. More than twenty years of experience guarantee authors the best possible service. To reach the goal of rapid publication at a low price the technique of photographic reproduction from a camera-ready manuscript was chosen. This process shifts the main responsibility for the technical quality considerably from the publisher to the authors. We therefore urge all authors and editors of proceedings to observe very carefully the essentials for the preparation of camera-ready manuscripts, which we will supply on request. This applies especially to the quality of figures and halftones submitted for publication. In addition, it might be useful to look at some of the volumes already published.

As a special service, we offer free of charge LATEX and TEX macro packages to format the text according to Springer-Verlag's quality requirements. We strongly recommend that you make use of this offer, since the result will be a book of considerably improved technical quality.

To avoid mistakes and time-consuming correspondence during the production period the conference editors should request special instructions from the publisher well before the beginning of the conference. Manuscripts not meeting the technical standard of the series will have to be returned for improvement.

For further information please contact Springer-Verlag, Physics Editorial Department V, Tiergartenstrasse 17, W-6900 Heidelberg, FRG

J. Heidmann   M. J. Klein (Eds.)

Bioastronomy
The Search for Extraterrestrial Life - The Exploration Broadens

Proceedings of the Third International Symposium on Bioastronomy Held at Val Cenis, Savoie, France, 18-23 June 1990

Springer-Verlag

Berlin Heidelberg New York
London Paris Tokyo
Hong Kong Barcelona
Budapest

Editors

Jean Heidmann
Observatoire de Paris
F-92195 Meudon, France

Michael J. Klein
Jet Propulsion Laboratory
California Institute of Technology
Pasadena, CA 91109, USA

ISBN 3-540-54752-5 Springer-Verlag Berlin Heidelberg New York
ISBN 0-387-54752-5 Springer-Verlag New York Berlin Heidelberg

Printing: Druckhaus Beltz, Hemsbach/Bergstr.
Bookbinding: J. Schäffer GmbH & Co. KG., Grünstadt
2158/3140-543210 - Printed on acid-free paper

LIFE IN THE COSMOS ...

Recent developments in the fields of radioastronomy, solar system space exploration and investigations of the origin of life suggest that life is likely to be the natural result of evolution in the cosmos. From this perspective it is reasonable to conclude that the appearance and evolution of life on the Earth may very well have occurred on other planets orbiting some of the 300 thousand million stars in our Milky Way Galaxy. Confronted with this tremendous possibility, astronomers find themselves on the front line of research because their domain of study is the cosmos.

... AND ITS EVOLUTION

The existence of extraterrestrial life does not necessarily imply that all life would have evolved to equal (or surpass) the level of intelligence and technology experienced by humans on Earth. The complex trail of evolution on the cosmic scale may be classified into five stages:

1- A cosmic stage that begins with the appearance of matter soon after the "Big Bang", proceeds to the formation of stars and planets several billion years later, and continues through the synthesis of chemical elements like carbon, the centerpiece for life as we know it.

2- An organic chemistry stage, with the formation of the first organic molecules that are the stepping stones towards our own life; these molecules have been discovered in interstellar space by radioastronomers and in comets by space probes.

3- A prebiotic chemistry stage, in which complex "building blocks" are produced, such as nitric bases that form the rungs of the DNA double helix ladder or amino acids, the essential elements of proteins; amino acids have been collected in meteorites that have fallen to Earth; some theories predict that a prebiotic chemistry could be proceeding even now on Titan, the large moon of the planet Saturn.

4- A primitive biological stage like the one of bacteria which dominated the first few billion years of the Earth; scientists engaged in space exploration hope to discover this level of biological activity, perhaps in a different form, buried in the frozen permafrost of Mars.

5- Finally, an "advanced" stage of life, perhaps more advanced than ours, is surely plausible; nothing in our investigations of the universe indicates that Homo sapiens are unique or that we occupy a special place in the cosmos; an effective approach to answer the question, "are we alone?", is to conduct comprehensive "SETI" (Search for Extraterrestrial Intelligence) observations that are designed to search for artificial radio signals produced by extraterrestrial civilizations.

BIOASTRONOMY...

Building on the discoveries of modern ground-based and space astronomy, the international community of scientists formed a new branch of astronomy, bioastronomy, which is dedicated to the study of the existence of life in the universe and the search for evidence of intelligent life. Echoing the noblest human aspirations, bioastronomy has become a very popular subject. Renowned institutions such as the International Academy of Astronautics and the National Academies of the United States and the Soviet Union were the first to support bioastronomical research. In 1982 the International Astronomical Union created a new commission (No. 51) devoted to bioastronomy. In 1989 the US National Aeronautics and Space Administration initiated the NASA Microwave Observing Project, and in 1990 the U.S. Congress voted to provide the funds required to design and build the special purpose digital processing systems capable of searching tens of millions of radio frequency channels.

... AND ITS THIRD SYMPOSIUM

Two international symposia devoted to bioastronomy have already occurred: the first was held in 1984 in Boston, Massachusetts (USA) and the second at Lake Balaton, Hungary, in 1987. Recognizing the work done in this field by French scientists, engineers and technicians, the international organizing committee for the third symposium accepted the invitation from France to hold the Third International Symposium on Bioastronomy in the beautiful region of Val Cenis in the French Alps. In June of 1990 the conference attendees were treated to the warm hospitality of the residents of the Val Cenis region and the enjoyable alpine atmosphere exemplified by hillside pastures, mountain streams, glaciers, and fields of wild flowers.

The scientific program was organized to maintain a broad perspective because the previous meetings had shown that the interaction of scientists from various disciplines was productive and stimulating. Consequently, the symposium was open to scientists from a variety of related fields: geology, climatology, biochemistry, origins of life, paleontology, ecology, neurology, sociology, intelligence and civilizations. The program was organized into sessions dealing with the five main stages of bioastronomy and a sixth session to encourage wider interdisciplinary connections. A special session on Post Detection Protocol was also held.

The symposium was attended by over one hundred scientists from around the world. Ninety-six abstracts were submitted and these were divided into invited, contributed or poster presentations. From the 87 manuscripts provided by the authors, the Editors, after consultation with referees, produced this volume of "Lecture Notes in Physics".

Jean Heidmann
Paris, FRANCE

Michael J. Klein
Pasadena, CA, USA

April 1991

ACKNOWLEDGMENTS

The editors acknowledge with pleasure the participation and sponsorship of organizations that made this symposium possible. We especially appreciate the monetary grants that enabled us to provide travel grants for several dozen participants and special grants to seven young scientists. We thank the following organizations:

- Centre National de la Recherche Scientifique, Paris, FRANCE
- Legal & General, Paris, FRANCE
- National Aeronautics and Space Administration, USA
- Observatoire de Paris, FRANCE
- SETI Institute, Mountain View, CA, USA
- The Alfred P. Sloan Foundation, New York, NY, USA

We also gladly acknowledge the District de Haute-Maurienne for its very efficient logistical support, Ciel et Espace for providing public relations services, and Maison de Val Cenis for its kind hospitality.

For the preparation of the manuscripts, we are grateful to Judy Hoeptner and Richard Chandlee from the SETI Project Office at the Jet Propulsion Laboratory (JPL) and the editorial personnel of the JPL Documentation Section.

We also wish to recognize the valuable contribution of several referees who wished to remain anonymous and the larger number of referees listed here:

L. H. Aller	M. M. Davis	P. Horowitz	G. S. Orton
H. D. Aller	K. Dose	M. A. Janssen	T. C. Owen
C. A. Beichman	G. C. Downs	J. M. Jenkins	S. T. Pierson
G. L. Berge	F. J. Dyson	K. Kaware	S. K. Pope
C. de Bergh	E. E. Epstein	K. I. Kellermann	H. Reeves
D. van den Bergh	B. R. Finney	B. N. Khare	D. Reiss
D. C. Berry	S. Firenstein	H. P. Klein	W. L. W. Sargent
A. Betz	A. G. Fraknoi	W. R. Kuhn	J. D. Scargle
J. J. Borucki	G. Gavoust	T. B. H. Kuiper	C. L. Seeger
R. N. Bracewell	E. Gerard	J. Lequeux	G. A. Seielstad
R. L. Carpenter	G. D. Gatewood	S. M. Levin	C. T. Stelzried
A. C. Clarke	D. W. Goldsmith	J. E. Lovelock	P. Thaddeus
R. C. Clauss	J. R. Hall	H. J. Melosh	A. Vidal-Madjar
Y. Coppens	J.P. Harrington	P. J. Morrison	J. A. Wood
R. D. Courtin	G. Helou	D. Muhleman	G. A. Zimmerman.
A. Dalgarno	E. Herbst	N. F. Ness	
J. A. Davidson	J. G. Hills	L. E. Orgel	

SPONSORING INTERNATIONAL COMMITTEE

President: Gyorgy Marx, Eotvos University, Budapest, Hungary

Vice President: Ronald D. Brown, Monash University, Clayton, Australia

Secretary: Michael D. Papagiannis, Boston University, USA

Members: Bruce Campbell, University of Victoria, Victoria, Canada
                    Frank D. Drake, University of California, Santa Cruz, USA
                    Samuel Gulkis, Jet Propulsion Laboratory, Pasadena, USA
                    Jean Heidmann, Observatoire de Paris, Meudon, France
                    Jun Jugaku, Tokai University, Hiratsuka, Japan
                    Nikolai Kardashev, Space Research Institute, Moscow, USSR
                    Vjachoslav I. Slysh, Space Research Institute, Moscow, USSR
                    Jill C. Tarter, NASA Ames Research Center, Moffett Field, USA

CO-SPONSORED by the International Academy of Astronautics

SCIENTIFIC ORGANIZING COMMITTEE

Chairman: Michael J. Klein, Jet Propulsion Laboratory, USA

Co Vice-Chair: Nikolai S. Kardashev, Space Research Institute, USSR
                    J. C. Tarter, NASA Ames Research Center, USA

Members: Ivan F. Almar, Konkoly Observatory, HUNGARY
                    Ronald Brown, Monash University, AUSTRALIA
                    Bruce Campbell, University of Victoria, CANADA
                    Paul A. Feldman, Herzberg Inst. of Astrophysics, CANADA
                    Jean Heidmann, Observatoire de Paris, FRANCE
                    Hirashi Hirabayashi, Inst. Space & Astronautical Science, JAPAN
                    Michael D. Papagiannis, Boston University, USA
                    Sir Martin J. Rees, Cambridge Institute of Astronomy, UK

LOCAL ORGANIZING COMMITTEE

President: Jean Heidmann, Paris Observatory

Secretary: Nicole Hallet, Meudon Observatory

Members: Francois Biraud, Meudon Observatory
                    Jean-Claude Ribes, Lyon Observatory

NASA Apollo 17/J. Heidmann

CONTENTS

II. ORGANIC AND PREBIOTIC EVOLUTION

III. PRIMITIVE EVOLUTION

IV. ADVANCED EVOLUTION - SEARCHING FOR EVIDENCE

V. ADVANCED EVOLUTION - POSSIBILITIES

OPENING REMARKS

George Marx
Department of Atomic Physics
Eotvos University, Budapest H-1088
and
Chairman of Sponsoring Committee
Third International Bioastronomy Symposium
Val Cenis, Savoie France

The great quest to find life out there is at least 400 years old, but it has been raised to the level of science just in the past three decades by devoted radioastronomers, and by planetologists, biologists, and space engineers. It is greatly to the merit of the founders and members of our Bioastronomy community*, and especially the leadership of Michael Papagiannis and Frank Drake, that this curiosity and search has been accepted by the scientific community. The Third International Symposium on Bioastronomy tries to answer old questions and to raise new ones. We are offered now new technological possibilities and we face new intellectual challenges. The exciting development of the late 1980's is that we learned to know interesting worlds in the neighborhood of the great planets within the Solar System, we have sighted protoplanetary dust disks around several nearby stars, and we have indications about great planets orbiting around alien stars. The Solar Family seems not to be alone in the Galaxy. We even speculate that there may be further steady sources of free energy, which are needed to create, nurture and develop life. This is one side to our understanding the role of life in the Universe.

As we open this Third International Bioastronomy Symposium, I would like to direct attention to a very peculiar astronomical object in the Solar System, worthy of specific interest; it is called Earth. This sphere of rock bore life 4 billion years ago, and most surprisingly it was able to sustain life without interruption since then. It has become understood now even by politicians, how precious this tiny piece of rock is. Global thinking and global responsibility is slowly diffusing from scientists to decision makers, to educators, to the young generation. We have begun to learn that this small planet may be vulnerable. The primordial Earth has been made habitable by simple pioneering aggregates of atoms which succeeded in copying themselves. It has been kept habitable by the biosphere, so creating time for evolution, up to intelligence. Now overpopulation, overexploitation, overpollution may upset this delicate balance between the forces of physics and those of biology. Newspapers report today about flood and drought, as they reported about war and peace not so long ago. Planetology is not a strange expensive hobby of a few scientists any more but it has become a common issue of mankind. This is how and why bioastronomy may become relevant for people.

---

* Professor Marx is currently the President of IAU Commission 51, Bioastronomy. Professors Papagiannis and Drake are past Presidents of Commission 51.

Let us pay attention to life out there <u>and</u> here on Earth as well!! At the General Assembly of the International Astronomical Union (1991) the Bioastronomy Commission intends to organize about the possible astronomical causes of mass extinctions on Earth.

On behalf of the International Sponsoring Committee, I express our thanks to Jean Heidmann (Chairman of the Local Organizing Committee) and Mike Klein (Chairman of the Scientific Organizing Committee) for helping to create this timely opportunity to discuss all these questions. The marvelous view of the Alps in Savoie reminds me of a previous similar seminar on the slope of the Caucasus where the moonlit snowwhite Mount Ararat was looking down at us about two decades ago. Let the earthly magnificence of Nature inspire us to understand her and to understand our relation to her.

I. COSMIC EVOLUTION

THE CREATION OF FREE ENERGY

George Marx
Department of Atomic Physics
Eötvös University, Budapest H-1088

The early Universe was hot and dense, with frequent collisions; therefore it was in thermal equilibrium, as indicated by the properties of the relic blackbody radiation, consequently the free energy of the Universe was zero. Our present world, however, is out of equilibrium, it is locally alive. Life needs a steady source of free energy. If one is interested in the search for life in the Universe, one has to look at places where free energy is being created or stored.

HEAT DEATH IN THE PAST

The creation of free energy is impossible in a closed system of fixed volume. In our world the galaxies accelerate due to gravity. By assuming a homogenous distribution of galaxies, for simplicity, the equation of motion of a galaxy at distance R is

$$d^2R/dt^2 = -G(4\pi G\rho R^3/3)/R^2,\tag{1}$$

where ρ is the mass density. The mass within the sphere of radius R is conserved:

$$M = 4\pi\rho R^3/3 = const,\tag{2}$$

consequently eq. (1) can be solved as

$$R(t) = (4.5GM)^{1/3}t^{2/3},$$

or taking eq. (2) as well into account,

$$\rho(t) = 1/6\pi Gt^2.$$

This is for a world with a dust of galaxies. A radiation dominated Universe behaves in a very similar way:

$$\rho(t) = 3/32\pi Gt^2.$$

It can be said that a non-empty Universe cannot be static, its density must have a singularity on the time axis. (This is true even for nonhomogenous distributions.) The singularity has been called Big Bang.

The present overall temperature of the Universe is T=2.73 K. Light can be considered to be a sine-wave drawn on a rubber sheet. As the size R of the sheet increases, the wavelength λ stretches, correspondingly T drops. This means that the Universe was hot and dense in the past. Frequent strong collisions produced thermal equilibrium within 10^{-40} seconds after the Big Bang. The Gibbs free energy of this extreme relativistic gas was zero. This has been proved by the isotropy (up to 10^{-4}) and by the Planck shape (up to 1%) of the relic microwave radiation [1].

RESURRECTION FROM THE HEAT DEATH

The radiation temperature is proportional to the average energy of quanta:

$$kT_{rad} \approx h\nu = hc/\lambda \sim 1/R(t).$$

For a nonrelativistic gas, taking the wavelength formula of de Broglie into account:

$$kT_{gas} \approx 0.5mv^2 = h^2/2m\lambda^2 \sim 1/R(t)^2.$$

While R increases 10-fold, say, radiation temperature drops to 10%, gas temperature drops to 1% of its former value. The different scaling behaviors can be seen from the corresponding entropy formulas as well:

$$S_{rad} = (16a\pi/3)(RT_{rad})^3, \qquad S_{gas} = 1.5N \, klog(R^2mT_{gas}).$$

For adiabatic noninteractive cooling, $T_{rad} \sim R^{-1}$ and $T_{gas} \sim R^{-2}$. When the neutralized nonrelativistic gas is decoupled from radiation, adiabatic expansion creates temperature difference, consequently free energy ([2], Figure 1). In reality, the temperature difference is increased by the expansion and decreased by collisions [3].

$$dT_{rad}/dt = -(T_{rad}/R)dR/dt - \alpha C_{rad}^{-1} \, \rho_{rad} \, \rho_{gas}(T_{rad} - T_{gas}),$$

$$dT_{gas}/dt = -2(T_{gas}/R)dR/dt + \alpha C_{gas}^{-1} \, \rho_{rad} \, \rho_{gas}(T_{rad} - T_{gas}).$$

The collision rate is proportional to radiation density ρ_{rad} and to gas density ρ_{gas}. In order to satisfy the energy condition $dE + pdV = O$, the corresponding heat capacities C appear in the equations. The heat transfer from the warmer radiation to the cooler gas is irreversible. The entropy increases (Figure 2).

Later, at lower temperatures and lower densities the coupling gradually faded away, thus the temperature difference (and the corresponding free energy) created by the expansion and by the different cooling rates survived. The neutralized gas cooled faster than radiation. Statistical density fluctuations (larger than the Jeans stability limit) became stabilized by their own gravity. Galaxies and stars were born.

LET THE SUN SHINE!

Let us consider a gas sphere (formed by gravity) in the sea of radiation. The kinetic energy of gas particles is proportional to the average temperature ($E_{kin} = 1.5Nk < T_{gas} >$). According to the virial theorem, the potential energy is $E_{pot} = -2E_{kin}$. The total energy is negative:

$$D = E_{kin} + E_{pot} = -1.5Nk <T_{gas}>.$$

The heat capacity of the star is negative:

$$C = dE/dT = -1.5Nk<0.$$

As space expands, radiation performs work, therefore it cools. The bound gas sphere is unaffected by the overall expansion, its temperature does not change. In this way the environment gets cooler than the star. The star gives away heat irreversibly. Its heat capacity is negative, therefore heat loss warms the star up. Temperature difference is created again [4]. The star becomes brighter, in its central region nuclear fusion begins (Figure 3).

In the early hot Universe composite nuclei could not survive: The fast cooling explains that the pre-stellar clouds were made mostly of hydrogen. In the star hydrogen nuclei collide:

$$H + H \rightarrow (^2He) \rightarrow H + H.$$

The ^2He nucleus is not stable, fast fusion is not possible. Weak decay enables a transition occasionally:

$$H + H \rightarrow (^2He) \rightarrow {}^2H + e^+ + \nu.$$

From this point on, strong nuclear fusions run fast: $^2H + {}^2H \rightarrow {}^4He$, but the time scale of the first weak interaction is billions of years (instead of microseconds), offering free energy (warm sunshine in a cold space) and time as well for biological evolution.

SOURCES OF FREE ENERGY

Our thesis can be formulated in the following way: *The Universe is dynamically unstable. It expands and cools. Different materials with differing scaling laws cool at different rates. When and where interactions become weak and slow, the thermalization is delayed. Free energy is created, transient disequilibria survive. The free energy offers a chance for life. Whether the life evolves to intelligence may depend on the duration of the transient disequilibrium. The slow radioactivity and gravity may produce delays long enough for biological evolution.* (Electric and nuclear forces work fast.) Search for extraterrestrial life and technology should be directed towards prolonged transient structures (trapped low-entropy fuels like nuclei, hydrogen, uranium, coal, wood). Possible examples:

Planets orbiting around a main sequence star, in which fusion has been slowed down by the *weak radioactive bottleneck* $H + H \rightarrow (^2He) \rightarrow {}^2H + e^+ + \nu$ from fraction of a second to billions of years (Earth). The circular planetary orbit is unstable thermodynamically, but its angular momentum is dispersed by *gravitational radiation* very slowly. On the interior planets of the Solar System an oxidized atmosphere (CO_2, H_2O) evolved. Organic material might have been supplied by comets. The exceptional elongated orbits of the comets are temporary phenomena; they are produced by occasional *gravitational perturbation.* Reduced cometary organics in the oxidized planetary atmosphere had become a chemical source of free energy.

Tidal friction on satellites of massive bodies may produce geothermal heating, liberating free energy from the *gravitational reservoir,* even far away from stars. The eccentric orbits (preventing the rotation period to be locked to orbiting period) may be sustained by a gravitational resonance among several bodies. (Volcanos on Io, liquid ocean on Europe or elsewhere [5].)

Supernova explosions at 10^{10} K produce unstable nuclei like U, Th, ^{40}K. When the ejected material condensates to rocks, slow *radioactive decays* liberate geothermal free energy on a time scale of 10^5 to 10^{10} years. This may support life independently of starshine. (Geothermally nurtured life forms, discovered in the Pacific Rift, 3000 m deep in the ocean [6]. In the early Solar System, giant comets might have liquid oceans, enriched inorganics, heated by the early intensive radioactivity [16].)

Further exotic sources of free energy may supply heat at even longer time scales. Cold nuclear fusion [7], proton decay [8], black hole decay [9] may nurture hypothetically very strange life forms, even on the crust of neutron stars [10].

RACE BETWEEN DECAY AND EVOLUTION

The precondition for the emergence of intelligent technology is that the weak transient should last longer than the biological evolution needs. This raises the second main question: what does the speed of evolution depend on?
- on the inherent laws of carbon biochemistry
- on the abundance of alien mutagen chemicals
- on outside (radioactive, ultraviolet) radiation level
- on the rate of astronomically triggered catastrophes [11-12-13]
- on the physically different rates of alternative information-creating structures (DNA, culture, computer, non-DNA-based biochemistry, life on the surface of brown dwarfs or neutron starts [10]).

Can technology emerge much faster than it happened on Earth? This is as important as to understand: how steady is the flow of free energy? (What makes the biosphere and climate of Earth so miraculously stable, in spite of increasing solar temperature and changing chemical composition of the atmosphere [15, 16]?) Without a wide-angle view, one may miss important potential sites of extraterrestrial life and technology.

References

1.   Sam Gulkis, Proc. of 3rd Symposium on Bioastronomy, Springer (1991)
2.   George Marx, Acta Physica Hungarica, <u>62</u> (1987) 139.
3.   George Marx, in H.P. Dürr Festschrift, Max Planck Inst. München, (1989) 197.

4.   George Marx and H. Sato, Journal of Modern Physics, 2 (1986) 133.
5.   R.T. Reynolds et al., Proc. of 2nd Symposium on Bioastronomy, Kluwer (1988) 21.
6.   J.B. Corliss, Proc. of 2nd Symposium on Bioastronomy, Kluwer (1988) 195.
7.   S.A. Jones et al., Nature, 338 (1989) 737.
8.   F.J. Dyson, Rev. Mod. Physics, 51 (1979) 447.
9.   S.W. Hawking, Phys. Rev., 13D (1976) 191.
10.   R.L. Forward, Dragon's Egg, Ballantine (1980).
11.   R.A. Muller, Proc. of 1st Symposium on Bioastronomy, Reidel (1984) 233.
12.   L.S. Marochnik, et al., Proc. of 2nd Symposium on Bioastronomy, Kluwer (1987) 49.
13.   B.A. Balazs, Proc. 2nd Symposium on Bioastronomy, Kluwer (1987) 61.
14.   G. Marx, Proc. 1st Symposium on Bioastronomy, Reidel (1984) 535.
15.   D. Schwartzmann, Proc. of 3rd Symposium on Bioastronomy, Springer (1991).
16.   W.M. Irvine et al., Proc. "Origin of Life". Reidel (1981) 27.

Discussion

D. BRIN: Atomic hydrogen is the principal "uphill" source of free energy used by life (as the source of sunlight). Would you credit the rapid expansion of the universe for the failure of hydrogen to equilibrate into larger, lower-energy atoms?

G. MARX: The expansion rate of the Universe is regulated by the dynamics, influenced by gravity. The survival of hydrogen (carrying free energy in a cold environment) is due to the rapidity of expansion (fast cooling from fusion temperatures to temperatures where Coulomb barrier prevents nuclear fusion) and due to the low probability of $H + H \rightarrow D + e^+ + \nu$ fusion, due to the weak coupling.

A. LEGER: How high are temperatures provided by tidal forces in the Solar System? Do you think that there is some chance that they provide conditions actually suitable for life (out of earth) and which are the best candidates?

G. MARX: There is a general agreement that the volcanic activity on Io, the nearest moon of Jupiter, is fed by the tidal forces of the giant planet (taking the eccentricity, too, into account). This means that tides can produce several hundreds of degrees of "geothermal" heat. This means pretty large temperature differences on Io. At the 2nd Bioastronomy Symposium (Hungary, 1987) there was a discussion that the ice-covered Jupiter moon Europa may have had oceans below the ice-cover, warmed by tides, and it could bear life as well. Similar effects are possible for other moons of the giant planet with eccentric orbits, even far away from the star.

D. BLAIR: Given that thermal disequilibrium is the key to the existence of evolutionary processes, is it not likely that the rate of evolution on a planet depends on the radiation temperature of the star - the rate may depend on both the "attempt frequency," set by cosmic rays and radioactivity, and a "process rate" set by the radiation temperature. Thus one might expect that advanced civilizations may develop more rapidly around shorter lived more massive stars.

G. MARX: This is a possibility indeed. But note that the precondition of spreading life on land (continents) was probably the O_2 in the atmosphere (created by sea algae). UV light from the Sun created ozone from oxygen and the ozone shield prevented UV light from reaching the surface and hurting land life. UV light and radioactivity influences mutation (and cancer) rate of simple organisms. Developed species have a system of enzyme defence which works up to a certain level of ionizing radiation. I think that to that end the mutation rate is an outcome of advantage in natural selection.

D. CUDABACK: When is the optimum time for development of life in the course of stellar evolution, based on availability of free energy?

G. MARX: For the creation of life, free energy is needed; e.g., sunlight in the region of kTs~1 ev, T~6000 K in an environment $0°C < T_e < 100°C$. But to my personal understanding, for development of life (and intelligence) the duration of the steady flow of free energy is the more restrictive (t$>10^9$ years), which can be afforded by main sequence stars. One has to think hard, how to overcome this restriction.

PARTICULATE DISKS AROUND MAIN SEQUENCE STARS

Dana E. Backman
NASA Ames Research Center, Moffett Field, CA 94035-1000

Francesco Paresce
Space Telescope Science Institute, 3700 San Martin, Baltimore, MD 21218

Paper Presented by D.E. Backman

The discovery of the so-called "Vega phenomenon" was perhaps the most important and unexpected result of the IRAS mission. Three nearby A main sequence stars, alpha Lyrae (Vega), alpha Piscis Austrinus (Fomalhaut), and beta Pictoris, were found to possess disks or clouds of solid particles emitting strongly in the far-infrared (IR). These have been spatially resolved by various means in the infrared, and in the case of beta Pictoris, also by visual and near-infrared coronagraphy. The grains around these stars have typical temperatures of 50-150 K, and the disks/clouds have fractional bolometric luminosities, L_{grains}/L_{star}, in the range $1x10^{-5}$ to $1x10^{-3}$. The spatial information has allowed the following conclusions: (1) the grains are substantially larger than interstellar grains, (2) the material probably lies in the stellar equatorial planes, (3) the disk material extends to distances of order 100-1000 AU from the stars, and (4) a zone about 20 AU in size around the central stars is relatively depleted of material. The total mass included in the microscopic particles detected by IRAS is of order $1x10^{-2}$ to $1x10^{-1}$ earth masses, although much more mass could be hidden in larger bodies.

Subsequent detailed surveys of IRAS data reveal 50-100 cases of main sequence stars of spectral classes A-K having far-IR excesses with similar temperature and fractional luminosity to the three prototypes. Roughly 40% of A stars and 15% of F, G, and K stars in subsets of the Gliese catalog (d < 22 pc) have excesses detectable by IRAS. This apparent trend by spectral class is due to a luminosity effect; a significant infrared excess around an A star requires less material than around lower luminosity stars at equal distances. However, the proportion of excesses with fractional luminosity $> 2x10^{-5}$ not show a trend with spectral class, i.e., that amount of material is roughly equally likely in nearby stars among types A, F, G, and K. There is further statistical evidence that most G stars have weak far-IR excesses compared to the expected photospheric flux. Thus, this phenomenon appears to be widespread, not limited to young stars, and may not necessarily be a sign of ongoing planet formation.

Possible connection of this phenomenon to the existence of planets is tantalizing. The scale of these clouds/disks corresponds most closely to the hypothetical "Kuiper disk" in our own solar system, a predicted remnant of the original planetesimal disk population expected to lie in or near the ecliptic just outside the planetary region, a transition between the planetary system and the Oort cloud. The necessity of larger bodies to resupply the observed small particles in the 3 prototype systems, however, is debatable. The central depletions could represent regions where accretion of planets has removed the small particles, or where grains have been destroyed by a mechanism such as ice sublimation, or they could be regions which were always deficient in material.

Discussion

G. MARX: If the formation of the planetary system is explained by the conservation of a (non-zero) angular momentum, we expect that the dust cloud of the Sun is a flat disk. What does observation indicate — is it a disk or a sphere?

D. BACKMAN: The Oort cloud comets, which come from $r > 10^4$ AU, arrive equally from all directions and so their reservoir is understood to be spherical. These comets are believed to have originally been part of the flat disk system, but were then scattered into large orbits by gravitational interaction with the planets or protoplanets. Once at great distances from the Sun, perturbations from passing stars and the galactic tide randomized the comet angular momenta and produce a spherical structure.

S. ISOBE: In your talk, you showed different values of transition zone temperature for three prototype stars. At the circumstellar conditions, in gas density of these stars, sublimation temperature does not vary for some definite type of grain composition. Is your temperature difference caused by the difference of dust composition among the stars?

D. BACKMAN: Difference of grain composition could be one explanation for the differences between the transition zone temperatures in the three resolved stellar disk/clouds (transition between sparsely and densely populated regions).

N. EVANS: What is the evidence for the Kuiper belt?

D. BACKMAN: Short period (p <20 yr) comet orbit inclinations appear to require a disk reservoir, unlike the long period (Oort cloud) comets. Also, the surface density of the present planetary system ("smearing" the planetary mass across the ecliptic) follows a radial power law, which when extrapolated beyond Neptune implies the possible existence of about 30 Earth-masses (about Neptune's mass), perhaps in the form of comet nuclei, between r=30 and 100 AU. Present IR and optical searches for small and large particle components of this region do not yet reach the sensitivity necessary to detect this.

BURSTS OF STAR FORMATION IN THE LOCAL GALACTIC DISK AND THEIR IMPLICATIONS FOR THE ORIGIN AND EVOLUTION OF LIFE AROUND THE SUN AND NEARBY STARS

A. A. Suchkov,
Rostov State University, Rostov-on-Don, USSR

It is widely recognized that a major driving mechanism for bioevolution, responsible perhaps also for the very origin of life, may be the hard cosmic ray radiation. This idea became well known when I. S. Shklovskij suggested that the extinction of the dinosaurs at the end of the Cretaceous era resulted from mutations caused by hard radiation from a supernova which exploded fortuitously at that time near the Solar System. In this paper I report some further speculations on how supernovae may have affected the bioevolution; I present some analysis of the recent history of star formation in the local galactic disk and argue that it may have a bearing on the history of life on the Earth, as well as on other life-suitable planets around nearby stars.

Cosmic rays are produced by supernovae remnants; hence their intensity and mutagenus efficiency depends on the galactic supernova rate. Since the latter is proportional to the current star formation rate, one may conclude that the rate of bioevolution is governed by the overall star formation rate in the galaxy. The history of life, both on individual life-suitable planets and in specific regions of the galaxy as a whole, could bear imprints of the history of star formation.

The effect of cosmic rays upon bioevolution is twofold. First, they bring forth through mutations new biological species. Second, also through mutations, they inhibit or even annihilate the already existing forms of life. The rate of both processes obviously depends on the intensity of the overall cosmic ray background, so one may expect to observe the most drastic changes in the life history to be correlated with just as drastic events in the star formation history.

I suggest that some of the most dramatic events which occurred in the life history of the Earth, such as the transition from Paleozoic to Mesozoic about 230 million years ago, might be caused by a burst of star formation in the local galactic disk reported by Barry (1988) to have commenced about 400 million years ago. (Barry's data suggest in fact 250 million years ago. As is well known, at the end of Paleozoic, within a rather short span of time, a large amount of biological species died out and simultaneously plenty of new species came into existence. A reason for this might well be the enhanced cosmic ray background from a multitude of supernovae which appeared a few million years after the onset of the burst. (The latter is the time required for a massive star to evolve before it explodes as a supernova.) The star formation activity may have had a similar effect upon possible life around nearby stars, accelerating or decelerating bioevolution, giving birth to life or destroying it.

It is to be noted, as argued by Rosa and Richter (1988), that supernova activity in galaxies is a recurrent phenomenon lasting tens of millions of years at a level nearly an order of magnitude higher than the average, and that the supernova rate is far below the average for 2 to 3 billion years between the peaks of activity. The current supernova rate in the galaxy is much lower than the peak values reported by these authors.

The fact that the most spectacular events in the life history of the Earth seem to coincide with a local star formation burst makes the latter issue very exciting for an astronomer studying the Galaxy. More than a decade ago, I suggested that star formation in the Galactic disk was not an even process but proceeded through two peaks separated by a period of inhibited star formation (Suchkov, 1977, 1981). The idea was initially stimulated by our finding that the metallicity distribution function of nearby stars appears to be bimodal (Marsakov and Suchkov, 1978); the same result was later obtained by some other authors (see Figure 1). We interpreted it as evidence for the two distinct epochs of star formation. The enhanced supernova rate at the end of the first epoch was supposed to rapidly increase metal abundance in the interstellar gas and simultaneously stop further star formation by heating the gas to high temperatures. New star formation was supposed to have started, after a long delay, in a medium substantially enriched with metals. This accounted for the relative deficiency of intermediate metallicity stars seen in Figure 1, as well as for the very existence of two distinct components: the old disk and the young disk.

An impressive result providing the evidence for uneven star formation, as well as the dating of its ups and downs, was obtained by Barry (1988). He concluded that about 3 billion years ago, the star formation rate in the local galactic disk reached a minimum and then, less than 400 million years ago, a burst of star formation started. Our recent study also indicates that very important changes in the history of star formation and metal enrichment have taken place 2 to 3 billion years ago (Marsakov et al., 1990). This can be seen from Figure 2 which displays the "two-dimensional" age-metallicity relation (AMR) for nearby F stars. Its lower envelope is defined by the age of the main sequence "turn-offs" revealed in the Hertspprung-Russell diagrams by groups of F stars of various metallicity as a sharp drop

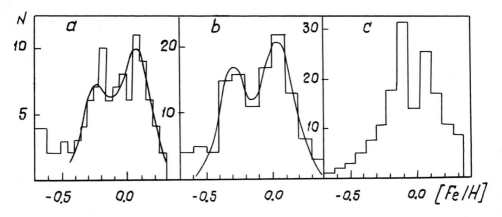

Figure 1. Metallicity distribution for nearby stars as obtained by: (a) Marsakov and Suchkov (1978), (b) Bartasiute and Dadurkevicius (1988), and (c) Knude (1989).

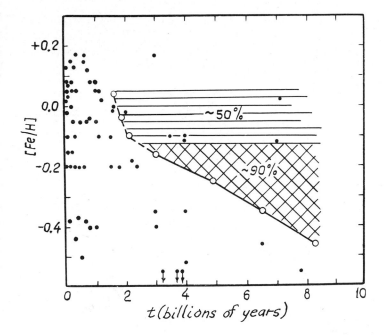

Figure 2. "Two-dimensional" age-metallicity relation for the local F stars; the filled circles represent the open cluster (Marsakov et al., 1990). The percentage of stars populating the dashed area is indicated for the two metallicity ranges.

in the number of stars at a certain effective temperature. I believe that Figure 2 shows the following: Prior to 2 to 3 billion years ago, the stars formed over a wide range of metallicity. As time goes on, the metallicity range gets narrower, perhaps because of mixing; such a behavior is implied by a gradual rise of the lower envelope of the AMR in Figure 2. However, the "turn-offs" for all groups of stars with [Fe/H] > – 0.15 indicate nearly the same age of about 2 to 3 billion years. At the same time, within this metallicity range, we observe a sharp increase in the portion of stars younger than those at the "turn-offs". This suggests that star formation has been stopped at 2 to 3 billion years ago, and after a long delay it has bursted anew.

If Barry (1988) is right, the time of the burst is about 250 million years ago. A hint as to when it started is also provided by open clusters. The well-known dominance of clusters younger than one billion years is commonly ascribed to cluster gradual decay. However, Figure 2 seems to point at another option: the majority of the open clusters may have formed simply in the latest burst of star formation. The puzzling difference between the loci of open clusters and field stars seen in Figure 2 has made us suggest that the burst was triggered by the infall of the external gas; e.g., that of the Magellanic Stream, so that metal-poor young clusters formed from external gaseous clumps, which avoided mixing and preserved their originally low metallicity.

In summary we conclude that presently there is good evidence in favor of a recent star formation burst in the local galactic disk which may have had a close relation to dramatic events in the life history on the Earth and possibly on other planets around nearby stars.

References

1. Barry, D.: 1988, Astrophys. J., 334, 436.
2. Bartasiute, S., and Dadurkevicius, V.: 1989, Astron. Tsirk., 1526, 8.
3. Knude, J.: 1989, Astron. Astrophys. Suppl. Ser., 81, 215.
4. Marsakov, V.A., and Suchkov, A.A.: 1978, Pis'ma Astron. Zh. (Soviet Astron. Lett.), 4, 500.
5. Marsakov, V.A., Suchkov, A.A., and Shevelev, Yu. G.: 1990, Astrophys. Space Sci., 172, 51.
6. Suchkov, A.A.: 1977, Pis'ma Astron. Zh. (Soviet Astron. Lett.), 3, 204.
7. Suchkov, A.A.: 1981, Astrophys. Space Sci., 77, 3.
8. Richter, O.-G., and Rosa, M.: 1988, Astron. Astrophys., 206, 219.

Discussion

C. MATTHEWS: Does your model predict the high iridium content found in the clay layers defining the Cretaceous/Tertiary extinction 65 million years ago?

A. SUCHKOV: My model certainly does not predict this, since it deals only with energetic particles from Supernova remnants. It predicts drastic changes in the life forms only, one or a few million years ago, and it has no relation to purely geological changes.

TESTING THEORIES OF STAR FORMATION

J. Neal Evans II

Department of Astronomy, University of Texas at Austin, Austin, TX 78712 USA

Summary

Recently developed models of star formation involve infall onto a disk which then accretes onto the star (Shu, Adams, and Lizano 1987, Ann. Rev. Astr. Ap., 25, 23). Since this disk is a likely candidate for the formation of planetary systems, this picture has considerable significance for bioastronomy. Current spatial resolution does not usually allow direct detection of the disks, so tests of the overall model are important. Various observational tests of these models will be presented, including far-infrared and molecular line observations.

The results of these tests so far provide qualified support for the models. For example, a high-resolution study of the Bok globule B335 in the 6 cm line of H_2CO finds a density distribution for the globule in good agreement with theoretical expectations: the density distribution appears to follow $n(r) \propto r^{-2}$ in the outer parts, but switch to $n(r) \propto r^{-1.5}$ inside a radius of about 0.03 pc, as expected for the infalling gas around a 1 $M_\odot$ star. High-resolution far-infrared observations of star forming regions in Taurus (L1551 IRS 5) and Orion (NGC 2071) are consistent with density gradients with $n(r) \propto r^{-1.5}$, again consistent with infalling gas; in addition, direct comparison of the data on L1551 IRS 5 to the predictions of the model of this source (Adams, Lada, and Shu 1987, Ap. J., 312, 788) indicate consistency.

Not all the data are consistent with the models. In the case of NGC 2071, the data require higher densities than would be predicted by a naive application of the ideas of Shu, Adams, and Lizano. A still more serious discrepancy occurs in the case of IRAS 16293-2422, a source in Ophiuchus. The far-infrared data on this source indicate a rather uniform density (if $n(r) \propto r^{-\alpha}$, $0 < \alpha < 0.5$), inconsistent with any simple models of star formation.

Discussion

C. LEVASSEUR-REGOURD: Could it be possible that the disk is not homogeneous, i.e. that the size distribution or the albedo of the dust grains is a function of r (as we recently derived for interplanetary dust, both from IRAS and visible observations)? Could such factors be taken into account in the models, and how would the models be sensitive to them?

N. EVANS: At the moment, the observations are too crude to constrain models of this complexity, but there are some indications that grains may be larger in the disks than in the general interstellar medium.

T. WILSON: For thin disks around central objects, geometry, i.e. face-on vs. edge-on disks, play a role. How do you take this into account?

N. EVANS: We do not model disks at present, but we do have plans to begin to include these in the models.

S. ISOBE: To give a good idea of distribution of gas and dust surrounding the star, it is good to show two-dimensional intensity distribution at some certain wavelengths. At radio and optical wavelengths, we are now reaching to milli-arcsec angular resolution. What is the future possibility to have high angular resolution at wavelength range of 20-200 μm, which gives good information on distribution of dust disk and planets?

N. EVANS: There are plans to place a larger (2.5-3m) telescope in a larger aircraft. In this way we could achieve ~5" resolution at (diffraction limit) at 50 μm and ~10" at 100 μm.

J. TARTER: Is the central mass in 4551 large enough to give in fall velocities that can be seen by using molecular line tracers?

N. EVANS: The current observations do not show unambiguous evidence of in fall (or rotation), but this may be possible in the future, by choosing appropriate tracers.

TOWARDS AN ESTIMATE OF THE FRACTION OF STARS WITH PLANETS FROM VELOCITIES OF HIGH PRECISION

B. Campbell, S. Yang, and A. W. Irwin
University of Victoria, Victoria, Canada

G.A.H. Walker
University of British Columbia, Vancouver, Canada

Paper Presented by B. Campbell

Summary

We have been measuring precise radial velocities for 20 solar-like stars in an attempt to detect reflex motion due to low mass, possibly planetary companions. This project is currently in its tenth year at the Canada-France-Hawaii telescope. The hydrogen fluoride technique that we utilize is now a proven method for obtaining velocities over long periods with a precision of order 10 meters per second.

We now believe that we understand what factors limit the precision attainable. We have also discovered a new phenomenon: stars with extreme levels of chromospheric activity show correlated velocity variations. This phenomenon does not interfere with our ability to detect long-term modulation. We have tested for such modulation, and have found a number of low-level variables. We will discuss the implication of these variations vis-a-vis planetary companions. We will also consider how such data might ultimately help to distinguish between low mass brown dwarfs and planets.

Discussion

W. SULLIVAN: What is your overall limit for the least massive planets that are definitely <u>not</u> in orbit around your stars?

B. CAMPBELL: By combining our data with astrometric information we know that the upper limit for any orbiting body is roughly 10 Jupiter masses.

D. LATHAM: Let me suggest a fourth "proof" that low-amplitude velocity variations are due to planetary-sized objects. If two (or more) strictly periodic variations can be demonstrated, with no correlation with other astrophysical parameters, then one can argue that there is a planetary <u>system</u>. In particular, one can argue against a system that contains two (or more) brown dwarf companions with the orbits viewed nearly face on, since such a system would be unstable. Note that a search for multiple periodicities requires many more observations spread over many cycles. Thus a long-term effort is necessary.

B. CAMPBELL: I agree that, as David Black has suggested, multiple bodies strongly imply the presence of a planetary system. Hence we are encouraged by our latest results for E Eri, which, if not a case of a "smoking gun," is perhaps a case where the trigger has been pulled.

L. DOYLE: It is interesting to note that the 11.2-year solar sunspot cycle is so close to the radial velocity variations expected by Jupiter's orbit. My question is, "do you presently fit your radial velocity variation observation for more than one planet?"

B. CAMPBELL: We test for significant fits of polynomials of degree 1, 2 and 3, as well as test for sinusoids in the residuals. So, yes, we do test for two components of motion, but only up to a sinusoid plus 3rd order polynomial (which is the fit for γ Cephei, a binary star).

M. PAPAGIANIS: I asked if there was a cooling of the hydrogen flouride tube to avoid line broadening.

B. CAMPBELL: No, in fact we heat the cell to 100°C to avoid polymerization of the HF. It is not thermal broadening which sets the line widths, but pressure broadening. So we are considering lowering the pressure in the cell.

J. TARTER: Does the detection of correlation of velocity with measures of chromospheric activity (such as the CAII line) suggest that we are at a natural limitation and should not try to push the technology to get better velocity accuracy - ultimately trying to get to the sub-meter/s range needed to detect earth masses?

B. CAMPBELL: Possibly, but we would like somewhat higher precision to better define the correlation with CAII, as well as to look for the effect of rotation plus spots. So an accuracy of 5 ms^{-1}, or perhaps 2ms^{-1} would be desirable for this. Below that we are probably up against a natural limit.

ON THE FEASIBILITY OF EXTRA-SOLAR PLANETARY DETECTION AT VERY LOW RADIO FREQUENCIES

Alain Lecacheux

Observatoire de Paris, Section de Meudon, France

ABSTRACT

The search for extra-solar planetary systems remains an important, continuing, astrophysical problem as well as a primary step in the quest for knowledge of the existence of life in the universe.

In addition to the methods currently in use for detecting planets, there is the possibility of revealing the existence of planets in other stellar systems by observing their natural, non-thermal low-frequency radiation; this was suggested a few decades ago, but has not been attained so far.

Five planets in the solar system, including the Earth, are now known to produce low-frequency radiations in the kilometer to decameter wavelength range. These radiations are mainly due to processes occurring in the auroral regions of the planetary magnetospheres. Thus they might be encountered every time that a planetary magnetized body interacts with a stellar wind. Similar radiation, probably due to the same processes, is observed to come from the Sun and a number of stars smaller and cooler than the Sun, those that are known to have a magnetic field.

The intensity of the planetary radiation is high enough to be detectable, in a reasonable range of distance, with a large but feasible low-frequency radio telescope, operated on the ground, on the Moon or in space.

Taking into account the observed properties of the planetary radio emissions in the solar system, and the scaling rules which can be inferred from the plasma physics, we discuss the feasibility of such a detection.

1. THE LOW FREQUENCY PLANETARY RADIATIONS IN THE SOLAR SYSTEM

An important result of the Voyager interplanetary mission is that it demonstrated that all the highly magnetized planets in the solar system produce powerful, low frequency radio emissions. They are the four giant planets (Jupiter, Saturn, Uranus and Neptune) and the Earth.

These emissions occur in the kilometer to decameter wavelength range (Fig. 1) and have a high contrast with respect to the radio emissions from the Sun: their intensity is several orders of magnitude above the "quiet sun" radiation (the thermal emission from the Sun and the solar corona), and are comparable in intensity to the "disturbed sun" emissions (the non thermal radio bursts occurring in the

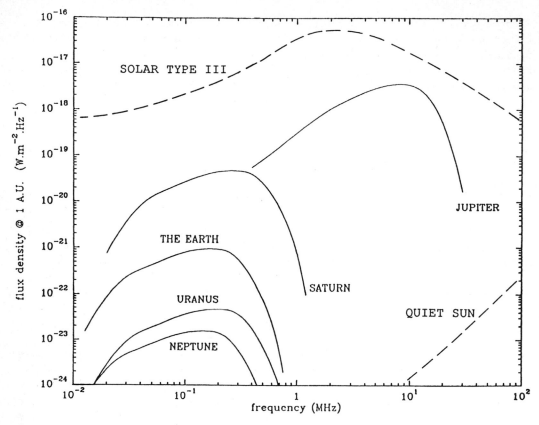

Figure 1. Planetary non thermal, low frequency radiations.

corona). Fig. 1 shows a simplified, quantitative view of the respective spectra of all these radio emissions; the displayed spectrum of each planetary radiation is an "averaged" peak flux spectrum, i.e., corresponds to the intensity versus the frequency reached during a few percent of the time over an observing period of several days. The actual instantaneous spectra are, of course, much more complex (for examples and a review, see [1]).

The spectrum of the jovian radiation is deduced from ground-based measurements obtained above 20 MHz [2]. Because of the presence of the terrestrial ionosphere, which has strong effects on the visibility of the incoming electromagnetic radiations below 20 MHz and is often opaque below 10 MHz, the 2 MHz to 15MHz frequency interval is not very well known. Below 2 MHz the frequency spectrum is better known from observations of RAE, IMP [2] and Voyager. The spectrum of the Saturnian radiation was recently computed by Galopeau et al. [3]. The spectra for the Earth, Uranus and Neptune were more simply drawn by assuming that their spectral shapes are the same as those of Saturn, and by scaling for the observed frequency ranges and peak flux densities. The lower and upper curves illustrate respectively the "quiet sun" spectrum [4] and the "active sun" (maximum recorded type III burst intensity [5]).

All these emissions are explained by a common scenario involving low energy (keV) particles precipitating into the planets' inner planetary magnetospheres (the "auroral zones"). At a given time, the radio source is distributed along a field line of the planetary magnetic field, each frequency being emitted close to the local electron gyrofrequency, at an altitude decreasing with the decreasing radio frequency. The active magnetic field lines are at high magnetic latitudes (about 75°) in the northern and the southern hemispheres of the planet. At the feet of these lines occur some of the visible and UV auroras due to the interaction of the particles flux with the dense upper atmosphere of the planet.

Among the main observational properties of these radiations, two are particularly important: the polarization is highly circular, and the overall properties of the emission (in particular its observed intensity) are deeply modulated at the planetary spin period.

2. THE VISIBILITY OF JUPITER RADIO SOURCE AT 1 PARSEC

In order to evaluate the possibility of detecting the low frequency radio emission produced by an extra-solar planet similar to Jupiter, we have plotted, in Fig. 2, the antenna temperature due to Jupiter if it was located at 1 parsec from the Earth: since such a radiosource would appear as a point source, the antenna temperature contribution depends only on the effective area of the antenna (its physical size, by assuming that the antenna efficiency is near unity). The three curves correspond to 1 km^2, 10 km^2 and 100 km^2 antennas respectively.

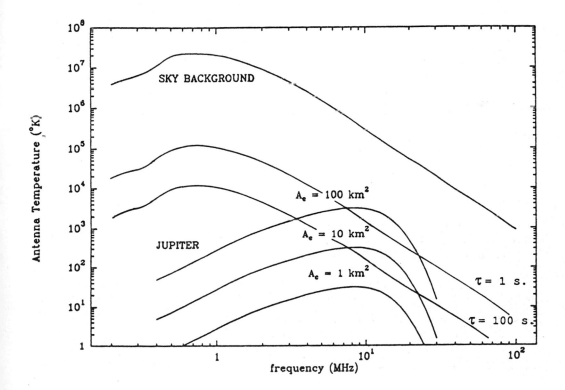

Figure 2. Jupiter's radiation observed at 1 parsec.

At these frequencies the main sensitivity limitation is the sky background, which is the result of the galactic and extragalactic synchrotron emissions. Fig. 2 also shows the equivalent antenna temperature due to the sky background as a function of the frequency. This temperature is proportional to the sky brightness average [6], divided by the square of the frequency, and does not depend on the antenna area. Even with quite a large 10 km x 10 km antenna, the source contribution is still modest. But it is in principle quite observable: the statistical noise fluctuations δT of the background signal T are given by the usual radiometer formula:

$$\delta T/T = (b\tau)^{-1/2}$$

where b in the detection bandwidth and τ the averaging time. The application of this formula shows that, in principle, the decameter emission of Jupiter, located at 1 parsec from the Earth, might be detected, with a signal to noise ratio of 5 and a detection bandwidth of 1 MHz, in less than 100 seconds. It is worth noticing that, at the wavelengths considered, a very large antenna can be constituted as a filled aperture array of simple, wire antennas. So, it certainly is technically feasible.

3. IS JUPITER AN EXCEPTIONALLY FAVORABLE CASE?

This favorable result was obtained by assuming that the jovian emission, the strongest planetary radiation in the solar system, is representative of a typical planetary radio emission. What can we say about the expectable intensity and frequency ranges for any extra-solar planet? This is indeed a very difficult question. The present state of the theory of the emission mechanism does not allow us to give a precise answer. Too many unknown factors are important in determining the production of the auroral radiations. In the case of the solar system planets, we have at most limited knowledge of several macroscopic quantities. While the structure of the planetary magnetic field and the plasma distribution around the planet are approximately known, the energetic particle distribution functions, the acceleration mechanism, and the origin of the particles are essentially unknown. Future, in situ observations will certainly be useful in order to understand the large differences encountered among the radiations of the different planets.

Table 1 summarizes several quantities of interest in comparing the different planetary radiations. For each of the five solar system bodies, the table gives its radius, its distance from the Sun, the value of the planetary magnetic dipole at the equator, the maximum emitted power (computed from Fig. 1 and assuming reasonable beaming properties) and the ratio to the corresponding terrestrial quantities. It appears that the solar system planetary radiations cover three orders of magnitude in frequency and more than six orders of magnitude in intensity.

The electromagnetic energy contained in the observed waves comes from an efficient conversion of the kinetic energy of energetic particles that have been accelerated in the planetary magnetospheres. It is known, mainly for the radiations of the Earth, Jupiter and Saturn, that the intensity variations of the radio emissions are correlated with the variations of the solar wind. This fact was used by Desch and Kaiser [7] to check that, more generally, the total power emitted by each solar system planet is approximately proportional to the solar wind power incident on its magnetosphere.

Table 1. Comparing planetary radiations.

	Earth	Jupiter	Saturn	Uranus	Neptune
R_P (km)	6378	71350	60400	23800	22200
R_P / R_E	1	11.3	9.4	3.8	3.4
D_P (AU)	1	5.2	9.5	19.2	30.1
B_{eq} (G)	0.31	4.3	0.21	0.23	0.13
B / B_E	1	13.6	0.68	0.74	0.42
P_r (W)	10^{7-8}	10^{10-11}	10^{8-9}	10^{6-7}	10^{6-7}
$P_r / P_{r,E}$	1	1000	10	0.1	0.05
n / n_E	1	154	6.4	2.8	1.4
P_i / P_E	1	80	3.4	0.22	0.07

Let

$$P_i = m_i \, (\epsilon_0/D^2) \, V^3 \, \pi \, L_0^2$$

be a measure of the power incident on the cross-sectional area of the planetary magnetosphere (D = distance of the planet from the Sun, L_0 = "radius" of the magnetosphere) due to the bulk motion (V = solar wind speed) of the solar wind plasma (m_i = proton mass, ϵ_0 = number density of particles in the solar wind at the Earth's orbit).

The radius L_0 may be expressed by writing the equilibrium condition between the kinetic ($\frac{1}{2}\epsilon V^2$) and the magnetic ($B^2/8\pi$) pressures at the front of the magnetosphere, i.e:

$$L_0 = (M^2 / 2\pi \, m_i \, (\epsilon_0/D^2) \, V^2)^{1/6}$$

where M is the magnetic dipolar moment of the planet. Then

$$P_i \approx (\epsilon_0^{2/3} \, V^{7/3}) \, (M^{2/3} / D^{4/3})$$

The solar wind input power is thus found to be proportional to the ratio $(M/D^2)^{2/3}$, so this ratio characterizes each planet. The relationship between P_r, the power radiated by the planet, and P_i, the solar wind input power, is displayed in Table 1 and Fig. 3.

In the solar system, the observed ratio P_r/P_i is about 10^{-6}, so that only a very small part of the solar wind energy appears to be converted into electromagnetic radiation. If this empirical relationship might be generalized, one could infer that the closer the planet from its star and the larger is its magnetic moment, the higher will be the intensity of the planetary low-frequency radio emission.

On the other hand, recent [8] theoretical work has shown that the inhomogeneity of the plasma in the vicinity of the planet plays an important role in the efficiency of the emission process. The cyclotron maser instability is a resonant process, occurring in low density, highly magnetized plasmas, in which the kinetic energy from a hot electron population, exhibiting an inversion of population in

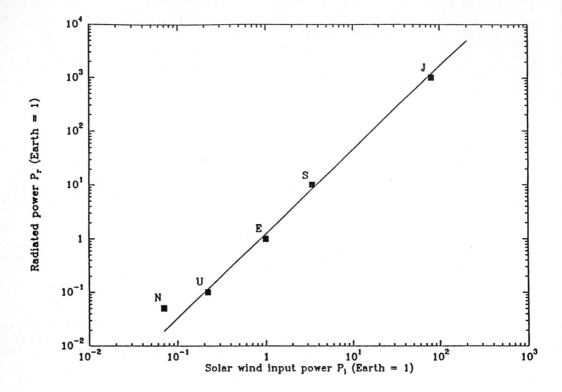

Figure 3. Solar wind input power and planetary radiation power.

perpendicular energy, is efficiently converted into electromagnetic energy. Briefly, if the medium is uniform, the amount of converted power depends on the source volume; if the medium is not homogeneous, the resonant condition is more or less satisfied at different points in the medium: this constrains the extent of the amplifying region and likely controls the amount of electromagnetic output power.

Since the plasma density must be very low, the main inhomogeneity is due to the planetary magnetic field variation. Let a planet be defined by its radius, R_P, and by $B_0 = 2 M/R_P^3$, the intensity of its magnetic field at the feet of the active field lines.

The scale height of the magnetic field is given by:

$$L_B = \mid \delta LogB/\delta R \mid -1$$

where R is the radial distance. In a dipolar magnetic field, $L_B \approx R/3$.

The emission frequency f is proportional to the local magnetic field intensity, so that: $f/f_0 \approx (R/R_P)^{-3}$. The "homogeneity" can then be defined by the parameter $\cap$ given by:

$$\cap = L_B f \approx B_0 R_P / (R/R_P)^2.$$

Here, R/R_P is the normalized altitude of the source. Clearly, as the "homogeneity" increases, so does the efficiency of the mechanism; this is true when the altitude decreases. On the other hand, the resonant process stops [8] when the cold plasma density becomes too large. This very crude argument explains why the spectrum of the planetary radiations exhibits a cutoff at some frequency higher than the frequency of the intensity maximum.

From Fig. 1 one can see that every spectrum of a solar system planet displays an intensity maximum at $f/f_0 \approx 0.3 - 0.4$. This maximum therefore comes from the altitude $R_{MAX} \approx 1.5\ R_P$, while the whole spectrum maps a broader altitude range (≈ 1 to $\approx 5 R_P$). The intensity maximum occurs at the frequency corresponding to $2(R_{MAX}/R_P)^{-3}$ times the equatorial magnetic field intensity at the surface of the planet; namely:

$$f_{MAX}\ (MHz)\ \approx 1.7\ B_{equ}\ (Gauss) = 1.7\ M\ R_P^{-3}$$

Taking $\cap$ at $R/R_P = 1.5$ to be representative of the mechanism efficiency for a given planet, we can now compare the solar system planets in terms of $\cap$. Table 1 and Fig. 4 show the relationship between the radiated power P_r and $\cap_{MAX}$. The expected relationship is well satisfied in the cases of the Earth, Jupiter and Saturn; the two other giant planets, Uranus and Neptune, significantly depart from it. This is likely because we have neglected important quantities that may vary from planet to planet, for example the hot particle distribution. This might also be related to the large departure of the magnetic fields of Uranus and Neptune from a simple dipole whose axis is approximately aligned with the rotation axis [9], leading to a more intricate topology of the source region and different behavior of the resonant process.

Notwithstanding Uranus and Neptune, it is likely that the power radiated by a planet is likely to be high when that planet is large and when it has a strong, well-ordered magnetic field at the surface.

In summary, in spite of the lack of theoretical results on the physics governing the absolute intensity of low frequency, planetary radiations, the previous discussion leads us to predict that an higher intensity will be radiated by a strongly magnetized, larger body, one that is close to its star. Even if Jupiter may be considered as one of the largest possible planets, close to the size of a brown dwarf, neither its magnetic field nor its distance to the Sun appear to be exceptional. Even stronger planetary radiation might be encountered, occurring at somewhat higher frequencies, leading to an easier search (lower brightness of the sky, lower level of manmade interferences) [10].

Now the expected probability of detecting such planetary signals should be discussed, taking into account the actual known stellar distributions in the solar neighborhood. The number density of stars, above the absolute visual magnitude of 14.3, is 0.06 pc^{-3} [11]. Among these stars, more than one half are multiple systems for which the existence of stable, planetary systems might be somewhat questionable. On the other hand, the stellar spectral classes and types do not, from the author's knowledge, clearly define the ability for stars of having planets. So, it appears that several tens of stars, in a radius of 10 pc, could be candidates for such a search.

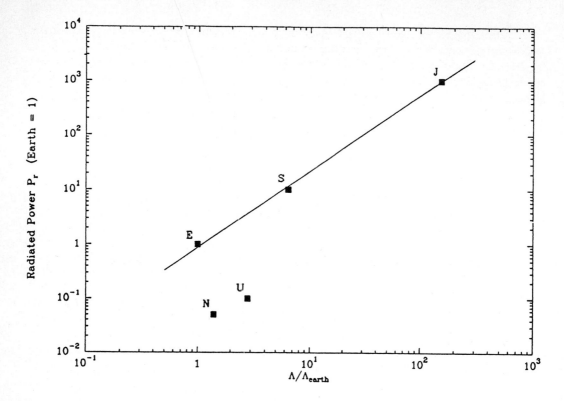

Figure 4. Mechanism efficiency of the solar system planets.

4. CONCLUSION

The arguments for and against the detection of extra solar planets by observing their low frequency radiation from their auroral zones are now summarized:

For:

- The planetary radiations have a large contrast when compared with intrinsic stellar emissions.

- This contrast is enhanced because the planetary radiations are highly polarized and deeply modulated by the planetary spin.
- In the case of very low signal to noise ratios, Fourier analysis techniques could help to detect faint planetary emissions.

- The method, in principle, is applicable to the detection of Earth-like planets, not only large (giant) planets.

- The method requires a very large antenna, typically 10 km x 10 km; but, at these frequencies, large antennas can be simple and (relatively) low in cost.

- Such an antenna could be used for other astrophysical purposes, for example, the study of low frequency stellar radio emissions.

Against:

- In the absence of theoretical predictions, and because the solar system "planetary statistics" does not give a clear estimate, a high intensity, Jupiter-like planetary radiation might be very unusual.

- There is presently no available radio telescope large enough to check the feasibility of such a detection. The broadband Clark Lake array is no longer in operation; the Nançay decameter array, with its effective area of 8000 m^2, is too small.

- From the ground, the terrestrial ionosphere (including the interference by signals reflected from it) precludes the detection of planetary emissions below about 25 MHz.

- The man-made interferences are particularly severe at decameter wavelengths. Even considering that modern technology and data processing methods could help to diminish their effects, it may be necessary to build such a radiotelescope on a quiet zone, such as the far side of the Moon.

- Even from the far side of the Moon, other natural, non thermal radiation of solar system objects (the solar and planetary radiations), will complicate the search.

Acknowledgements

The author would like to thank Ph. Zarka and G. Dulk for their helpful suggestions and comments.

References

1. Genova, F., Zarka, P., and Lecacheux, A., Jupiter Decametric radiation, in Time variable phenomena in the Jovian system, edited by M.J.S. Bolton, Proceedings of a workshop held Aug. 1987 at Flagstaff, Arizona, 1988
2. Desch, M.D. and Carr, T.D., Decametric and hectometric observations of Jupiter from RAE-1 satellite, Astrophys. J., 194, L57, 1974
3. Galopeau, P., Zarka, Ph., and Le Quéau, D., Theoretical Model of Saturn's Kilometric Radiation Spectrum, J. Geophys. Res., 94, 8739, 1989
4. Boischot, A., and Denisse, J.F., Solar Radio Astronomy, Adv. in Electronics and Electron Phys., 20, 147, 1964
5. Lecacheux, A., Steinberg, J.L., Hoang, S., and Dulk, G.A., Characteristics of type III bursts in the solar wind from simultaneous observations on board ISEE-3 and Voyager, Astron. Astrophys., 217, 237, 1989
6. Brown, L.W., The galactic radio spectrum between 130 and 2600 kHz, Astrophys. J., 180, 359, 1973
7. Desch, M.D., and Kaiser, M.L., Predictions for Uranus from a radiometric Bode's law, Nature, 310, 755, 1984
8. Le Quéau, D., Planetary Radio Emissions from High Magnetic Latitudes: the "Cyclotron Maser" Theory, in Planetary Radio Emissions, edited by H.O Rucker, S.J. Bauer and B.M.-Pedersen, Proceedings of the 2nd International Workshop held Sept. 1987 at Graz, Austria, 1988

9. Ness, N.F., The magnetic environment of the known radio planets, in Planetary Radio Emissions, edited by H.O Rucker, S.J. Bauer and B.M.-Pedersen, Proceedings of the 2nd International Workshop held Sept. 1987 at Graz, Austria, 1988

10. Winglee, R.M., Dulk, G.A., and Bastian, T.S., A Search for Cyclotron Maser Radiation from Substellar and Planet-like Companions of Nearby Stars, Astrophys. J. (Letters), 309, L59−L62, 1986

11. Allen, C.W., Astrophysical Quantities, University of London, The Athlone Press, p 237, 1963

Discussion

W. SULLIVAN: In 1975 Bill Erickson and I used the Clark Lake array to search for Jupiter-like decametric bursts at ~25 MHz from extra-solar planets around ~10 nearby stars. Our sensitivity was such that we could detect bursts only if they were approximately greater than 10^3 times as luminous as Jupiter bursts. We spent several hours monitoring each star and saw nothing that we could identify as a planetary burst, but our observations were greatly hampered by man-made radio interference (also burstlike!). (For an abstract, see Sullivan, Erickson and Spangler 1976, Bull. Amer. Astron. Soc.)

A. LECACHEUX: A number of attempts were indeed done by using the existing earth-based low frequency radio telescopes. New attempts should use, in my opinion, sophisticated data processing methods, in order to better discriminate between interferences and planetary low frequency emissions whose temporal and spectral behaviors are known (polarization, burstiness, etc.)

M. HARRIS: If your method selects planets with strong magnetic fields, these are planets on which life is likely to survive, since the field will protect against energetic particles.

A. LECACHEUX: I would not especially talk about life, but I agree.

D. BLAIR: Would the sensitivity for searching for planetary radio emissions be increased by using VLBI techniques, especially to reduce the sensitivity to terrestrial radio interference?

A. LECACHEUX: The radio telescope must be, due to the needed effective area, a filled aperture, phased array of elementary antennas. The interferometric technics will be naturally used. In addition polarization capability may help to reduce the level of terrestrial radio interferences.

BIRAUD: A 10 x 10 km phased array will already have a very narrow beam, giving a good protection against man-made interference.

A. LECACHEUX: The best protection should be, as mentioned in the talk, to build the telescope on the far side of the moon.

THE ESO MICROVARIABILITY KEY PROGRAM AND THE DETECTION
OF EXTRASOLAR PLANETS AND BROWN DWARFS

Nikolaus Vogt

Astrophysics Group, Physics Faculty

Casilla 6014, Santiago, Chile

Summary

A few months ago the European Southern Observatory, La Silla, Chile, approved a proposal for a "Key Program" called "Microvariability of Bright Stars." This project pretends to monitor several thousands of bright stars ($V \leq 6^m$) by means of photoelectric photometery with a fully-automatic 50-cm telescope using one broadband filter (Johnson V). One thousand stars per night will be observed with an accuracy of $\pm 0^m003$; each star will be measured *each* night at least once. The main aim is to search in a large, unbiased sample of stars of all spectral types and luminosities for variations with amplitudes $\geq 0^m01$ and time scales of $\geq$ one day.

As a by-product, this project includes the possibility to detect large planets and brown dwarfs by means of eclipses. Constraints on size and number of detected planets are given, taking into account the relative radii of planets and stars, the separation, inclination and periods of planetary orbits, as well as the expected eclipse durations and amplitudes. The photometric accuracy implies that the planetary radii must be of the order of 10% of the stellar radius; i.e., we refer to Jupiter-sized and larger bodies. Only planets around main-sequence stars will be detectable, not those of giants and supergiants. If a total of about 2000 main sequence stars are monitored each night during one year (or every n[th] night during n years), we expect statistically one eclipse event if all stars would have planets at an orbital distance of about 400 stellar radii (a $\approx$ 2.0 AU); 10 eclipse events if all stars would have planets at a distance of 130 R_* (a $\approx$ 0.6 AU); and 100 eclipse events if all stars would have planets at a distance of 40 R_* (a $\approx$ 0.2 AU). Statistically relevant information can only be obtained for those types of planetary systems for which a considerable number of events is expected if they are frequent. Therefore, the rate of occurrence of systems with giant planets and brown dwarfs (or upper limits for this rate) will be determined for orbital distances smaller than 100 stellar radii (about 0.5 AU) in a statisically complete manner.

Acknowledgement

This research was supported by "Fondo Nacional de Ciencia y Tecnología" in Chile (grants FONDECYT 369-88 and 481-89).

Discussion

B. CAMPBELL: While I suspect you may become frustrated in a search for eclipses, may I impress upon you the value of continuing such a project. In particular, many of us would like to know if there are light variations at the few x $0^m.001$ level in stars due to effects other than low mass companions. However, are you able to observe relatively bright stars, say of magnitude 1 or 2, such as we are observing in our high precision radial velocity program?

N. VOGT: Yes. For this purpose we will apply a neutral density filter whose density will be controlled when observing stars of intermediate brightness without and with the filter. This way even the brightest stars (1^m) will be monitored which is normally very difficult to achieve.

P. BOYCE: If you only use one filter you will not be able to determine the extinction well enough to achieve the photometric accuracy which you expect to reach ($<0^m.01$).

N. VOGT: No! The extinction will be determined by the large sample of about 1000 stars measured by each night, always near meridian at air masses between 1.0 and 1.5. Since only stars with known colors are observed it is easy to derive the color term of the extinction coefficient, even without a color measurement.

A. LEGER: When are you going to start this program?

N. VOGT: As soon as the automatic mode at the ESO 50-cm telescope is installed and operational. This should be completed in early 1991.

L. DOYLE: You can get your detected photometric variations down to 0.01 magnitude, Earth-sized planets, if you use two or more filters. Also, if you can get the stellar rotation periods, you can pre-determine the stellar rotation axis inclination - say to within approximately $+10^\circ$ from edge-on, and lower the number of stars you have to monitor.

N. VOGT: This would be very helpful. Indeed, we may use two filters in case that the observing procedure (pointing and centering) is sufficiently rapid. The present idea, however, is to give priority to a large sample of stars monitored, even when this is possible only in one filter.

A PROPOSAL FOR THE SEARCH
OF EXTRASOLAR PLANETS BY OCCULTATION

Jean Schneider and Michel Chevreton
CNRS, Observatoire de Paris
92195 Meudon, France

Summary

We intend to search for extrasolar planets by looking for the occultations they provoke on their parent stars. Because normal (solar-type) stars give only very weak occultations with low occultation probabilities, we will investigate a class of stars more suited to this method, the dM Dwarfs. We expect show that:

1) the occultation factor is significantly enhanced

2) the probability of occultation is much large.

We therefore propose to monitor a sample of 250 dMs during several months using a photometric telescope with a specially designed photometer.

This project has been submitted as a routine program to the solar 90 cm telescope THEMIS, on which we propose to permanently install the photometer. The photometer can be adapted to any photometric telescope.

Editor's Note:

The subject matter of this paper can be found in Proceedings of the 24th ESLAB Symposium on the Formation of Stars and Planets and the Evolution of the Solar System, Friedrichshafen, ed. B. Battrick, ESA SP-315, 1990, pp. 67-71.

NEAR-TERM PROSPECTS FOR EXTRA-SOLAR PLANET DETECTION: THE ASTROMETRIC IMAGING TELESCOPE

Richard J. Terrile
Jet Propulsion Laboratory, Pasadena, CA USA

Eugene H. Levy
University of Arizona, Tucson, AZ USA

George D. Gatewood
Allegheny Observatory, Pittsburgh, PA USA

Paper Presented by R. J. Terrile

Summary

The Astrometric Imaging Telescope (AIT) is a 1.5 to 2 meter diameter space-based telescope designed to carry out a comprehensive program of direct and indirect extra-solar planet detection. The telescope consists of two separate instruments, an astrometric experiment to measure the reflex motion of the parent stars and an imaging coronagraph to directly image planets and the circumstellar region.

The astrometric technique utilizes the Multichannel Astrometric Photometer (MAP) which passes a Ronchi ruling over a field of stars and measures the centroid of the stars in two orthogonal observations. Used above the Earth's atmosphere this will be about two orders of magnitude more accurate than any existing astrometric instrument and will achieve an accuracy of about 10 microarc-seconds. This will allow detection and study of Uranus-size or larger planets in Jovian orbits around several hundred nearby stars. The astrometric study of a parent star of a planetary system will lead to an accurate determination of its distance. The distance is inversely proportional to the magnitude of the annual parallactic motion. The same study yields the major characteristics of the individual bodies within the planetary system. The periods of the orbits are obtained from the analysis of the motions of the central star. They are directly related to the distances between the individual planets in the system and the system's sun. The distances in turn determine the thermal radiation level, or effective temperature, at the planet's orbit. With sufficient precision and time, the analysis of the apparent motion of the target star will also yield the eccentricities and the relative inclinations of the orbits of each of the planetary bodies. Assuming the mass of the primary star can be accurately estimated, the study will also yield the mass of each planet.

The imaging instrument relies on a high efficiency coronagraph to suppress the diffraction wings of the bright parent star by a factor of 1000. Laboratory tests on the coronagraph have demonstrated its high efficiency at concentrating diffracted light into an area where it can be removed by the apodization of an image of the telescope pupil. However, in order to utilize this high efficiency, the scattered light floor of the telescope must also be a factor of 1000 below the diffraction wings. This requirement puts strong constraints on the mid-spatial frequency errors of the primary mirror. Laboratory experiments in mirror fabrication have demonstrated, with sub-scale mirrors, the required flight quality optics. Space-borne, the AIT will directly image planetary systems with the capability of detecting Jupiter-sized planets in hours around the nearby stars. Its sensitivity to faint material near bright stars exceeds that of the Hubble Space Telescope (HST) by 5 stellar magnitudes and makes the instrument ideally suited for the detection and study of the circumstellar material associated with planetary system formation.

Discussion

T.WILSON: Could you contrast the capability of your system, as compared to the orbiting 8-m infrared telescope and orbiting interferometer, in regard to the detection limits for earth-sun-like systems (i.e., how distant would the systems be, to what magnitude, what is the expected number of stars, ...)?

R.TERRILE: Our system will get down to detection limits of Neptune mass planets around the nearest 100 or so stars. It also is designed to utilize current technology and could be started today. Eight-meter diameter space-based telescopes will have greater capability but are much further away technically from being ready. The same is true for interferometers. Earth detections will clearly be a goal for second generation planet detection instrumentation.

F.DRAKE: Did you consider using apodization with the primary mirror in the coronographic images?

R.TERRILE: We have found that the focal plane apodization used in our coronagraph is far more efficient than pupil plane apodization. This is always true if there is one central bright object in the field of view with faint target objects around. We do have a pupil plane apodization capability which we use for crowded field imaging.

A SEARCH FOR T-TAURI STARS BASED ON THE IRAS POINT SOURCE CATALOG

J.C. Gregorio Hetem and J.R.D. Lepine
Instituto Astronomico e Geofisico, Universidade de São Paulo

G.R. Quast and C.A.O. Torres
Laboratorio Nacional de Astrofisica Itajuba, Minas Gerais

R. de la Reza
Observatorio Nacional Rio de Janeiro

Paper Presented by J.R.D. Lepine

ABSTRACT

T-Tauri stars represent the class of pre-main sequence stars which evolve to solar-type main-sequence stars. It is of interest to investigate the occurrence of circumstellar disks, which are believed to give origin to planets at this stage of evolution.

We present the preliminary results of a survey for new T-Tauri stars that we have undertaken, starting from a list of candidates selected in the IRAS Point Source Catalog. Spectra of the objects are obtained with the 1.6 m telescope at Brazopolis, MG, Brazil. Forty new T-Tauri stars, and a number of other interesting objects, have been detected. In a few cases the presence of a pre-planetary circumstellar disk can be inferred from an analysis of the relative velocity of forbidden lines with respect to the stellar rest velocity, but further observations are required to obtain better statistics on the presence of circumstellar disks.

INTRODUCTION

The spatial distribution and mass distribution of pre-main sequence (PMS) stars are important clues to understanding the process of star formation and the evolution of star-forming molecular clouds, from the hierarchical fragmentation to the dispersion of the cloud remnants. In particular, the investigation of the spatial distribution of a complete sample of T-Tauri could shed some light on the puzzling question of the existence of isolated T-Tauri at a considerable distance from any molecular cloud, of which TW Hya is the prototype (Rucinski and Krauter, 1983; de la Reza et al., 1989). The study of a complete sample of T-Tauri is potentially also of interest for estimating the occurrence of planets around stars, since the presence of pre-planetary disks around young stars seems to be more easily detected than the planets themselves in later stages of evolution.

About 800 PMS stars are optically identified at present, the large majority being T-Tauri stars, and most of which are listed in a catalog by Herbig and Bell (1988). These objects have been discovered either by objective-prism or by grating surveys, or because they were attractive candidates for spectroscopic observations for some reason such as the proximity to a T-association, variability or X-ray emission. The sample of known T-Tauri is therefore not complete to any given magnitude, except for limited areas of the sky.

In the present work we present the first results of a systematic search for T-Tauri stars that we have undertaken, based on the IRAS Point Source Catalog. Our method takes advantage of the almost complete sky coverage of the IRAS Survey, of the fact that the T-Tauri stars emit a large fraction of their energy in the infrared, and of the low interstellar extinction at these wavelengths. The candidate sources were selected from the PSC, with colors in the same range of the known T-Tauri, and were observed with a Coudé spectrograph. Up to the present the search has led to the discovery of 40 new T-Tauri and a number of other interesting objects. Except in a few cases, the information contained in the spectral range that we observe is not sufficient to reveal the presence of a circumstellar disk, and complementary observations are being made.

SOURCE SELECTION AND OBSERVATIONS

The candidate sources were selected in the following color range:

$$0.95 < (F25/F12) < 3.40$$
$$0.50 < (F60/F25) < 3.30.$$

The box so defined in the color diagram is similar to that used by Beichman (1986) to search for T-Tauri stars in Orion, and used by Harris et al. (1988) in a similar search in Taurus. We did not use the 100 μm flux in our selection, since the signal in this band is often dominated by the emission from cirrus or from dust clouds. Only sources with flux qualities 2 or 3 (i.e., excluding upper limits) at 12, 25, and 60 μm were chosen. Our list of candidates contains 888 objects, considering only sources situated to the South of declination $+30°$.

Spectra showing H alpha and Li I in the region 655-670 nm have been obtained with a CCD camera at the Coudé focus of the 1.6-m telescope of the Laboratorio Nacional de Astrofisica, in Brazopolis, Minas Gerais. The data reduction was performed with the Vax computer of the Instituto Astronomico e Geofisico, University of São Paulo.

RESULTS AND DISCUSSION

Up to the present, the telescope has been pointed towards 203 IRAS sources, less than 1/4 of the total number of candidates. In 57 cases, no visible counterpart bright enough to obtain a spectrum was found at the pointed position. In almost all the remaining cases, our goal to obtain spectra on at least three different nights for each object has been reached.

The results for a sample of 109 spectra that have been reduced can be summarized as follows:

New T-Tauri discovered 22

Suspected T-Tauri 14

Previously known T-Tauri 28

H alpha absorption and Li absorption 7

H alpha absorption but no Li 22

H alpha emission but no Li 8

No lines in the spectrum 8

We can therefore expect that about 100 new T-Tauri stars will be discovered by this survey, when completed. Although this number does not represent a substantial increase in the total number of known T-Tauri, our search covers a large region of the sky, which is allowing us to detect some T-Tauri stars distant from known star-forming regions. Some examples of these are Cod -298887 and Hen 600 (A and B), of which we have revealed the T-Tauri nature (de la Reza et al., 1989); these objects are close to TW Hya, the prototype of "isolated" T-Tauri. In addition we are constituting a large sample of T-Tauri stars observed in the same conditions.

One line of evidence for the presence of pre-planetary disks around T-Tauri stars comes from IRAS and near-infrared photometry as discussed for instance by Cohen et al. (1989). Another line of evidence, that we could in principle access with our data, comes from an analysis of the profile of forbidden lines (Edwards et al., 1987). These lines, which originate in the expanding gas at large distance from the star, are blueshifted because the redshifted part of the line is occulted by the disk. Only in a few cases have we detected forbidden lines; one example is CT Cha, a previously known T-Tauri. Its spectrum is shown in Figure 1; the 6731 Å (SII) emission line is about 100 kms^{-2} blueshifted with respect to the velocity of the Li I line at 6708 Å, which is well above the uncertainties. We are also performing UBVRI photometry of the detected T-Tauri stars in order to verify if the energy distribution reveals the presence of a disk. We expect to be able to estimate from this study the fraction of T-Tauri stars of our sample for which there is evidence for the presence of a pre-planetary disk.

Figure 1:
The spectrum of IRAS 11027-7611 (CT Cha). The Li 6708 Å and (S II) 6731 Å lines are indicated.

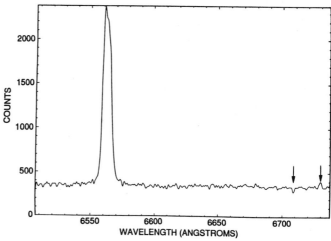

References

Beichman, C.A., 1986, in "Light on Dark Matter", ed. F.P. Israel (Dordrecht: Reidel), 279.

Cohen, M., Emerson, J.P., Beichman, C.A., 1989, Astrophys J., 339, 445.

de la Reza, R., Torres, C.A.O., Quast, G., Castilho, B.V., and Vieira, G.L., 1989, Astrophys J. 343, L61.

Edwards, S., Cabrit, S., Strom, S.E., Heyer, I., Strom, K.M., Anderson, E., 1987, Astrophys J. 321, 473.

Harris, S., Clegg, P., and Hughes, J., 1988, Mon. Not. R. Astr. Soc. 235, 441.

Herbig, G.H., and Bell, K.R., 1988, Lick Obs. Bull. 1111.

Rucinski, S.M., and Krautter, J., 1983, Astron. Astrophys. 121, 217.

HOW MANY SINGLE STARS AMONG SOLAR-TYPE STARS?

A. Duquennoy and M. Mayor
Geneva Observatory, Switzerland

Paper Presented by A. Duquennoy

INTRODUCTION

In the context of establishing probabilities for the existence of extraterrestrial life, several potentially important parameters (in relation to the stability of planetary orbits or to the surface temperature of the planets), can be derived from this kind of study such as the frequency of single stars and the distribution of the mass ratios between the secondary and the primary components in binary stellar systems. It is a well known fact that a large fraction (more than 50%) of solar-type stars appears as binary systems. The distribution of the mass-ratios ($q = M2/M1$), mostly due to observing biases, was still recently believed to be increasing with q and peaking at q=1. With such a distribution, the fraction of companions with substellar masses (brown dwarfs) was not of prime importance in establishing the fraction of single stars.

Our high-precision survey of an unbiased sample of solar-type stars has completely reversed the shape of the mass-ratio distribution, the maximum now being observed close to q=0 . This result is an important constraint on stellar formation processes, but also opens the question of the importance of sub-stellar companions on the real fraction of single stars. The existing cross-correlation spectrometers and other Doppler-shift instruments already allow significant possibilities to detect companions with masses M2sini lower than a tenth of solar mass. And in fact many objects with well documented orbital elements have been discovered in this domain of low-mass companions.

The mass-decade from 0.01 to 0.10 solar masses is an exciting domain covering the transition of stars to giant planets. The brown dwarfs also appear in this domain as probably the lower limit for the mass of stars formed theoretically by fragmentation processes. Long-term, systematic surveys with now fiberfeed spectrometers or accelerometers are expected to provide significant progresses in the coming decade.

The distinction between a giant planet or a low-mass brown-dwarf is less given by the physical properties of these objects than by their formation processes. The orbital eccentricity is probably the key-parameter to set the limit between the two kinds of objects.

1. THE CORAVEL SURVEY OF SOLAR-TYPE STARS IN THE SOLAR NEIGHBORHOOD

In view to reanalyse the duplicity of solar-type stars, we selected an unbiased distance-limited sample. With a limiting distance to the sun of 22 parsecs (pi(trig) >0.045"), we constructed a sample of 164 solar-type primary stars from the catalogue of nearby stars (Gliese 1969).

A long-term spectroscopic (Doppler shift) survey of this sample has been performed with the radial velocity spectrometer CORAVEL mounted on the 1-m swiss telescope at the Haute-Provence Observatory (OHP). CORAVEL uses the technique of cross-correlation (see Baranne, Mayor and Poncet 1979) to derive high precision (0.1-0.3 km/s) velocities. Its precision is not as high as that of other spectrometers such as those used by Campbell or by McMillan and their collaborators (see e.g. the Second Bioastronomy Meeting, 1988). However CORAVEL has other advantages such as the rapidity of the data acquisition and the much fainter stellar luminosity accessible. For example, a precision of 1 km/s is obtained for a 11-mag star in 1 mn, and a precision of 0.1 km/s is obtained for a 7-mag star in 10 mn. This is of extreme importance for the detection of very low mass secondaries, which is favoured around low-mass (usually faint) primary stars.

Another advantage is that the OHP CORAVEL spectrometer is permanently mounted on its dedicated 1m-telescope since 1977. This allows a combination of continuous, long-term and high precision radial-velocity surveys of well-defined and significant samples of nearby solar-type stars, such as the sample presented here.

We obtained for this sample about 4200 radial velocities spanning over 13 years with an average precision of 0.25 km/s.

2. LIMITS IN DETECTABLE (SUB-)STELLAR SECONDARY MASSES

Being given the precision in velocity determination for each measurement, the number of measurements and the dates of observations throughout the 13 years for each primary star, it is possible to simulate the detection with CORAVEL of a secondary mass orbiting with arbitrary orbital elements around a primary star showing no a priori variations. In the case of our nearby solar-type sample, the mean precision is about 0.25 km/s and the mean number of measures for such primary stars is about 11.

Fig. 1 gives the detection probabilities for a given secondary mass M2 orbiting at a given orbital period P, according to a variability criterion described by Duquennoy, Mayor and Halbwachs (1991b). It shows in particular the ability of CORAVEL to detect one fourth of the secondaries with

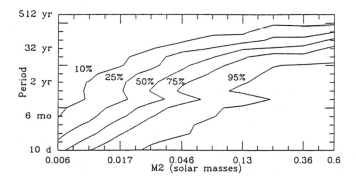

Figure 1. An average 11 radial velocity measurements distributed in 10 years, each one with a precision of 0.25 km/s, allow the detection of only a fraction of low-mass companions to solar-type stars. The detected fraction is a function of the orbital period P and of the mass M2 of the companion.

M2 = 6 Jupiter masses and P = 1 month (provided such a case exists in reality!), or with M2 = 20 Jupiter masses and P = 8 yrs. It can also detect 90% of all the secondaries with P <1 yr and M2 = 0.08 solar masses (or 80 Jupiter masses), the latter being the transition mass between stars and brown dwarfs.

In a near future, we can imagine that new generation detectors such as accelerometers or optic fiber spectrographs with CCDs, able to achieve velocity precisions down to 0.02 km/s or less, for a wide range of stellar luminosities, will bring far below the secondary-mass limit detection, down to real planets.

3. RESULTS ON THE NEARBY SOLAR-TYPE STARS SAMPLE

In addition to our spectroscopic survey which is limited to binary systems with orbital periods less than about 30 yrs, we collected all the informations concerning the systems with longer periods detected by visual techniques. With due allowance made to detection biases, as described in details by Duquennoy and Mayor (DM, 1991a), and according to the simulations quoted in the previous section, we derive the corrected distribution of the mass ratios $q = M2/M1$ (Fig. 2). Note that for solar-type stars, M1 = 1 solar mass.

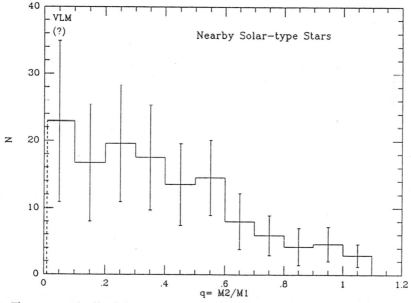

Figure 2. The mass-ratio distribution of solar-type binaries.

We see in particular that the distribution appears as a monotonous function rising towards the small values of q, at least down to $q = 0.2$. Furthermore, this result seems independent of the orbital period (see DM). We also show that about 23 very low mass companions are expected in the histogram bin q=0.01-0.10 for the whole range of orbital periods (log P= −1 to 10, with P in days). This is roughly the number of objects observed in each of the next two bins of q. Thus, the probable number of brown dwarfs in binary stellar systems appears neither high, neither low, but simply similar to the number of secondaries with masses around 0.2 solar masses.

In other terms, the proportion of objects with M2= 0.01 to 0.10 solar masses around nearby solar-type stars possibly amounts to 14% of the primaries. It can be compared with the proportion that can be estimated around another kind of extremely well measured stars with CORAVEL which are the IAU velocity-standard stars. This proportion is found to be about 11% (see DM), in good agreement with the number above.

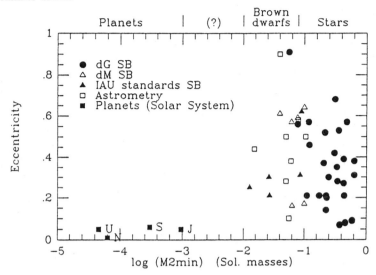

Figure 3. The stellar formation processes (still poorly known) produce binaries with notable eccentricities. Probably planetary orbits are much more circular (as shown by the four most massive planets or our solar system). The orbital eccentricity of objects (still to be discovered) with masses between 1 and 10 Jupiters seems to be the key-parameter to distinguish between brown dwarfs and giant planets.

Another interesting result concerns the comparison of the orbital eccentricities between the three following samples (see Fig. 3):

i)   for the nearby solar-type stars with secondary masses M2 >0.08 solar masses, all eccentricities associated with orbital periods above a certain cut-off value are widely displayed among non-zero values. The cut-off period is interpreted as the circularization period (about 11 days in our sample) due to tidal effects between the two components of the binary system. Below this cut-off period, the observed eccentricities are no longer representative of the original eccentricity distribution.

ii) for a sample of 19 primaries with probable brown dwarf secondaries (M2 or M2sini = 0.01 to 0.08 solar masses) taken among IAU velocity standards, M-dwarf primaries and astrometric binaries (see DM), the eccentricities also appear significantly non-zero (e-mean = 0.33).

iii) for the most massive planets of our solar system, all the eccentricities are close to zero.

This "visually provocative" drawing may be interpreted as a probable difference existing in the formation processes between stars and brown dwarfs on one hand, and planets on another hand. We recall that the mass of 0.01 solar mass is the present theoretical lower limit for an object formed by fragmentation processes, while fission processes could form planetary-size objects (see Boss 1987). No object has been firmly detected in the mass range 0.010-0.001 solar masses yet. Fig.3 indicates that the orbital elements (still to be determined) for such secondaries probably could be used to distinguish between the brown dwarfs and the giant planets.

4. CONCLUSION

Finally, the question of the number of true single stars may be aborded. In our sample of nearby solar-type stars, we observe 44% of the primaries with at least one stellar companion. Corrected from detection biases and taking into account the very low mass secondaries down to 0.01 solar masses, this number rises to 71%. Consequently, less than 30% of the primaries may be real single stars.

However, concerning the stability of the hypothetic planetary orbits and in the context of the search for extraterrestrial life, the long period systems (say with P >100 yrs) may be considered as two independent single systems. As they represent about half of the total of the binary systems, we may expect actually about two thirds of the solar-type stars which could accept stable planetary orbits.

References

Baranne, A., Mayor, M., Poncet, J.-L. 1979, Vistas Astron. 23, 279
Boss, A.P. 1987, Theory of Collapse and Protostar Formation, Eds D. J. Hollenbach and H. A. Thronson, Jr., Reidel Publ.
Duquennoy, A., Mayor, M. 1991a, Astron. Astrophys. (in preparation)
Duquennoy, A., Mayor, M. , Halbwachs, J.-L. 1991b, Astron. Astrophys. Suppl. Ser. (in preparation)
Gliese, W. 1969, Veroll. Astron. Rechen Inst. Heidelberg, No. 22
Marx, G. (ed.) 1988, Bioastronomy-The Next Steps, Kluwer Academic Publishers

Discussion

J. STRELNITSKI: Defining the "solar-type" stars, did you take into account the age of the star, as e.g., Soderblom did composing his "short" list of SETI stars?

A. DUQUENNOY: No. Our solar-type sample is defined by all stars in the Gliese (1969) catalogue with spectral types F7V to G9V, trigonometric parallax above 0.045" and declination above −15°, independently of their age. If the "SETI" stars of Soderblom are defined by stars of age above 3 Gyr, then our sample of old disk stars contains statistically about 70% of such stars.

S. ISOBE: Your velocity resolution is 0.25 km/s and that by Campbell is much higher. What different mass of planets can you detect compared with him?

A. DUQUENNOY: Our detection of planets is very limited, since to detect a 6-Jupiter mass companion we need it to be in a short period orbit and the primary (solar-type star) must be measured several hundred times! However, we can easily detect objects in the range of brown dwarfs, with orbital periods up to a few years, and around primary stars much fainter than Campbell does, increasing significantly the statistics on such objects. In that sense, our detection limits are complementary of those of Campbell.

HABITABLE PLANETARY ORBITS AROUND α CENTAURI AND OTHER BINARIES

D. Benest
O.C.A. Observatoire de Nice, B.P. 139, F-06003 Nice, France

INTRODUCTION

Cosmogonical theories as well as recent observations (e.g. the I.R.A.S. observations of dust-disks around nearby young stars like Vega or Fomalhaut, or the probable preplanetary disk around β Pictoris) allow us to expect the existence of planets not only around single stars — such as the Sun—, but also in double — and perhaps even multiple — star systems, which are established to be the most numerous, at least in the Solar Neighborhood (see the detailed bibliography given by Paprotny et al., 1980, 1983, 1984).

We are then faced with the following dynamical problem: do long-term, stable orbits exist for planets in double star systems? The earliest paper I know of that refers explicitly to a "Planetary Orbit in a Binary" appeared in 1907 (Pavanini); more recently, more and more numerical and semi-numerical studies have been published in this field (see e.g. Benest, 1988b, for a recent bibliography).

My aim here is to try to begin to answer the following question: do such planetary orbits, around one of the two components of fairly wide binaries, exist inside the so-called "habitable zone", as defined by Hart (1979)?

For this purpose, numerical simulations are made within the frame of the elliptic plane restricted three-body problem, to search for the existence of stable orbits for planets (of negligible mass) around one of the two components in double stars (the two massive bodies). A systematic exploration of the phase space of the initial conditions is undertaken, for given values of the two parameters μ and ϵ (μ is the reduced mass of the considered component of the two stars, and ϵ is the eccentricity of their orbit around each other). A planetary orbit is called stable when there has been neither escape nor collision (with one of the two stars) during at least 100 revolutions of the binary. Comparisons with analogous studies (Dvorak, 1982, 1984, 1986, 1988; Rabl and Dvorak, 1988; Dvorak, Froeschlé and Froeschlé, 1989) have shown that this time span is sufficient. The set of initial conditions corresponding to stable orbits and its extension are thus determined. (See details in Benest, 1988b.)

In the first stage, a systematic exploration of the circular case ($\epsilon = 0$) of the restricted three-body problem showed that stable planetary orbits exist at large distances from each star (Benest, 1974, 1975, 1976). But known binaries generally have fairly high eccentricities; we need therefore to study the more realistic elliptic case ($\epsilon > 0$).

The first real binary systems explored have been α Centauri ($\mu = 0.45 - 0.55$, $\epsilon = 0.52$) (Benest, 1988a) and Sirius ($\mu = 1/3 - 2/3$, $\epsilon = 0.592$) (Benest, 1989); recently, we began to investigate η Coronae Borealis ($\mu = 0.45 - 0.55$, $\epsilon = 0.28$). Up to now, the results for these typical elliptic cases confirm what we obtained in the circular case: that large stable planetary orbits exist up to distances from each star of the order of more than half the binary's periastron separation. I will speak now more about habitable planetary orbits in the α Centauri system.

THE α CENTAURI SYSTEM

In the α Centauri system, among all the stable planetary orbits, quasi-circular stable planetary orbits were found up to a little more than 4 AU (Astromical Unit) around α Cen A and up to a little more than 3 AU around α Cen B. Figure 1 presents a sample of such quasi-circular stable planetary orbits, chosen to follow approximately the so-called Bode's Law. Around each star, the selected planetary orbits correspond roughly to a Mercury-like (A1 and B1), a Venus-like (A2 and B2), an Earth-like (A3 and B3) and a Mars-like orbit (A4 and B4), together with two asteroid-like orbits (A5-A6 and B5-B6).

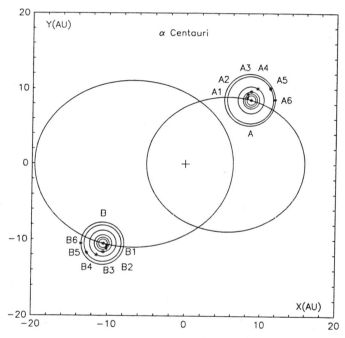

Figure 1: A sample of stable planetary orbits around α Cen A and α Cen B.

Following Hart (1979), the Habitable Zones around α Cen A (spectral type G_o) and around α Cen B (spectral type K1) are about 1.2 AU and 0.7 AU respectively. From our results, there exist therefore quasi-circular stable planetary orbits in the habitable zone around each component of the α Centauri system; we may call these orbits "Habitable Orbits".

Among these habitable orbits, we may select the orbit A3 around α Cen A (Fig. 2) and the orbit B2 around α Cen B; although we previously called B2 a "Venus-like" orbit, it is in fact inside the habitable zone around α Cen B, as this star is less luminous than the Sun.

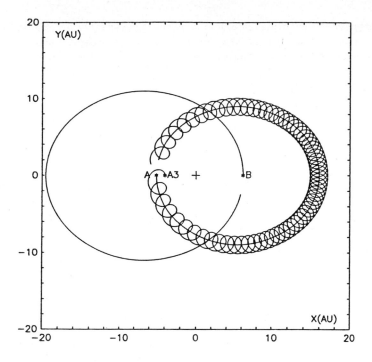

Figure 2: An example of habitable planetary orbit around α Centauri A.

CONCLUSION

This study has established from a dynamical point of view the existence of stable orbits for planets around each star in a binary system. Of course, it does not imply that there are planets on such orbits; this is a separate cosmogonical question: is the formation of such a planet possible, and under which conditions, during or immediately after the formation of the binary system?

However, a consequence of these results is the possible existence of planetary companions in perhaps very numerous multiple stellar systems, increasing the possible number of habitable planets in the Universe, if such quasi-circular stable planetary orbits are found around other numerous binaries.

References

Benest, D.: 1974/1975/1976, "Effects of the Mass Ratio on the Existence of Retrograde Satellites in the Circular Plane Restricted Problem I/II/III." Astron. Astrophys, 32, 39-46 / 45, 353-363 / 53, 231-236.

Benest, D.: 1988a, "Planetary Orbits in the Elliptic Rectricted Problem I. The α Centauri System." Astron. Astrophys 206, 143-146.

Benest, D.: 1988b, "Stable Planetary Orbits around One Component in Nearby Binary Stars." Celes. Mech. 43, 47-53.

Benest, D.: 1989, "Planetary Orbits in the Elliptic Rectricted Problem II. The Sirius System." Astron. Astrophys. 223, 361-364.

Dvorak, R.: 1982, "Planetenbahnen in Doppelsternsystemen." Sitz. Österreich. Akad. Wiss., Math.-Naturw. Klasse, Abt. II 191, 423.

Dvorak, R.: 1984, "Numerical Experiment on Planetary Orbits in Double Stars." Celes. Mech. 34, 369-378.

Dvorak, R.: 1986, "Critical Orbits in the Elliptic Restricted Three-Body Problem." Astron. Astrophys. 167, 379-386.

Dvorak, R.: 1988, "Orbites Critiques dans le Problème Restreint Circulaire et Elliptique (Orbites Planètaires dans des Binaires)." in Dévelopements Récents en Planétologie Dynamique, D.Benest, C.Froeschlé, eds. (Obs. Nice), pp. 177-186.

Dvorak, R., Froeschlé, Ch., Froeschlé, C.: 1989, "Stability of Outer Planetary Orbits (P-Types) in Binaries." Astron. Astrophys. 226, 335-342.

Hart, M.: 1979, "Habitable Zones about Main Sequence Stars." Icarus 37, 351-357.

Paprotny, Z., et al.: 1980/1983/1984, "Insterstellar Travel and Communication Bibliography I/II(update 1982)/III(update 1984). (section 05.04: Planets of Multiple Star Systems)", J. Brit. Interplanet. Soc. 33, 201-248/36, 311-329/37, 502-512.

Pavanini, G.: 1907, "Sopra una nuova Categoriadi Soluzioni Periodiche nel Problema dei 3 Corpi." Annali di Matematica, Serie III, 13, 179-202.

Rabl, G., Dvorak, R.: 1988, "Satellite-type Planetary Orbits in Double Stars: A Numerical Approach." Astron. Astrophys. 191, 385-391.

Discussion

M. PAPAGIANIS: Would the star manage to keep the larger, farther out planets as well it would an Earth-like planet?

D. BENEST: Yes, up to distances of the order of half the binary's periastron.

D. BRIN: I have three quick questions. First, were your simulations two-dimensional, with planets orbiting in the plane of the binary pair motion? Second, with only 100 stellar orbits, can you definitely exclude so-called "chaotic" perturbations after longer periods? And finally, I wonder about the stability of such systems during the proto-planetary period. Might disks of planetismals be more susceptible to disruption in a binary system than already-formed planets?

D. BENEST: 1. Yes, my numerical simulation is done within the two-dimensional restricted 3-body model. 2. As far as my numerical simulations show, chaotic orbits appear to have an influence only at the border of the stability zone, but not for quasi-circular orbits, which lie well inside the stability zone. 3. I do not speak from the cosmogomical point of view, then I can't tell you anything about the stability of an accretion disk in a binary system.

J. TARTER: Does your work represent a significant difference from earlier work by Don Black and collaborators that indicated the possibility of stable planetary orbits as long as the orbital radius was <1/3 of the binary separation or >3 times the binary separation?

D. BENEST: My work is consistent with previous, other works cited; although I found stable planetary orbits up to 1/2 the binary periastron separation, quasi-circular planetary orbits were found only up to 1/4 to 1/3 the binary periostran separation. I did not make studies about outer planetary orbits; R. Dvorak did, and his results are consistent with the paper you cited.

THE SEARCH FOR PROTOSTARS

T. L. Wilson

Max-Planck-Institut f. R., Auf dem Huegel 69, D53 Bonn 1, F.R.G.

ABSTRACT

A summary of the basic properties of molecular clouds is given. For a selection of nearby clouds, search stategies for protostars are summarized. A critical discussion of low-mass candidates in nearby molecular clouds is presented.

INTRODUCTION

Not all stars were formed at the time of galaxy formation. The usual argument, based on stellar evolution theory, is that the lifetime of massive stars, such as those ionizing the Orion Nebula have been producing energy for less than 10^6 years. Even shorter timescales have been observed in the Orion-KL nebula. The proper motions of water vapor masers (Genzel et al. 1981), the ratio of the size and radial velocity of the bipolar CO outflow (see e.g. Wilson et al. 1986) and the deuterium-to-hydrogen ratio in NH_3 (Walmsley et al. 1987) are consistent with a time scale of less than 10^4 years. At that time, a fairly massive star embedded in this molecular cloud "switched on," that is, started producing large amounts of energy, presumably by hydrogen burning and this has altered the kinetic temperature of molecular gas over a region of size 10^{17} cm. Downes et al. (1981) have identified this star with the infrared source IRc2.

The birth and death of stars is less well understood than their evolution during mid-life, that is, their evolution on the main sequence (see Renzini and Fusi Pecci 1988). It would be of great interest to find an instance where stellar birth is occurring, that is a protostar. One dimensional collapse calculations have been carried out by many authors (see Larson 1974). More realistic calculations should include rotational and magnetic effects. These calculations are very involved and the final configurations can be very sensitive to initial conditions (see e.g. Tscharnuter 1985). It would be of great interest to observe the kinetic temperatures, H_2 densities and gas motions in a region where collapse of a single object occurs, in order to estimate the boundary conditions and relative sizes of rotational and turbulent motions. Tscharnuter (1985), Yorke (1985) and Larson (1974) show calculations where scale sizes are of order 10^{15} cm.

Before considering the question of searches for protostars, it is important to define what one means by the term. As pointed out by Wynn-Williams (1982), depending on the definition, either thousands of protostars have been found already, or none have been found. The strict definition of a protostar is "A collapsing object in which the energy is generated by contraction." In terms of this definition, no protostar has yet been found. Before discussing the searches for protostars, one must first know where to look.

THE BIRTHPLACES OF STARS

Until about 1970, it was generally thought that stars formed from HI clouds. With the discovery of complex molecules (see Rank et al. 1971) and the direct detection of molecular hydrogen via ultraviolet absorption lines (Carruthers 1970), it became clear that clouds of molecular hydrogen were common. The first measurements showed that these clouds were cooler and denser than HI clouds (see e.g. Penzias 1975). Molecular hydrogen is dissociated by 11 ev photons, which are plentiful in the interstellar radiation field. For an H_2 cloud to survive, the column density must exceed 10^{20} cm^{-2} and the local density must exeed 100 cm^{-3} (see e.g. the discussion in Solomon and Rivolo 1987). Because of large extinction, the interiors of molecular clouds can be investigated only in the infrared or radio wavelength ranges. Since the H_2 molecule itself does not radiate for kinetic temperatures less than about 1000K (see the discussion of Drapatz 1985), the properties of molecular clouds must be deduced from measurements of molecules having permanent dipole moments, such as CO, CS, H_2CO, HC_3N or NH_3. The spatial distribution of such molecules appears not to be the same. In part these differences may be caused by excitation effects, or may be related to dissociation energies of the species. However in a number of instances there are large differences which are puzzling (see Swade 1989, Olano et al. 1988).

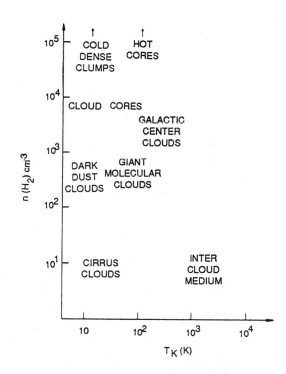

Figure 1: A classification of molecular clouds, on the basis of density and kinetic temperature. There may be structure on a finer scale in these objects

Molecular clouds show a variety of kinetic temperatures and densities. In Figure 1, we show a classification. These clouds are not relaxed objects. In nearly all cases, the linewidths exceed the sonic, or thermal, width. This is usually referred to as turbulence. Wilson and Walmsley (1989) have reviewed the causes of such behavior. There are two schools of thought: The turbulence is caused by the interaction with embedded stars (Norman and Silk 1980) and the motions are transferred from small to large scales, or the turbulence begins on large scales and trickles down to smaller scales (Larson 1981). Supersonic turbulence decays in a characteristic time which is short compared to most other time scales. Thus some replenishment of turbulence or other support mechanism is needed (see the discussion in Shu et al. 1987). A related observation is the following: the rate of star formation is much less than the ratio of molecular cloud mass and free-fall time (see e.g. the discussion in Zuckerman and Palmer 1974). On the large scale, these clouds are generally thought to be close to virial equilibrium. Thus there must be support against collapse. This support might be due to magnetic fields or embedded energy sources. The general opinion is that the number of embedded stars seems to be too small for such support. The strength of the magnetic field has been measured in only a small percentage of the clouds and because of geometry, a well ordered field cannot support a cloud against collapse in all directions. Across field lines there is support, but along field lines, there is no support. It may be argued that Alfven waves could provide support along field lines, but these waves must travel from inside to outside the cloud. This presupposes embedded sources. Myers and Goodman (1988) argue that the magnetic energy is comparable to the gravitational energy. This is a reasonable assumption, but has not been tested in many cases.

All molecular clouds have spatial structure on a finer scale. For example in Dark Dust Clouds, there are cloud cores which have H_2 densities which may be 100 times larger than the envelope. One could imagine that such structure is the result of a collapse process which will lead inexorably to the formation of a star. One fundamental limit to simple collapse is the differential angular momentum imparted by the rotation of the galaxy (see e.g. the discussion in Tscharnuter 1985).

WHERE TO SEARCH FOR PROTOSTARS

Lada (1987), Myers (1987) and Downes (1987) have reviewed observations related to star formation processes. As is generally accepted, in warmer Giant Molecular Clouds or GMCs, stars with higher masses are formed. In the cooler Dark Dust Clouds, lower mass stars are formed. The angular resolution of radio telescopes is limited by diffraction. For single radio telescope, the best angular resolution obtained is presently about 8 in. For aperture synthesis radio telescopes, higher angular resolutions are possible, but the limit for thermally excited gas seen in emission is sensitivity. At present this limit is about 1 in. for centimeter wavelength spectral line observations with the Very Large Array of the U.S. National Radio Astronomy Observatory. Thus to obtain the finest linear scales, one should concentrate on the closest molecular clouds. Dame et al. (1987) have collected a list of nearby molecular clouds. To this should be added the nearest molecular cloud, MBM12 which is 65 pc from the sun (see Pound et al. 1989). We show a version of their plot in Figure 2. In this compilation, one must differentiate between GMCs, where larger mass stars are produced and dark dust clouds, where only low mass stars are to be found. In Figure 3 we show linear and angular scales for the distances to MBM12, the Taurus and Ophiuchi dark clouds and the Orion molecular cloud, which resembles a GMC.

In the following, we list a few regions where photostars might be found. This list is not meant to be complete. The high density clumps in the Orion Molecular Cloud are promising, since the H_2 density probably exceeds 10^6 cm^{-3} and the kinetic temperature is about 30 K. There is no direct evidence of collapse, but rather of rotation (Harris et al. 1983, but see Wilson and Johnston 1989). For the maxima in the dust map of NGC2024, Mezger et al. (1988) have argued that collapse is likely, since the kinetic temperature is low and the density is very large. If the molecular cloud is in front of a continuum source, such as is the case for W3(OH), one could argue that infall is occuring (see Wilson et al. 1978, Reid et al. 1980, but also Welch and Marr 1988). This seems also to be true for the galactic

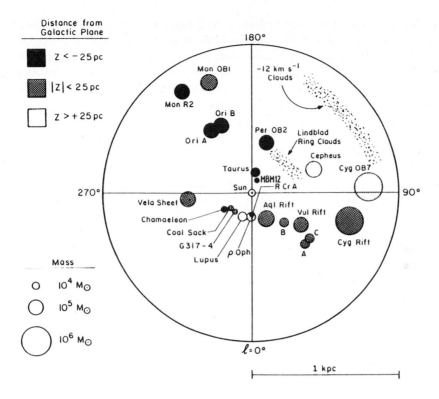

Figure 2: A schematic of the positions and identification of molecular clouds near the Sun (from Dame et al. 1987). To this has been added the location of MBM12 (see Pound et al. 1989).

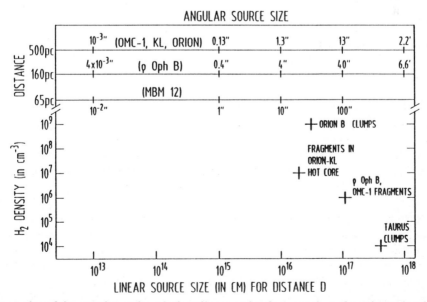

Figure 3: A plot of the angular and equivalent linear scales for a number of nearby molecular clouds (see also Fig. 2). On this, we have plotted a number of objects for which densities have been estimated. (See text for appropriate references.)

source G10.6-0.4 (Keto et al. 1988). As these molecular clouds are falling into a hot ionized medium, collapse may be initiated or enhanced, but this is not completely clear (Hunter et al. 1986). Also those regions where massive stars have already formed are stirred up, so that fine details of the motion cannot be followed. Searches in dark dust clouds may be more rewarding since gas motions in these regions are more quiescent. These searches will not lead to the discovery of high mass protostars, since those formed in such regions are less than a few solar masses.

THE IDENTIFICATION PROCESS

The most sophisticated approach to find a protostar using spectral lines is the method applied by Walker et al. (1986) to the cool object IRAS 1629A. This source is located in the Rho Ophiuchi dust cloud. These authors measured transitions of CS and the rarer isotope $C^{34}S$. The line shapes of these lines were quite different (see Figure 4 taken from Menten et al. 1987). The interpretation depends crucially on the behavior of optically thick and optically thin lines. The sharp negative-going feature in the center of the CS line is an indication that warmer gas is being absorbed by cooler gas along the line of sight. This is referred to as self-absorption. That the rarer isotope has a different lineshape is proof that self-absorption is affecting the shape of the normal isotope line. The more highly excited transition of CS shows a larger asymmetry than seen in the lower excitation line. This line arises from higher lying energy levels. Thus the line is formed in warmer material closer to the center of the cloud. The crucial fact here is that the blue shifted wing of the more highly excited CS line is more intense than the red shifted wing. The blue shifted photons arise in gas approaching us. This material is closer to the warmer center of the cloud. Because of the sense of the gas motions, we see the hottest blue

Figure 4: A collection of CS line profiles for the object IRAS1629A (taken from Menten et al. 1986). The differing shapes of the CS and $C^{34}S$ profiles are caused by self-absorption. The CS J=3-2 line has a significantly different shape from the CS J=2-1 line. (See also the J=5-4 data of Walker et al. 1986.)

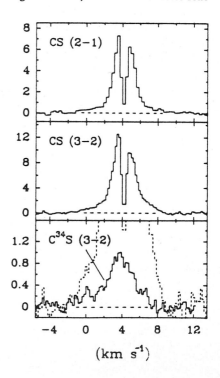

shifted material. For the red shifted gas, there is cooler material moving with a slightly lower velocity. Due to the thermal motions and turbulence in the gas, this cooler material absorbs photons from the warmer gas. In an expanding cloud, the roles of red and blue shifted maxima are reversed. In both cases, the relative intensities of red and blue peaks should be spatially symmetric about the cloud center. Using such analyses, one can differentiate between expansion and contraction. Higher resolution mapping by Menten et al. (1988) showed that the relative intensities of the red and blue shifted peaks changed systematically over the map. Their conclusion was that rotational motions or perhaps a bipolar outflow were more likely than collapse. In addition, they concluded that the narrow absorption feature is probably caused by unrelated foreground gas. The conclusion of this discussion is that IRAS1629A is not a collapsing object, but that the analysis used by Walker et al. (1986) is the most realistic method for deducing collapse motions in molecular clouds. Such investigations require the highest possible angular resolution.

References

Carruthers, G.: Astrophys.J. 161, L81.

Dame, T.M., Ungerchts, H., Cohen,R.S., de Geus, E.J., Grenier, I.A., May, J., Murphy, D., Nyman, L.-A., and Thaddeus, P.: 1987 Astrophys.J. 322, 706.

Downes, D., Genzel, R., Becklin, E.E., and Wynn-Williams, C.G.: 1981 Astrophys.J. 244, 869 Downes, D.: 1987 in Star Forming Regions, ed. M.Peimbert, J. Jugaku, Reidel, Dordrecht, p. 93.

Drapatz, S.: 1985 in Birth and Infancy of Stars, ed. R. Lucas, A. Omont, R. Stora, North-Holland, Amsterdam, p. 803.

Genzel, R., Reid, M.J., Moran, J.M., and Downes, D.: 1981 Astrophys.J. 244, 884.

Harris, A., Townes, C.H., Matsakis, D., and Palmer, P.: 1983 Astrophys.J., 265, L63.

Hunter, J.H., Sanford, M.T., Whitaker, R.W., and Klein, R.I.: 1986 Astrophys.J. 305, 309.

Keto, E.R., Ho, P.T.P., and Haschick, A.D.: 1988 Astrophys.J. 318, 712.

Larson, R.B.: 1974 Fund. Cosmic Phys. 1, 1.

Larson, R.B.: 1981 Mon.Not.Roy.Astron.Soc. 194, 809.

Lada, C.J.: 1987 in Star Forming Regions, ed. M.Peimbert, J. Jugaku, Reidel, Dordrecht, p. 1.

Mezger, P.G., Chini, R., Kreysa, E., Wink, J., and Salter, C.J.: 1988 Astron.Astrophys. 191, 44.

Myers, P.C.:1987 in Star Forming Regions, ed. M.Peimbert, J. Jugaku, Reidel, Dordrecht, p. 33.

Myers, P., and Goodman, A.: 1988 Astrophys.J. 326, L27.

Norman, C., and Silk, J.: 1980 Astrophys.J. 238, 158.

Olano, C., Walmsley, C.M., and Wilson, T.L.: 1988 Astron.Astrophys. 196, 194.

Penzias, A.A.: 1975 in Atomic and Molecular Physics and the Interstellar Matter, ed. R. Balian, P. Encrenaz, J. Lequeux, North Holland, Amsterdam, p. 373.

Pound, M.W., Bania, T.M., and Wilson, R.W.:Astrophys.J. 351, 165.

Zuckerman, B., and Palmer, P.: 1974 Ann.Rev.Astron.Astrophys. 12, 279.

Rank, D.M., Townes, C.H., and Welch, W.J.: 1971 Science 174, 1083.

Reid, M.J., Haschick, A.D., Burke, B.F., Moran, J.M., Johnston, K.J., and Swenson, G.W.: 1980 Astrophys. J. 239, 89.

Renzini, A., and Fusi Pecci, A.:1988 Ann.Rev.Astron.Astrophys. 26,199.

Shu, F., Adams, F.C., and Lizano, S.:1987 Ann.Rev.Astron.Astrophys. 25, 23.

Solomon, P., and Rivolo, A.R.: 1987 in The Galaxy, ed. G. Gilmore, B. Carswell, Reidel.

Swade, D.: 1989 Astrophys.J. 345, 828.

Tscharnuter, W.M.: 1985 in Birth and Infancy of Stars, ed. R. Lucas, A. Omont, R. Stora, North-Holland, Amsterdam, p. 601.

Walmsley, C.M., Hermsen, W., Henkel, C., Mauersberger, R., and Wilson, T.L.: 1988 Astron.Astrophys. 172, 311.

Walker, C.K, Lada, C.J., Young, E.T., Maloney, P.R., and Wilking, B.A.: 1986 Astrophys.J. 309, L47.

Welch, W.J., and Marr, J.: 1987 Astrophys.J. 317, L21.

Wilson, T.L., Serabyn, E., and Henkel, C.: 1986 Astron.Astrophys. 167, L17.

Wilson, T.L., and Walmsley, C.M.: 1989 Astron.Astrophys.Reviews 1, 141.

Wilson, T.L., Bieging, J.H., and Downes, D.: 1978 Astron.Astrophys. 63, 1.

Wilson, T.L., and Johnston, K.J.: 1989 Astrophys.J., 340, 894.

Wynn-Williams, C.G.: 1982 Ann. Rev. Astron.Astrophys. 20, 587.

Yorke, H.W.: 1985 in Birth and Infancy of Stars, ed. R. Lucas, A. Omont, R. Stora, North-Holland, Amsterdam, p. 645.

Discussion

A. SUCHKOV: Can you make a short summary of what are the major differences in the chemistry of the Earth and molecular clouds which you have mentioned in the beginning of your talk? Which may be the implications of these differences for the chemistry of planets around stars forming in molecular clouds and perhaps for other characteristics of these planets?

T. WILSON: As will be discussed by W. M. Irvine and R. D. Brown, the interstellar molecules are formed by ion-molecule reactions, at least in dark dust clouds. Some grain surface reactions are possible, but are proven only for the case of H_2. The longer chain molecules are unsaturated and hydrogen-rich. For star formation, the chemistry is crucial for promoting the cooling of collapsing clouds, and, if magnetic fields are present, allowing ions to slip by neutral (ambipolar) diffusion. The final states of planetary atmospheres are a result of final states of solar-system developments, but this is probably unrelated to molecular cloud chemistry.

S. ISOBE: On the contraction of gas clouds. Although you mentioned mostly at the phase of protostars, I would like to make a comment on YY Ori type stars (one of T Tau stars) which show inverse P Cygni type absorption lines and are at the pre-main sequence phase. From our observations, DRTau (one of YY Ori) shows variation of the inverse P Cygni-type absorption lines in a time period shorter than one day. This gives an infalling cloud with a size several times Jupiter's radius.

T.WILSON: I agree that observations indicated infall for some pre-main sequence stars. These are not protostars under my definition, since a substantial part of the energy production arises from nuclear reactions. In Ann. Rev. Astron. Astrophys., 20, 587 (1982) Wynn-Williams discusses the definition of "protostars."

J. TARTER: Tom - You need not wait until your 70th birthday and the mm array! Hat Creek will expand to nine elements and longer baseline within two years. Caltech will expand shortly thereafter and the Japanese are planning expansion for their mm array - in addition the IRAM interferometer is now ready to start. There is much useful work to be done in the near future.

T. WILSON: These interferometer systems are excellent, but may not allow measurements down to scales of 10^{14}-10^{15} cm in thermally or quasi-thermally excited lines. In Astron. Astrophys. Reviews 1, 141, (1989) Malcolm Walmsley and I give some estimates needed for future systems. These requirements will be difficult to meet.

THE HABITABILITY OF MARS-LIKE PLANETS AROUND MAIN SEQUENCE STARS

Wanda L. Davis and Laurance R. Doyle
SETI Institute

Dana E. Backman and Christopher P. McKay
NASA Ames Research Center, Moffett Field, CA 94035-1000

Paper Presented by W. L. Davis

ABSTRACT

We have developed a model to investigate the duration of conditions necessary for the origin of life on a Mars-like planet. We investigate various star-planet distances and a range of stellar spectral types in order to provide guidance for the Search for Extraterrestrial Intelligence (SETI). We consider planets suitable for the origin of life if they contain liquid water habitats for time periods comparable to the maximum time required for the origin of life on Earth. A Mars-like planet will differ from an Earth-like planet primarily because it will have no plate tectonic activity and therefore no long-term recycling of CO_2. We find that there is sufficient time for the origin of life on Mars-like planets around F, G, K, and M type stars. In addition, we find that Mars-like planets with an initial insolation less than the present insolation on Mars have the greatest potential for the origin of life.

1.   INTRODUCTION

Understanding the origin of life is of fundamental importance to the organization of a search for extraterrestrial intelligence (SETI). Investigating the range of stellar spectral types that can potentially support habitable planets, determining appropriate ranges of star-planet distance and calculating the duration of conditions suitable for the origin of life on these planets will be useful in guiding SETI efforts. Earth is the only planet that we know with life. Unfortunately, the early record of the origin and evolution of life has been erased by geological processes. Nonetheless, models based on Earth-like planets have yielded considerable insight into the question of habitable zones around main sequence stars (Whitmire et al., 1991).

Mars may also hold clues to the prevalence of habitable planets. Many of Mars' planetary features, especially the fluvial features, indicate that Mars and Earth may have had similar conditions early in their histories and life may have originated on Mars as it did on Earth (McKay and Stoker, 1989). Fortuitously, on Mars the record of these early events may still be preserved. If Earth and Mars had similar early environments and life arose on both planets, this would increase our confidence in the ubiquitousness of life under favorable planetary conditions.

In this paper we develop a model to investigate the suitability of Mars-like planets around main sequence stars as sites for the origin of life. Since life on Earth requires liquid water, we will use liquid water as the definitive characteristic of a planet suitable for the origin of life. If the duration of liquid water habitats on Mars-like planets is comparable to the time required for the origin of life on Earth, this would be of biological significance. A Mars-like planet will differ from an Earth-like planet primarily because it is smaller in size and thus will have no plate tectonics to allow for the recycling of CO_2 from carbonates.

Our model is adapted from the work of McKay and Davis (1991). Based upon a silicate weathering model (Pollack et al., 1987), McKay and Davis (1991) concluded that liquid water habitats could have persisted on Mars for up to 700 $\pm$ 300 million years _after_ mean temperatures fell below 0°C. This is comparable to the amount of time for life to inhabit Earth (Dott and Batten, 1988). In adapting this model we assumed that all Mars-like planets would have CO_2 atmospheres. Generally, the planet cools as the amount of CO_2 is reduced.

We examine the history of a Mars-like planet located such that its initial conditions are just like Mars' were 3.8 billion years ago, independent of specific stellar spectral type. We consider this condition plausible because terrestrial planets might coalesce around other stars in temperature zones similar to those zones where terrestrial planets formed in our solar nebula rather than at similar absolute radii. There is some evidence from theoretical models to support this (Lin and Papaloisou, 1985). Thus the stellar evolutionary rate is the variable factor in this model. Our results suggest that stars other than our sun have the potential for planets with persistent liquid water habitats. In addition, we have done a more general investigation of the effect on Mars-like planets of a range of stellar evolution rates for a range of initial insolation values (L_*/D^2), where L_* is the stellar luminosity in solar units, and D is the planetary orbital radius in astronomical units, AU. These results seem to suggest that the rate of stellar evolution would have minimal impact on the duration of liquid water habitats for F to M type stars.

2. MARS CLIMATE MODEL

Models of the climate on early Mars must take into account the decreased luminosity of the early sun and changes in the sun's insolation due to variations in orbital eccentricity, obliquity and latitude. It has been suggested that the early atmosphere of Mars was composed of CO_2 in sufficient quantities to warm the surface above freezing despite the lower luminosity of the early sun (Moroz and Mukhin, 1977). This is similar to suggestions for warming the early Earth (Owen et al., 1979; Kasting and Ackerman, 1986). Pollack et al. (1987) have developed a model for the formation of carbonates and loss of atmospheric CO_2 due to weathering of silicate materials as a function of the surface temperature and atmospheric pressure of CO_2. In addition they present results from a 1-D radiative convective model for the surface temperature increased by atmospheric CO_2 for several values of the solar luminosity.

McKay and Nedell (1988) have suggested that ice-covered lakes on early Mars could have contained liquid water long after the mean temperatures on the surface were below 0°C. Ice-covered lakes can exist when the mean annual temperature is below freezing as long as the yearly temperature _maximum_ is above freezing. Snowmelt and groundwater flowing into the lakes during these periods of peak temperature would have carried atmospheric CO_2 and cations leached from the surface materials. Dissolution resulting from freezing at the underside of the ice cover would result in the concentration of the ions and gases in the water column below the ice cover. As long as there is a source of ice to provide meltwater for the lakes, these liquid water habitats continue. A similar phenomenon is observed in the Antarctic lakes in the McMurdo Dry Valleys (Wharton et al., 1986,

1987). There, liquid water habitats are found under perennial ice covers and are a major site for microbial life in the valleys (Parker et al., 1982).

McKay and Davis (1991) computed the history of atmospheric CO_2 pressure, surface temperature and degree-days above freezing on early Mars based on the weathering parameterization of Pollack et al. (1987). Degree-days above freezing is defined as the integral of temperature versus time for all times that the temperature is above 0°C. Their model assumes that the yearly temperature profile is sinusoidal with an average value given by the mean annual temperature. The maximum (peak) temperatures reflect increased solar insolation---by a factor of 1.89---due to variations in eccentricity, obliquity, and latitude.

3.   STELLAR MODEL

In order to determine the duration of liquid water habitats for specific stellar spectral types, the stellar evolution model presented here follows the classic calculations of Iben (1967). It is assumed that proton-proton reactions dominate and that stars all start with population I abundances (Hydrogen = 93%, Helium = 6.9%, metals = 0.1%). It is assumed, in general, that convection will not play a significant role in energy transport for stars near 1 solar mass. The rate of hydrogen burning is assumed to increase linearly with the stars' age on the main sequence due to the increased core pressure necessary to sustain the denser helium layer being produced.

The model begins 0.7 billion years after the star's arrival on the main sequence (corresponding to 3.8 Gyr ago for our solar system), after active recycling of CO_2 on the Mars-like planet either by impacts or volcanic activity has ended (Pollack et al., 1987; Carr 1989). The star-planet distance is set to give the planet the same post-bombardment insolation that Mars had 3.8 Gyr ago. At this time we assume the CO_2 pressure is about 5 bars implying that temperatures are just at freezing. The history of pressure and temperature for a Mars-like planet, given the rate of evolution for specific stellar types is calculated. The rate of evolution of the star is defined as the ΔL_* per 4.5 Gyr in present solar units ($L_\odot$). For example, our sun would have a post-bombardment solar insolation of 0.75 $L_\odot$ and a rate of brightening of 0.3 $L_\odot$ per 4.5 Gyr. The global average temperature (Pollack et al., 1987) is used to compute the CO_2 loss rate; the decrease in CO_2 lowers the pressure and temperature which in turn decreases the CO_2 loss rate. Finally, based on the model of McKay and Davis (1991) the number of degree-days above freezing is determined. The duration of liquid water habitats for a Mars-like planet around these specific spectral types is presented below.

In addition we investigated a wider set of cases, with ΔL_* over the range of 0.0 to 1.0 solar luminosities per 4.5 Gyr and values of the insolation on a Mars-like planet ranging from 0.5 to 3 times the insolation on the present Mars. The insolation depends on the stellar luminosity and the star-planet distance. Our range of stellar evolution rates corresponds to stellar types later than about F9; we did not model cases where the star evolves off the main sequence in less than about 1 billion years. Note that stars that do evolve off the main sequence in less than 1 billion years represent less than about 3% of the stars in the galaxy (Allen, 1976). Thus we consider the vast majority of main sequence spectral types represented in our galaxy. The duration of liquid water habitats for a Mars-like planet around general spectral types is presented below.

4.   RESULTS AND DISCUSSION

Figure 1 shows the reduction of CO_2 atmospheric pressure over time for several main sequence stellar types. The length of the line denotes the duration of liquid water. For some spectral types there is between 0.5 and 1 bar of CO_2 remaining after the liquid water disappears. Figure 2 shows the corresponding temperature histories. For rapidly evolving stars the brightening of the star compensates at first for the loss of atmospheric pressure resulting in an initial warming of the planet with time. For example, the A5 curve shows that the rate of brightening for this type star provides an increase in temperature even though the planet is losing its atmosphere. In all cases the atmosphere eventually disappears and the temperature drops to the black body temperature. Note that A and early F type stars leave the main sequence so quickly that they have liquid water habitats for a period of time much shorter than the maximum time required for the origin of life on Earth---given the incomplete fossil record on Earth. Provided that the planet has the same post-bombardment insolation that Mars had 3.8 Gyr ago, the stars with slower rates of evolution permit a longer duration time for liquid water habitats. The F, G, K and M stars have liquid water habitats for about 600 to 800 million years <u>after</u> mean annual temperatures are below freezing. This would suggest that there is sufficient time for the origin of life on a Mars-like planet around stars of these types.

Figure 3 illustrates the duration of liquid water habitats as a function of 1) the initial effective insolation on a Mars-like planet around a zero-age main sequence star normalized to current Mars conditions $(1.52/D_{AU})^2 (L_*/L_\odot)$ and 2) the rate of increase in luminosity due to hydrogen conversion to helium on the main sequence. These results show that for an initial planetary insolation less than 1 the duration for a liquid water habitat is about 500 million to 1 billion years. For initial planetary insolations larger than about 1 the rate of atmospheric reduction is so rapid that the change in stellar

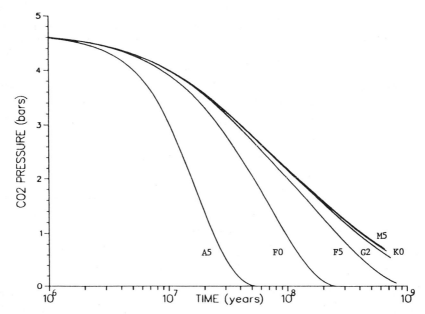

Figure 1: Atmospheric CO_2 pressure versus time on Mars-like planets for several spectral types. The length of the line denotes the duration of liquid water. Note that for some spectral types there is between 0.5 and 1 bar of CO_2 remaining after the liquid water disappears.

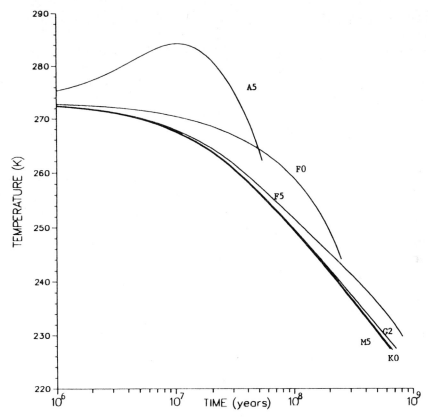

Figure 2: Temperature versus time for several spectral types. Note that in A-type stars the evolution of the star dominates the greenhouse effect. Thus, although the planet is losing its atmosphere, it temporarily increases in temperature because of the brightening of the star.

luminosity has virtually no effect over the range of rates considered here. Thus, these results would suggest that planet-star combinations resulting in an insolation less than about 1 have the greatest potential for maintaining liquid water habitats sufficient for the origin of life on Mars-like planets.

5.   SUMMARY

Recently there has been a reassessment of exobiological habitats (see e.g. Doyle and McKay, 1991; Whitmire et al., 1991). This has been motivated in part, by the renewed interest in the search for extraterrestrial intelligence. The current studies of the potential habitable zone around main sequence stars seem to suggest that our previous limits (e.g., Hart, 1979) may be extended. In addition, we have shown that planets other than Earth-like have the potential for the origin of life. These studies broaden our perspective about the likelihood of stellar systems favorable to life.

Acknowledgments

The authors would like to thank K. Zahnle, J. Kath, and B. Hogan for their helpful discussions. This research was supported by NASA Cooperative Agreement 2-336.

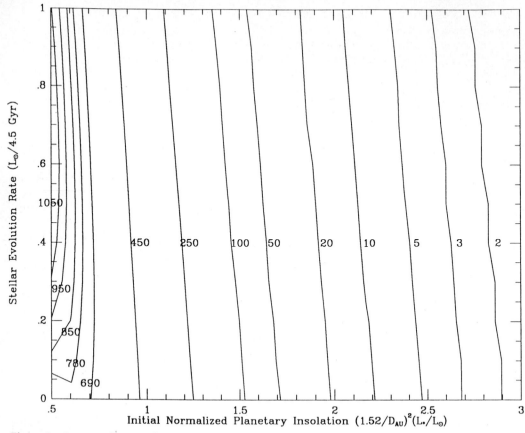

Figure 3: Contour diagram illustrating the duration of liquid water habitats (in millions of years) as a function of the initial planetary insolation in units of the current Mars insolation $(1.52/D_{AU})^2 (L_*/L_\odot)$ and the rate of increase in stellar luminosity due to hydrogen conversion to helium on the main sequence (in units of $L_\odot/4.5$ Gyr).

References

Allen, C.W, 1976, *Astrophysical Quantities* (Humanities Press Inc., New Jersey), 249.

Carr, M.H., Recharge of the early atmosphere of Mars by impact-induced release of CO_2, *Icarus* **79**, 311-327, 1989.

Cess, R.D., V. Ramanathan, and T. Owen, The martian paleoclimate and enhanced carbon dioxide, *Icarus* **411**, 159-165, 1980.

Dott, Jr., R. H., and R. L. Batten, 1988, *Evolution of the Earth* (McGraw-Hill, New York).

Doyle, L., and C.P. McKay, Exobiological Habitats: An overview, (this volume), 1991.

Hart, M.H., Habitable zones about main sequence stars, *Icarus* **37**, 351-357, 1979.

Iben, I., Stellar evolution VI. Evolution from main-sequence to the red-giant branch for stars of mass 1 $M_\odot$, 1.25 $M_\odot$, and 1.5 $M_\odot$. *Astrophysical Journal* **147**, 624, 1967.

Kasting, J.F., and T.P. Ackerman, Climatic consequences of very high CO_2 levels in Earth's early atmosphere, *Science* **234**, 1383-1385, 1986.

Lin, D.N.C., and J. Papaloizou, On the dynamical origin of the solar system, in *Protostars & Planets II*, D. Black and M.S. Matthews, eds., 981-1072, 1985.

McKay, C.P., G.A. Clow, R.A. Wharton, Jr., and S.W. Squyres, Thickness of ice on perennially frozen lakes, *Nature* **313**, 561-562, 1985.

McKay, C.P., and W.L. Davis, Duration of liquid water habitats on early Mars, *Icarus*, in press 1991.

McKay, C.P., and S.S. Nedell, Are there carbonate deposits in Valles Marineris, Mars?, *Icarus* **73**, 142-148, 1988.

McKay, C.P., and C.R. Stoker, The early environment and its evolution on Mars: implications for life, *Reviews of Geophysics* **27**, 189-214, 1989.

Moroz, V.I., and L.M. Mukhin, Early evolutionary stages in the atmosphere and climate of the terrestrial planets, *Cosmic Research* **15**, 774-791, 1977.

Owen, T., R.D. Cess, and V. Ramanathan, Enhanced CO_2 greenhouse to compensate for reduced solar luminosity on early Earth, *Nature* **277**, 640-642, 1979.

Parker, B.C., G.M. Simmons, Jr., K.G. Seaburg, D.D. Cathey, and F.C.T. Allnut, Comparative ecology of plankton communities in seven Antarctic oasis lakes, *J. Plankton Res.* **4**, 271-286, 1982.

Pollack, J.B., J.F. Kasting, S.M. Richardson, and K. Poliakoff, The case for a wet, warm climate on early Mars, *Icarus* **71**, 203-224, 1987.

Wharton, Jr., R.A., C.P. McKay, G.M. Simmons, Jr., and B.C. Parker, Oxygen budget of a perennially ice-covered Antarctic dry valley lake, *Limnol. Oceanogr.* **31**, 437-443, 1986.

Wharton, Jr., R.A., C.P. McKay, R.L. Mancinelli, and G.M. Simmons, Jr., Perennial N_2 supersaturation in an Antarctic lake, *Nature* **325**, 343-345, 1987.

Whitmire, D.P., R.T. Reynolds, and J. F. Kasting, Habitable Zones for Earth-like Planets around main sequence stars, (this volume), 1991.

PLANETARY ACCRETION DEBRIS AND IR EXCESSES IN OPEN STELLAR CLUSTERS

D. E. Backman, J. R. Stauffer, and F. C. Witteborn
NASA Ames Research Center
Moffett Field, CA 94035-1000

Poster Paper

We have initiated an infrared survey of 550 A-G main sequence stars in clusters with ages less than 1 billion years. This survey consists primarily of coadded IRAS survey photometry. The goals are: (1) to identify infrared excesses which might be attributable to protoplanetary accretion/collisional debris, and (2) to search for a possible trend with stellar age.

LITHIUM ABUNDANCES AS A PROBE OF THE EARLY EVOLUTION OF SOLAR-TYPE STARS

Eduardo L. Martin
Institute d'Astrophysique de Paris, 98bis Bd Arago, F-75014 Paris
Instituto de Astrofisica de Canarias, E-38200 La Laguna,Tenerife, Spain

Gibor Basri
University of California at Berkeley, CA 94720, USA

Claude Bertout
Institute d'Astrophysique de Paris, 98bis Bd Arago, F-75014 Paris

Paper Presented by E. L. Martin

Poster Paper

Lithium is presently a rare element on the Sun. Nevertheless its abundance in the presolar nebula was two orders of magnitude higher than solar as inferred from meteorites (Nichiporuk and Moore, 1974 Geo. Cosmochim., Acta 38). The depletion of lithium on the Sun presumably started when the star was only a few million years old, central temperature was rising because of contraction, light elements like lithium were easily destroyed deep in the interior and rapid mixing with surface material was ensured by convection. We are interested in examining observationally the relation of lithium depletion and other characteristics of low-mass pre-main sequence evolution.

Associated with star-forming regions, there are many very young solar type stars (M $\leq$2M$\odot$ and age $\leq$2 x 10^7 yr), generally known as T Tauri stars (TTS). In these stars lithium shows up as a strong photospheric line at λ 6707Å and is taken as a spectral signature of youth (Herbig, 1962 Adv. Astr. Astroph.,1). Our aim is to compute lithium abundances on the basis of high quality data which is now becoming available. The most interesting points will be: (1) confronting observed lithium abundances to theoretical isochrones and evolutionary models (cf. Profitt and Michaud 1989, Ap.J., 346), (2) comparing populations of young clusters of different ages and comparing stars very similar in all properties but one (mass, age or rotational velocity), and (3) trying to differentiate the evolution of very young stars from the evolution of their circumstellar material.

Proper study of the lithium resonance line must take into account variations due to non-photospheric continuum emission (possibly disk-star boundary layer) and thermal inhomogeneities similar to large sunspots and plages. Allowing for these effects leads to measurements of the true line strength of LiIλ6707, which in turn can be used to estimate the photospheric Li abundance. The major drawback when passing from line strengths to chemical abundances comes from uncertainties in the atmospheric parameters like effective temperature, surface gravity, microturbulent velocity and metallicity appropriate for TTS. Thus, the study of lithium abundances encourages basic research on the properties of TTS atmospheres. Full discussion of our results will be presented elsewhere.

LOW-AMPLITUDE STELLAR VELOCITY VARIATIONS: OTHER POSSIBILITIES

David W. Latham
Harvard-Smithsonian Center for Astrophysics, USA

Guillermo Torres
Harvard-Smithsonian Center for Astrophysics, USA and
Universidad Nacional de Cordoba, Argentina

Tsevi Mazeh
Wise Observatory, Tel Aviv University, Israel

Paper Presented by D. W. Latham

Poster Paper

Orbital motion due to a low-mass companion is not the only possible source for low-amplitude radial-velocity variation in a star. Other possibilities include stellar rotation combined with surface features, or pulsation.

The periodic velocity variation with amplitude 0.57 km s^{-1} for the solar-type dwarf star HD 114762 is almost certainly due to the presence of an unseen low-mass companion. The sharp peak in the power spectrum, shown in Fig. 1, suggests that the period has been strictly constant over the 56 cycles observed. No photometric variations have been observed for this star, supporting strongly the argument against pulsation or rotation with surface features as the source of the velocity variations. The period of 84 days implies a mass of 12 Jupiters if the orbit is viewed edge-on. The orbital solution based on 401 CfA velocities, shown in Fig. 2, has an eccentricity of 0.26, about five times the estimated error. We find no evidence for a second periodicity in these data.

Many of the late-type giants for which we have extensive velocity observations show quasi-periodic velocity variations at the level of 1 km s^{-1} or so. A good example is the M5 giant FS Comae. Intensive monitoring of its velocity over the past few months clearly demonstrates the variation with a roughly 55-day cycle, as shown in Fig. 3. However, the messy power spectrum, shown in Fig. 4, suggests that the velocity variation has not been strictly periodic over the 28 cycles observed. Furthermore, the star exhibits photometric variations, shown in Fig. 5, which are consistent with radial pulsation as the source for the velocity variations.

Watch out for giants when using low-amplitude velocity variations to search for evidence of low-mass companions! In many cases the velocity variation is probably not due to orbital motion.

Get simultaneous photometry! Light variations are a smoking gun for shooting down orbital motion. Simultaneous curves for velocity, color, and brightness can help distinguish between orbital motion, pulsation, and rotation with surface features. This is an ideal application for Automatic Photoelectric Telescopes.

Live a long time! Many cycles need to be monitored to check for strict periodicities. Star spots rarely last more than a few cycles, and even pulsation can drift in period. Orbital motion should be rock steady by comparison.

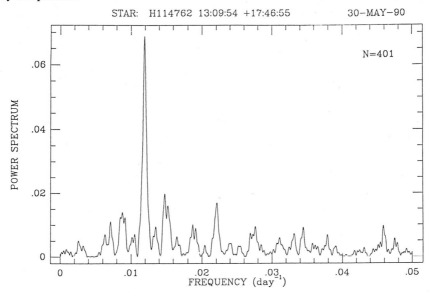

STAR: H114762 13:09:54 +17:46:55   30−MAY−90

Fig. 1. The sharp peak at 84 days in the power spectrum for the 401 CfA velocities of HD 114762 suggests that the variation has been strictly periodic over the 56 cycles observed.

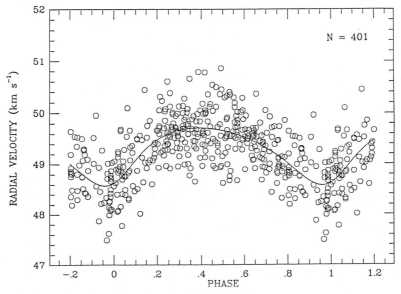

Fig. 2. The orbital solution for HD 114762 has an amplitude of 569 m s^{-1}. If the orbit is viewed edge-on, the mass of the unseen companion is 12 Jupiters.

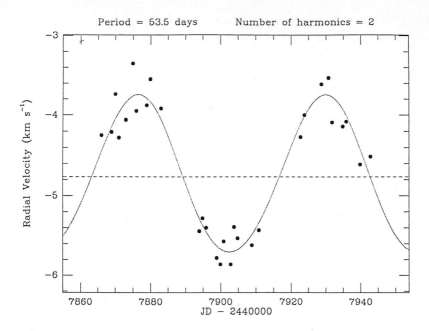

Fig. 3. Intensive monitoring of the velocity of FS Comae over the past few months clearly shows the variation with roughly 55-day cycle.

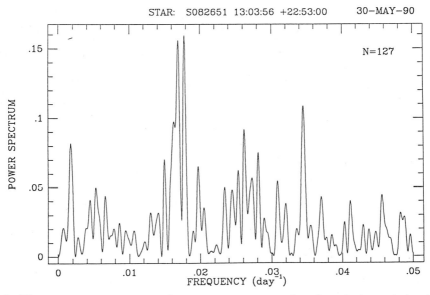

Fig. 4. The messy power spectrum for FS Comae suggests that the velocity variation has not been strictly periodic over the 28 cycles observed.

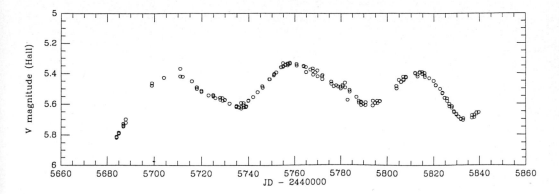

Fig. 5. The light curve for FS Comae is consistent with radial pulsation as the source of the velocity variations. The photometry was obtained for D. Hall with one of the Automatic Photoelectric Telescopes operated by the Fairborn Observatory at the Smithsonian's Whipple Observatory, Mt. Hopkins, Arizona.

II. ORGANIC AND PREBIOTIC EVOLUTION

SOME RECENT DEVELOPMENTS IN INTERSTELLAR CHEMISTRY

William M. Irvine
Five College Radio Astronomy Observatory
University of Massachusetts, USA

Åke Hjalmarson
Onsala Space Observatory, Chalmers University of Technology
Sweden

Masatoshi Ohishi
Department of Physics, Toyama University
Japan

Paper Presented by W. M. Irvine

ABSTRACT

The list of securely identified interstellar molecules has grown to about ninety, with molecular weights up to almost twice that of glycine. We review recent detections of new interstellar molecular species, which include the second phosphorus-containing molecule and new series of heterogeneously terminated carbon chain molecules. New observations allow us to refine estimates of the distribution of the "biogenic" elements (H, C, O, N, S, P) among various chemical reservoirs in interstellar clouds. Differences in relative chemical abundances among different clouds and even within given molecular clouds are providing new information on chemical processes, including exchange of material between the gas phase and grain mantles.

I.   INTRODUCTION

The dense interstellar clouds are the regions where stars, and presumably planets, form. The chemistry and physics of these regions is inextricably linked as they evolve toward star formation, since trace molecular constituents of the clouds provide crucial cooling agents which abet contraction, and chemical processes control the degree of ionization and hence the coupling to local magnetic fields. The properties and abundance of carbon play crucial roles in these processes. There is an increasing amount of evidence that the link between these interstellar molecules and the solar system itself is more than ancestral; isotopic ratios indicate that "primitive" solar system objects such as carbonaecous meteorites and comets contain interstellar material which has survived in molecular form (Zinner, 1988). These relatively unprocessed solar system objects may themselves be the major source of volatiles for the terrestrial planets, and some complex organic molecules have certainly been delivered to the Earth by meteorite and cometary impacts (Anders, 1989). The study of interstellar chemistry is thus important to bioastronomy for a number of reasons.

The complexity of the chemistry of dense molecular clouds is being increasingly revealed, with some ninety molecular species currently identified in the gas phase, containing the elements H, C, N, O, S, Si, Cl, P, Na, Al, K, and F (we include species found in extended, expanding circumstellar envelopes but not molecules observed in stellar photospheres). The securely identified molecules include species with up to thirteen atoms and weights up to 147 AMU, among which are many examples which would be unstable by terrestrial standards (radicals, positive ions, and highly unsaturated species; e.g., Irvine, Ohishi, and Kaifu, 1990; Irvine, 1990a). We review here some recent progress in the field, citing for convenience primarily our own previous reviews, rather than attempting an exhaustive survey.

II.   RECENTLY DETECTED INTERSTELLAR MOLECULES

The second phosphorus-containing interstellar species, CP, has recently been detected, in the expanding envelope of the evolved star IRC+10216. Thus CP joins PN, previously observed in the massive star-forming region Sgr B2 near the Galactic Center (Guélin et al., 1990).

Somewhat earlier (Yamamoto et al., 1987a), the second cyclic hydrocarbon was observed, when c-C_3H was detected in the cold, dark cloud TMC-1. A linear isomer of C_3H had previously been found in this source.

A variety of carbon chains have been known for some time in the interstellar medium, with their abundance being particularly large in quiescent clouds in Taurus such as TMC-1 (Irvine, 1990b). All of the hydrocarbon radicals C_nH (n = 0,..., 6) have now been detected. The cyanopolyynes, $H(C \equiv C)_{2n}CN$(n = 0,..., 5) and some of their methylated relatives are also well known. More recently, members of the sulfur-terminated series C_nS (n = 1, 2, 3) have been observed (Saito et al., 1987; Yamamoto et al., 1987b), as have the silicon-terminated species CSi, cyclic-C_2Si, and C_4Si (e.g., Ohishi et al., 1989; Cernicharo et al., 1989). The oxygen-terminated series is also present, with CO and C_3O well established, and recent tentative identifications made for C_2O and C_5O (Turner, 1990; Ohishi, 1990b). The linear species C_3 and C_5 have been observed by infrared absorption in the envelope of the star IRC+10216 (Bernath, Hinkle, and Keady, 1989), and thus join C_2, which is known from ultraviolet observations of diffuse and translucent clouds. Very recently the cumulenes $H_2C=C=C$ and $H_2C=C=C=C$ have been identified in both the laboratory and in space (Cernicharo et al., 1990b; Killian et al., 1990; S. Saito, private communication).

The heaviest non-linear radical identified to date, CH_2CN, has been found in both warm and cold molecular clouds. The abundance appears to exhibit a temperature dependence, moreover, such that the abundance ratio CH_2CN/CH_3CN is greater than 1 in cold clouds and decreases significantly in warmer sources (Irvine et al., 1988; Turner et al., 1990).

Some of the strongest evidence for the preservation of interstellar molecular material in the present Solar System comes from the very large isotopic fractionation observed for hydrogen in certain fractions of carbonaceous chondrites. On the interstellar side, a recent example of such fractionation is provided by the small cyclic hydrocarbon C_3H_2. Comparisons of the normal isotopic form, the ^{13}C form and the deuterated form indicate that the C_3HD/C_3H_2 ratio is some ten thousand times the cosmic D/H ratio. Although this pushes theories to the limit, it can probably be understood in terms of gas phase reactions in very cold interstellar clouds (Bell et al., 1988).

III.   RESERVOIRS FOR THE VOLATILE ELEMENTS IN INTERSTELLAR CLOUDS

A fundamental question in interstellar chemistry is the distribution for each element among possible chemical reservoirs. For hydrogen, there is no question that the major reservoir in dense interstellar clouds is molecular hydrogen, H_2. For the other volatile elements, however, the situation is much less clear. This problem has been considered recently by Irvine and Knacke (1989), to whom we refer for a discussion of the difficulties involved in determining these distributions, as well as the results as of that date. We shall concentrate here on some more recent information. It is of interest that, although in a chemical sense the interstellar medium is very strongly reducing, the kinetics at low temperature make is difficult to hydrogenate the species that are present. As a result, many of the major trace constituents (after H_2) are quite unsaturated or are oxidized.

In the case of carbon, for example, the principal gas phase constituent is CO, which contains approximately 10% of the cosmic abundance of carbon. Symmetric hydrocarbons such as CH_4 and HC_2H have no permanent electric dipole moment and are thus very difficult to observe in emission. Recent measurements against background embedded infrared sources are consistent with the theoretical models which predict that the abundance of both these hydrocarbons should be much less than that of CO (e.g., Lacy et al., 1989). Carbon dioxide likewise has no electric dipole moment, but its abundance may be inferred from that of its protonated ion, HOCO+. In general throughout the Galactic disk, including the neighborhood of the Sun, the abundance of CO_2 appears to be much less than that of CO. Although the result is somewhat model-dependent, recent observations suggest that in molecular clouds near the Galactic Center the abundance of CO_2 may be much higher than elsewhere in the Galaxy (Minh et al., 1990a), perhaps as a result of cosmic ray processing of icy grain mantles. Atomic carbon (CI) has proven to be much more abundant than predicted in dense interstellar clouds, but still appears to contain not more than 10% or 20% of the carbon present in CO.

In contrast, there is a variety of evidence indicating that complex organic compounds form a major constituent of the interstellar grains. Spectral features range from the ultraviolet to the infrared, and include both emission and absorption lines. The precise chemical form for this organic material is still a matter of dispute, but suggestions include polycyclic aromatic hydrocarbon molecules, hydrogenated amorphous carbon, graphite, and material analogous to that referred to by geologists as kerogen. Solid CO and CO_2 are also present. In sum, it appears that the bulk of the carbon in dense interstellar clouds forms part of the grains.

There is some evidence that the principal nitrogen reservoir is molecular nitrogen, N_2. The difficulty in making an unequivocal determination lies partly in the lack of a permanent electric dipole moment for this homonuclear molecule. As for CO_2, the abundance of N_2 must be estimated from that of its protonated ion. N_2H+ emission appears to be optically thick in many sources, however, complicating the abundance determinations. The rare ^{15}N isotope has been observed in one cloud, where the results support the theoretical conclusion that most nitrogen is present as N_2. Chemical models predict that the next most important nitrogen repositories will be atomic nitrogen and NO. No relevant observations yet exist for atomic nitrogen. In the case of nitric oxide (NO), results have recently become available on the abundance in both quiescent molecular clouds and regions of active star formation. The results support the theoretical models, indicating that in quiescent clouds $NO/N_2 << 1$ (McGonagle et al., 1990; Ziurys et al., 1990). Although ammonia is widely present in interstellar clouds, and the ammonia abundance is significantly enhanced in regions heated by active star formation, the data indicate that $NH_3/N_2 << 1$. Another widespread nitrogen-containing interstellar molecule is HCN, but it does not appear to contain a major fraction of the cosmic abundance of nitrogen.

Determination of the major reservoirs for oxygen is plagued by the difficulty in observing three possible major chemical forms, O_2, H_2O, and atomic oxygen. In the case of O_2, there are no electric dipole transitions, but strong magnetic dipole transitions exist at radio frequencies and, in fact, make the terrestrial atmosphere opaque. Searches for the rare ^{18}O isotope have been carried out, but only upper limits have been set. Measurement of the abundance of O_2 will apparently have to wait for the deployment of space-borne equipment such as that proposed for the Submillimeter Wave Satellite (SWAS). Emission lines from interstellar water have been observed with ground-based radio telescopes and from the Kuiper Airborne Observatory, but these are from fairly high-lying energy levels and contain substantial maser emission, making abundance determinations difficult. A tentative detection of $H_2\,^{18}O$ has been made, and there may be a thermal component of some of the maser lines in warmer molecular clouds; these data are consistent with theoretical predictions that $H_2O/CO < 1$ (Cernicharno et al., 1990a). Accurate measurements of the water abundance must wait, as in the case of O_2, for appropriate satellite measurements. In hot gas, such as the presumably shock-heated regions in Orion, the fine structure transition of atomic oxygen at 63 μm has been observed, and the abundance of atomic oxygen may be comparable to that of CO. No such information is available for cold clouds, although theoretical models predict that atomic oxygen could have a few times the abundance by number of CO. Carbon monoxide accounts, however, only for 6% or so of the cosmic abundance of oxygen in dense clouds.

More of the oxygen is almost certainly contained in the interstellar grains. Between 10% and 20% is probably in refractory cores as silicates, while another 10% to 15% in dense clouds may be present as H_2O ice. Adding this complement to that in the gas phase, however, still leaves about half the oxygen unaccounted for, at least according to some inventories.

Although, as indicated above, two phosphorus-containing molecules have now been detected astronomically, neither represents a major reservoir for phosphorus in dense interstellar clouds. Theoretical predictions suggest that the major gas phase repository may be atomic phosphorus. The principal phosphorus-containing molecule is predicted to be PO, but thus far observations have failed to confirm this prediction. Note that phosphine (PH_3) is not readily formed under the low temperature, low density conditions in interstellar clouds. Most phosphorus may be contained in the grains.

A number of sulfur-containing molecules have been observed in molecular clouds, including SO, SO_2, and CS, but none of these appear to be major sulfur reservoirs in comparison with the cosmic abundance. Recently, extensive observations of H_2S have been carried out for both cold, quiescent clouds and warmer regions where star formation is taking place (e.g., Figure 1). Note that there are suggestions that H_2S may be formed primarily on grain surfaces, because the reactions of both atomic sulfur and S+ with H_2 are endothermic. The observed abundances have the interesting result that, although H_2S is detected in cold, quiescent clouds, its abundance is enhanced perhaps one thousand-fold in regions such as the Orion "hot core", where temperatures and densities are considerably higher. It is suggested that this may result from evaporation of icy grain mantles, a mechanism that has also been put forward to explain the strongly enhanced abundances of NH_3, CH_3OH, and HDO (Minh et al., 1990b). Even in this case, however, the bulk of the sulfur is probably present in or on the grains.

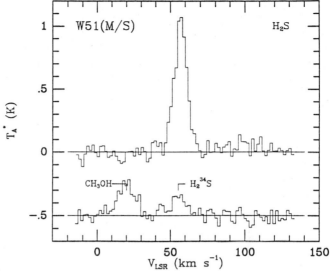

Figure 1: The 1(1,0)-1(0,1) transitions of hydrogen sulfide and its rare 34-S isotopic species observed towards the W51 molecular cloud with the FCRAO 14m telescope (Minh, 1990).

IV.   ABUNDANCE DIFFERENCES AND GRADIENTS

Since astronomy is an observational rather than an experimental science, chemical theories may be tested by comparing their predictions for regions with different characteristics, including physical properties such as temperature and density as well as other factors such as age or initial conditions. It is becoming increasingly well documented, for example, that striking differences in chemical abundances exist between quiescent molecular clouds and regions of massive star formation, presumably as a result of the effects of intense radiation fields and shock waves on both the gas and embedded grains in the hotter regions.   The best studied example is probably the core of the Orion molecular cloud, where some four regions, each exhibiting a characteristic chemistry, have been identified (Johansson et al., 1984; Irvine et al., 1987; Blake et al., 1987).   Such data are consistent with phenomena such as evaporation of icy grain mantles and perhaps sputtering of the refractory grain components in the vicinity of star formation, and the conversion from ion-molecule chemistry to a more thermodynamically controlled chemistry in warmer, denser regions. Nonetheless, even in the relatively nearby Orion region, perplexing puzzles remain, including anomalously high abundances for certain isomers relative to others (Millar and Herbst, 1990).

Significant gradients in chemical abundance also appear to be present in the giant molecular cloud near the Galactic Center, Sgr B2. Following some pioneer studies by Goldsmith et al. (1987), spectral surveys of multiple positions in Sgr B2 are currently underway at the Swedish-European Sub-millimeter Telescope in Chile under the direction of a group at the Onsala Space Observatory (Sweden), and likewise at the Nobeyama Radio Observatory in Japan. The SEST observations are in the 1mm wave-length band, while those at Nobeyama include data in the 7mm as well as the 3mm bands. Some sulfur species such as SO peak at the so-called M position, other molecules like CH_3CN in high energy transitions are strongest at the N position, while a number of molecules exhibit their strongest emission for low energy lines at the LP position (Ohishi, 1990a; Hjalmarson, 1990; cf. Fig. 2). Much more data needs to be analyzed in order to interpret these results in terms of local physical and chemical conditions.

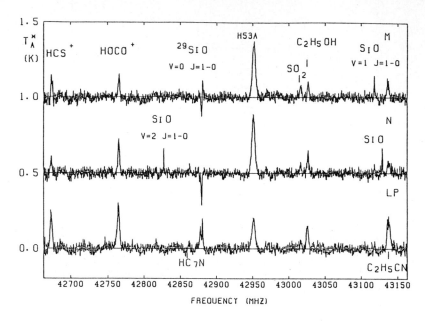

Figure 2: Spectra in the 7mm band at three locations (M, N, and LP) in the Sgr B2 molecular cloud taken with the Nobeyama Radio Observatory 45m telescope (Ohishi, 1990a).

In fact, abundance gradients exist even in quiescent clouds which appear to lack embedded stars or protostars. Well studied examples include TMC-1, where there appears to exist a gradient in the relative abundance of saturated molecules like ammonia and H_2S relative to that of the cyanopolyynes (e.g., Minh, Irvine, and Ziurys, 1989); and L134N where Swade (1989) has shown that quite different distributions exist for a number of molecules that have been mapped. In particular, classic "density tracers" used by molecular astrophysicists, such as NH_3 and HCO+, do not appear to define the same regions, suggesting the existence of real chemical differences. The trend is also apparent from a comparison of the distributions of sulfur-containing molecules such as SO_2 and SO with that of NH_3 or HCO+. Although Swade speculated that such differences might result from differential depletion of oxygen relative to carbon onto the grains, perhaps as H_2O ice, it is fair to say that these chemical gradients in cold clouds are not understood at present.

To end on a hopeful note, observatories such as FCRAO are now bringing into operation receiver systems able to observe multiple pixels simultaneously (Erickson et al., 1990). This will allow images to be built up of molecular clouds in the emission from several molecular transitions much more rapidly than heretofore, providing a data base which can be used to try to solve these problems.

This research was supported in part by NASA grant NAGW-436, and NSF grant INT-8915252 under the U.S.-Japan Cooperative Science Program.

References

Anders, E. 1989, Nature, 342, 255.

Bell, M.B., Avery, L.W., Matthews, H.E., Feldman, P.A., Watson, J.K.G., Madden, S.C., and Irvine, W.M. 1988, Astrophys. J., 326, 924.

Bernath, P.F., Hinkle, K.H., and Keady, J.J. 1989, Science, 244, 562.

Blake, G.A., Sutton, E.C., Masson, C.R., and Phillips, T.G. 1987, Astrophys. J., 315, 621.

Cernicharo, J., Gottlieb, C.A., Guélin, M., Thaddeus, P., and Vrtilek, J.M. 1989, Astrophys. J. (Letters), 341, L25.

Cernicharo, J., Thum, C., Hein, H., John, D., Garcia, P., and Mattioco, F. 1990a, Astron. Astrophys., 231, L15.

Cernicharo, J., Gottlieb, C.A., Guélin, M., Killian, T.C., Paubert, G., Thaddeus, P., and Vrtilek, J.M. 1991, Astrophys. J. (Letters), 368, L39.

Erickson, N.R., Goldsmith, P.F., Predmore, C.R., Novak, G.A., and Viscuso, P.J. 1990, in IEEE MTT-S Internat. Microwave Symp., in press.

Goldsmith, P.F., Snell, R.L., Hasegawa, T., and Ukita, N. 1987, Astrophys. J., 314, 525.

Guélin, M., Cernicharo, J., Paubert, G., and Turner, B.E. 1990, Astron. Astrophys., 230, L9.

Hjalmarson, Å. 1990, private communication.

Irvine, W.M., Goldsmith, P.F., and Hjalmarson, Å. 1987, in Interstellar Processes, D. Hollenbach and H. Thronson, eds. Dordrecht: D. Reidel, 561.

Irvine, W.M. et al. 1988, Astrophys. J. (Letters), 334, L107.

Irvine, W.M. and Knacke, R.F. 1989, in Origin and Evolution of Planetary and Satellite Atmospheres, S.K. Atreya, J.B. Pollack, and M.S. Matthews, eds. Tucson: U. Arizona, 3.

Irvine, W.M. 1990a, in Chemistry in Space, J.M. Greenberg and V. Pironello, eds. Dordrecht: Reidel, in press.

Irvine, W.M. 1990b, in Chemistry and Spectroscopy of Interstellar Molecules, N. Kaifu, ed. Tokyo University Press, in press.

Irvine, W.M., Ohishi, M., and Kaifu, N. 1990, Icarus, in press.

Johansson, L.E.B. et al. 1984, Astron. Astrophys., 130, 227.

Killian, T.C., Vrtilek, J.M., Gottlieb, C.A., Gottlieb, E.W., and Thaddeus, P. 1990, Astrophys. J. (Letters), 365, L89.

Lacy, J.H., Evans, N.J., Achtermann, J.M., Bruce, D.E., Arens, J.F., and Carr, J.S. 1989, Astrophys. J., 342, L43.

McGonagle, D., Ziurys, L.M., Irvine, W.M., and Minh, Y.C. 1990, Astrophys. J., 359, 121.

Millar, T.J. and Herbst, E. 1990, Astrophys. J., in press.

Minh, Y.C., Irvine, W.M., and Ziurys, L.M. 1989, Astrophys. J. (Letters), 345, L63.

Minh, Y.C., Brewer, M.K., Irvine, W.M., Friberg, P., and Johansson, L.E.B. 1990a, Astron. Astrophys., in press.

Minh, Y.C., Ziurys, L.M., Irvine, W.M., and McGonagle, D. 1990b, Astrophys. J., 360, 136.

Minh, Y.C. 1990, Ph.D. Dissertation, University of Massachusetts.

Ohishi, M. et al. 1989, Astrophys. J., 345, L83.

Ohishi, M. 1990a, in Chemistry and Spectroscopy of Interstellar Molecules, N. Kaifu, ed. Tokyo University Press, in press.

Ohishi, M. 1990b, in Submillimeter Astronomy, G.D. Watt and A.S. Webster, eds. Dordrecht: Kluwer, 113.

Saito, S., Kawaguchi, K., Yamamoto, S., Ohishi, M., Suzuki, H., and Kaifu, N. 1987, Astrophys. J. (Letters), 317, L115.

Swade, D.A. 1989, Astrophys. J., 345, 828.

Turner, B.E., Friberg, P., Irvine, W.M., Saito, S., and Yamamoto, S. 1990, Astrophys. J., 355, 546.

Turner, B.E. 1990, in Chemistry and Spectroscopy of Interstellar Molecules, N. Kaifu, ed. Tokyo University Press, in press.

Yamamoto, S., Saito, S., Ohishi, M., Suzuki, H., Ishikawa, S.I., Kaifu, N., and Murakami, A. 1987a, Astrophys. J. (Letters), 322, L55.

Yamamoto, S., Saito, S., Kawaguchi, K., Kaifu, N., Suzuki, H., and Ohishi, M. 1987b, Astrophys. J. (Letters), 317, L119.

Zinner, E. 1988, in Meteorites and the Early Solar System, J.F. Kerridge and M.S. Matthews, eds., Tucson, U. Arizona, 956.

Ziurys, L.M., McGonagle, D., and Minh, Y.C. 1990, Astrophys. J., in press.

Discussion

A. SUCHOV: Is there any evidence for systematic differences in the chemistry of the Galactic disk gas clouds and the high-velocity clouds falling down onto the disk?

W. IRVINE: My impressions are that the high velocity clouds may be a heterogeneous set of objects and that the densities are typically too low to expect much formation of molecules. The question of the metallicity of these clouds is clearly of basic importance for theories of their origin, and deep searches for molecules might be useful in this regard. Cf. IAU Colloq. 120 (LNP 350, Springer-Verlag).

L. DOYLE: What about detecting molecular chirality?

W. IRVINE: The simplest potentially chiral molecules that are related to known interstellar species would be CH_3CHDOH and CH_3CHDCN. CH_3CH_2OH has been detected in two or three warm molecular clouds where ratios of deuterated-to-normal species are not high. CH_3CH_2CN is somewhat more widespread; for example, it is detected in the Orion "hot core", the closest region of massive star formation to the Sun. However, even if the CH_3CHDCN/CH_3CH_2CN ratio were enhanced as much as the HDO/H_2O ratio (e.g.; ~100X), it would be very difficult to detect CH_3CHDCN. Even after detection, it would be very much more difficult to demonstrate a non-racemic mixture.

R. BROWN: When comparing observed interstellar abundance contours it may sometimes be difficult to make adequate allowance for the differing excitation requirements for different lines of different interstellar species. Thus comparing SiO with other species may be uncertain if any SiO masering is involved.

W. IRVINE: I agree that to accurately measure abundances is difficult. Multi-transition maps are needed for each molecule to really determine the excitation and to estimate transition optical depths for each species. In the data I showed for SiO, however, the emission was presumably thermal and not maser emission (the transition mapped was $v = O$, $J = 2$-1, whereas the strong maser lines occur in excited vibrational states).

C. MATTHEWS: What is the origin of interstellar dust?

W. IRVINE: The cores of the dust grains are thought to be formed in environments such as the envelopes of evolved stars, novae, and supernovae. In dense interstellar clouds these cores acquire "icy" mantles by accretion of volatiles, which may then be further precessed by irradiation by cosmic rays or UV photons.

A. LEGER: A place for grains to grow is dense cold clouds where gaseous species condense on grains nuclei and form icy mantles.

W. IRVINE: I agree, see question by Matthews.

T. WILSON: You mention "evolution" in trying to explain the differences in the maps of different species. Could you give some specifics? Can you rule out excitation effects in regard to the L134 differences?

W. IRVINE: I meant evolution primarily as a possible explanation for differences among different clouds, although one could imagine different regions in the same cloud evolving at different rates because (e.g.) of initial density or temperature gradients.

Swade has suggested that differences in the gas phase O/C ratio in L134N could result from differential depletion of H_2O ice as a function of local temperature.

THE FORMATION OF LONG CHAINS OF CARBON ATOMS IN SPACE

Ronald D. Brown, Dinah M. Cragg, and Ryan P. A. Bettens
Monash University, Victoria, Australia

Paper Presented by R. D. Brown

INTRODUCTION

Theoretical models of the chemistry of interstellar molecular clouds have been based mainly on gas-phase reactions proceeding without activation energy, i.e., at the collision rate for the reactants. The majority of reactions included in such models have been between positive ions (atomic or molecular) and uncharged molecules (Herbst and Klemperer 1973, Watson 1973). However some other types of reactions have been included, particularly formation of molecular hydrogen from atoms on grain surfaces. Although several alternative processes have been proposed to account for the formation of interstellar molecules, catalytic formation on grain surfaces being a popular alternative (for a recent gas-grain model incorporating both gas-phase and grain-surface chemistry, see Brown and Charnley 1990 and references therein), the gas-phase ion-molecule scheme has proved very successful in accounting for the observed abundances of all small molecular species. It is so successful that it leaves little room for other processes to make significant contributions of molecules to molecular clouds.

One or two noticeable uncertainties remain however. There are difficulties in accounting for the appreciable abundances of molecules containing long strings of carbon atoms, as in the cyanopolyacetylenes, HC_nN, and the hydrocarbons, C_nH. In some recent work (Brown, Cragg, and Bettens 1990; Herbst and Leung 1989; Herbst and Leung 1990; Millar, Leung, and Herbst 1987) it was possible to produce enough or even over-produce complex molecules in comparison with observation for times earlier than the time taken to reach "steady state" ($>10^7$ yr) with various reasonable, but quite uncertain, assumptions about the values of critical rate constants. Here we shall summarize previous deliberations on the problem of long carbon chains and outline recent investigations of whether some new gas-phase processes, involving "hot-ions" reactions, could explain the observed abundances.

THE ION-MOLECULE SCHEME

So far as the build-up of carbon chains is concerned, the processes of insertion, condensation, and radiative association are of prime concern. Examples are:

insertion: this can involve either C or C^+:

$$C^+ + CH_2 \longrightarrow C_2H^+ + H$$

$$C + CH_5^+ \longrightarrow C_2H_3^+ + H_2$$

condensation:

$$CH_3^+ + CH_2 \rightarrow C_2H_3^+ + H_2$$

radiative association:

$$C_nH_m^+ + C_iH_j \rightarrow C_{n+i}H_{m+j}^+ + h\nu$$

Other processes that affect the buildup of hydrocarbon chains are electron recombination reactions and hydrogen-atom abstraction reactions (HAARs).

electron recombination:

$$C_2H_3^+ + e \rightarrow C_2H_2 + H$$

hydrogen-atom abstraction reaction:

$$C_2H_2 + H_2 \rightarrow C_2H_3^+ + H$$

Processes of these kinds when included in the general gas-phase model for interstellar chemistry, lead to predictions of observable amounts of hydrocarbons with short chains of C atoms (see Fig. 1a). But chains of five or six carbons are not predicted to result in observable abundances, unless rather arbitrary assumptions about selected neutral-neutral reactions are made, because there is evidence that the HAARs that are important for the formation of such species involve activation energy, or are endothermic, thus not proceeding at rates high enough to generate observable abundances.

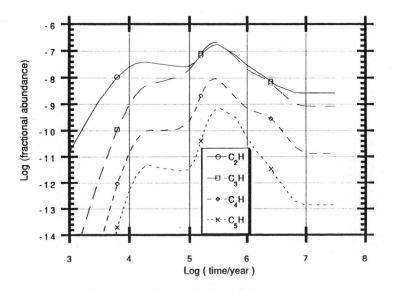

Fig. 1a

Some additional processes have been proposed as responsible for buildup of carbon chains, e.g., Huntress (1977) included reactions such as:

$$C_2H_2^+ + CH_4 \longrightarrow C_3H_{5+} + H$$

$$C_2H_2^+ + C_2H_2^+ \longrightarrow C_4H_{3+} + H$$

and Suzuki (1983) proposed atomic carbon ion insertion reactions:

$$C^+ + C_n \longrightarrow C_{n+1}^+ + hv$$

but these reactions are not efficient in producing observable abundances of long chain molecules because the reactants are present in such low abundances that too little product is produced. Suzuki's reactions are also rendered ineffective because of competing reactions with atomic O (one kind of neutral-neutral reaction the rate of which under interstellar conditions is quite uncertain).

We are therefore forced to conclude that such long chain carbon molecules are produced by processes that are not gas-phase chemical reactions (i.e., they may come from processes involving dust grains, or from fragmentation of carbonaceous microparticles of dust) or else the uncertain processes previously mentioned, discussed by Millar, Leung, and Herbst (1987) and Herbst and Leung (1990), proceed at an appropriate rate.

There have been suggestions (Herbst, Adams, and Smith 1984) that if the hydrocarbon ion involved in a HAAR is "hot," i.e., has been generated with considerable kinetic energy, this may be sufficient to overcome the endothermicity. However, this proposition had not previously been incorporated in gas-phase models. When we proceeded to do so (Brown, Cragg, and Bettens 1990), we found that such "hot-ion" processes could account (See Fig. 1b) for observed abundances of the larger hydrocarbon ions (and some other long chain compounds), but only if the activation energies

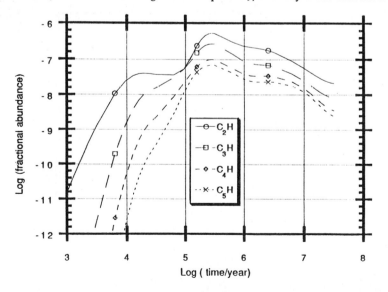

Fig. 1b

of the HAARs are lower than current estimates based on laboratory data (4 - 8 kJ mol^{-1}). It should also be noted that tunneling might be responsible for the HAARs proceeding at an efficient rate. This could mean that relatively large barriers at room temperature might be tunneled through at low temperatures. The corresponding rate constants would have to be within about two powers of ten of the collision rates.

The present position of the theory of interstellar chemistry for long chain molecules is therefore that the only suggestion so far advanced that might explain their observed abundances involves "hot-ion" reactions or allowing selected neutral-neutral reactions to proceed at the requisite rates. If hot-ion reactions are responsible, then it means that current estimates of the activation energies for the corresponding HAARs are too high. Unless one of these possibilities proves to be the underlying cause of chain building, it appears that we may have to look for other types of processes, presumably involving interstellar grains, to explain our observance of long chain molecules in space.

References

Brown, P. D., and Charnley, S. B., 1990. Mon. Not. R. As. Soc, 244, 432.
Brown, R. D., Cragg, D. M., and Bettens, R. P. A., 1990. Mon. Not. R. As. Soc, 245, 623.
Herbst, E., Adams, N. G., and Smith, D., 1984. Ap. J., 285, 618.
Herbst, E., and Klemperer, W., 1973. Ap. J., 185, 505.
Herbst, E., and Leung, C. M., 1990. Astrophys. J. Supp., 69, 271.
Herbst, E., and Leung, C. M., 1990. Astr. Astrophys., 223, 177.
Huntress, W. T., 1977. Ap. J. Supp., 33, 495.
Millar, T. J., Leung, C. M., and Herbst, E., 1987. Astr. Astrophys., 183. 109.
Suzuki, H., 1983. Ap. J., 272, 579.
Watson, W. D., 1973. Ap. J. Lett., 183, L17.

Discussion

W. IRVINE: How large are the carbon chains (cyanopolyynes) included in your present model, and what abundance ratios does the model predict for successive members of the series?

R. BROWN: Our model calculations, including "hot ion" reactions but not attempting to find "best" values for the cross sections of the hydrogen-atom abstraction reactions, gives [HC$_5$N]/HC$_3$N] $\approx$8 x 10^{-2}. Values closer to the experimental ratio of $\sim$ 0.3 could no doubt be obtained by adjustment of rate constants (still keeping to the plausible range of values). However we do not think it worth further forcing of the model to improve agreement with observations until we are more certain of the actual importance of "hot-ion" reactions. Our personal studies are still speculative.

J. TARTER: Given the observation of C$_2$ and C$_5$ in circumstellar regions and the laboratory determination of IR frequencies for C$_7$ (and C$_9$?) - might we not expect to <u>observe</u> the relative abundances of these clusters and decide whether these basic structures could be playing a role in the cyanopolyacetylene chemistry?

R. BROWN: Additional observation of relative abundances of C species will certainly provide a more stringent list of current chemical models of molecular clouds. Whether our suggestion that the buildup of chains of C atoms depends upon the effectiveness of "hot" molecular ions in overcoming energy barriers to hydrogenation reaction (so-called hydrogen-atom abstraction reactions, HAARS) is correct will depend both on more observations and more laboratory measurements on cross sections for HAARS at varying temperatures. At present there seems to be no other plausible suggestion as to how such C chains are so efficiently built up in clouds such as TMC 1. Although it has been speculated that such chains may come in some manner by the breakup of larger molecules on dust grains, there seems no way of testing such a suggestion quantitatively. If we were sure of both the chemical composition and the surface structure of grains, it might be possible to do laboratory experiments to see whether C_n are formed by heating on photolysis. But this does not appear to be a feasible experiment at present.

HYDROGEN CYANIDE POLYMERIZATION:
A PREFERRED COSMOCHEMICAL PATHWAY

Clifford N. Matthews
Department of Chemistry, University of Illinois at Chicago,
Chicago, IL 60680, USA

Summary

In the presence of a base such as ammonia, liquid HCN polymerizes spontaneously at room temperature to a brown-black solid from which a yellow-brown powder can be extracted by water and further hydrolyzed to yield α-amino acids. Two types of structural units appear to be present in these polymeric products, stable ladder polymers with conjugated -C=N- bonds, and polyamidines, readily converted by water to polypeptides.

Several kinds of investigations, including electric discharge experiments which produce HCN from methane and ammonia, give results consistent with the hypothesis that the original polypeptides on Earth were synthesized directly from such HCN polymers and water without the intervening formation of α-amino acids. In the absence of water - on land - the intermediate polyamidines could have been the original condensing agents directing the synthesisis of nucleosides and nucleotides from available sugars, phosphates and nitrogen bases. Most significant would have been the parallel synthesis of polypeptides and polynucleotides arising from the dehydrating action of these polyamidines on nucleotides.

The expected predominance on cometary nuclei of frozen volatiles such as methane, ammonia, and water subjected to high energy sources makes them ideal sites for the formation and condensed-phase polymerization of HCN. Dust emanating from the nucleus, contributing to the coma and tail, would also arise partly from the polymer. Results of the recent Halley missions support this view, particularly the detection of cyanide radicals, HCN itself, and particles consisting only of H, C and N. HCN polymerization could account, too, for the dark material detected on some asteroids and for much of the yellow-brown-orange coloration of Jupiter and Saturn, as well as for the orange haze high in Titan's stratosphere.

In sum, laboratory and extraterrestrial studies increasingly suggest that hydrogen cyanide polymerization is a truly universal process that accounts not only for the past synthesis of protein ancestors on Earth but also for chemistry proceeding elsewhere today within our solar system, on satellites around other stars, and in the dusty molecular clouds of the Milky Way and other spiral galaxies. The existence of this preferred pathway adds greatly to the probability that life is widespread in the universe.

Discussion

F. RAULIN: Cliff, I would like to reduce (slightly) your enthusiasm. Although I agree that HCN must have played an important role in prebiotic chemistry, <u>pure HCN</u> polymerization is not so easy to get. First, in aqueous solution, HCN polymerization requires high HCN concentration ($\geq 10^{-2m}$). Second, HCN is not alone in planetary or cometary environments and pure HCN chemistry seems unlikely. Now, the coupling of HCN and HCHO chemistry, for instance, may change drastically the figure. Third, although I like very much your idea of directly getting the polypeptides (or their precursors) from HCN without the need of condensation of amino acids, this possibility has never been fully demonstrated, nor has the structure of the HCN oligomers been fully elucidated.

C. MATTHEWS: Francois: HCN liquid (b.p. 25°C) plus a trace of ammonia readily polymerizes to a black-brown solid possessing a water-extractable portion that can be hydrolyzed to yield many of the α-amino acids found in proteins today. [See Nature <u>215</u>, 1230 (1967)]. On comets, frozen methane and ammonia subjected to high energy sources would yield liquid HCN which would readily polymerize in the presence or absence of water to give a black crust. On Titan (no H_2O in the atmosphere) gas-phase reactions of clouds of HCN (the molecules attracted by H-bonding) would yield HCN polymer, accounting for the stratospheric orange haze. You are right to say that the postulated, polyamidined structure of the polymer has yet to be fully demonstrated. However, we have obtained many kinds of supportive evidence, including solid-state NMR studies showing unequivocally the presence of peptide bonds [See Origins of Life <u>14</u>, 243 (1984)] and are currently using pyrolysis and supercritical fluid extraction techniques in our continuing investigations.

Essentially we emphasize that (1) HCN is ubiquitous in reducing environments (a well established fact); (2) HCN polymers are readily formed in the presence of a base (not so well known); (3) among the polymers are polyamidines easily converted by water to polypeptides (more controversial but potentially of great significance).

R. BROWN: One problem in relating HCN polymers to proteins of biological importance is that the latter have all their chiral centers in the L-series. One needs to explain how such chiral purity could be obtained by hydrolysis of HCN polymer.

C. MATTHEWS: So far, amino acids obtained by hydrolysis of HCN polymers have been racemic. However, we have changed the question! Instead of asking why L-amino acids are preferred in nature, we ask why polyamidines (HCN polymers) or the resulting polypeptides prefer to form right-handed helices instead of left-handed helices. It seems, though, that both kinds form equally in our laboratory experiments. Perhaps this happened on the primitive Earth as well, and led to the origin of both left-handed and right-handed life. The present form then won out over the other, early in the course of evolution. This may have been the first war!

R. E. DAVIES: (1) I am surprised that you didn't mention that HCN can polymerize to form adenine. (2) We have mechanisms of getting rid of D-amino acids. We all have some D-amino acid oxidase, and all older people have D-amino acids especially at the end of polypeptide chains. To make alpha-helices you need either all L- or all D-. The L-amino acids won out early on.

C. MATTHEWS: Indeed, Oró showed that adenine (empirical formula $H_5C_5N_5$) forms directly from HCN in ammoniacal solution. HCN and acetylene give rise to many nitrogen heterocycles. So HCN is important for the origin both of proteins and nucleic acids. We have further proposed [see Origins of Life, 16, p. 500 (1986)] that the polyamidines present in HCN polymers were the original dehydrating agents of prebiotic chemistry, linking together nitrogen heterocycles, sugars and phosphates to yield nucleosides, nucleotides and nucleic acids. Regarding your interesting chirality comment, note that the cyanide model leads to the formation of right- or left-handed helices without the need for first synthesizing α-amino acids.

PAHs: VERY ABUNDANT ORGANIC MOLECULES IN THE INTERSTELLAR MEDIUM

A. Léger, L. d'Hendecourt, L. Verstraete, and C. Joblin
Groupe de Physique de Solides
Tour 23 - Universite Paris 7-75251 Paris 05

Paper Presented by A. Léger

ABSTRACT

The presence of Polycyclic Aromatic Hydrocarbons (PAHs) in the interstellar gas has been proposed recently on the basis of a suggestive resemblance between the set of observed IR emission lines (3.3, 6.2, 7.7, 8.6, 11.3 μm) and the vibrational spectra of these molecules.

Their abundance is derived from the intensity of their IR emission: they contain 10% of the cosmic carbon. It raises them to the level of the most abundant free organic molecules in space far ahead molecules like HCN, HC_3N, CH_3OH,...detected at radio wavelengths.

1.   OBSERVATIONAL EVIDENCE FOR VERY SMALL GRAINS

Information on interstellar matter can be obtained by absorption measurements, in front of stars, or emission measurements. In fact, these two pieces of information are quite complementary.

The first data available were absorption measurements leading to the determination of the extinction curve. Standard dust models that were constructed to account for it proposed the presence of interstellar grains made of silicates, graphite, and ices in dense and dark regions. On the other hand, radio wavelength data showed the presence of small molecules of CO, HCN, HC_3N, NH_3 ... . Unfortunately, when IR emission data became available, they were in conflict with the prediction of these models (see Puget and Léger, 1989, for a review). Photometric measurements, including IRAS data, show a hot emission ($2 \leq \lambda \leq 60$ μm) that could not be explained by grains whose temperature was expected less than 30K. In addition, in the brightest regions, the spectrum of this emission could be measured and results in a set of emission bands at 3.3, 6.2, 7.7, 8.6, and 11.3 μm that were not identified although observed since 1973 (for a review: Allamandola, 1984).

Our understanding of the problem greatly improved when Andriesse (1978) and Sellgren (1984) showed that thermal emission by Very Small Grains (~50 atoms) during the transient heating following the absorption of individual uv photons can explain color temperature of ~ 10^3K in spatial regions where their mean temperature could not be above 60K.

2. NATURE OF VERY SMALL GRAINS: PAH MOLECULES

Léger and Puget (1984) discussed the nature of these grains and, considering their necessary stability against sublimation when heated, they eliminated ices and silicates as candidates, but retained graphitic materials. More specifically, they proposed planar molecules of Polycyclic Aromatic Hydrocarbons (PAHs) (Fig. 1).

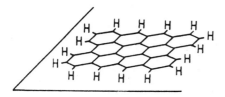

Figure 1: An example of PAH.

Then, using the benefit of having a precise model, they were able to calculate the expected emission I_λ from such species as:

$$I_\lambda = B_\lambda \, (T) \, \sigma_\lambda^{\text{abs}} \,,$$

where $B_\lambda \, (T)$ is the Planck function and $\sigma_\lambda^{\text{abs}}$ is the absorption cross section. Using the cross section of the biggest PAH whose IR spectra was measured in the laboratory, coronene ($C_{24} H_{12}$), they obtained its emission spectrum. The similarity between it and the set of observed emission bands strongly suggested that a mixture of PAHs could be the carrier of the emissions. The measurement in the laboratory of spectra from other large compact PAH molecules and a detailed treatment of the emission process have confirmed this view (Fig. 2)

3. ABUNDANCE

The abundance of PAHs in the Interstellar Medium directly results from the observed ratio of near-IR emission, F_{PAH}, to the far-IR emission because of grains, F_{grain}, if the respective absorption cross sections σ are measured or estimated, in the radiation field $u(\lambda)$, one reads:

$$\frac{F_{PAH}}{F_{grain}} = \frac{\int u(\lambda)\sigma_{PAH}(\lambda) \, N_{c,PAH} \, d\lambda}{\int u(\lambda)\sigma_{H,gr} \, (\lambda) \, N_H \, d\lambda}$$

where $N_{c,PAH}$ is the number of carbon atoms contained in PAH species, N_H and $\sigma_{H,gr}$ the number of H nuclei and the grain cross section per H nucleus, respectively.

To account for observations, about 10% of the interstellar carbon has to be in PAH molecules, making these species the *most abundant free organic molecules* known in the interstellar medium, far ahead of the other organic molecules detected in the radio wavelengths as seen in Table 1.

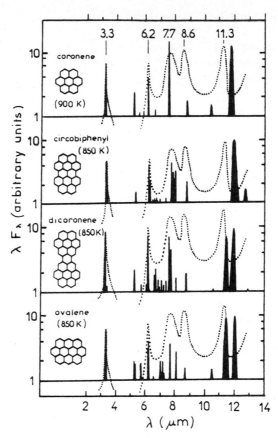

Figure 2: Emission spectra of several compact PAHs calculated from their absorption spectra measured in the laboratory (dark lines). The spectrum of the reflection nebula NGC 2023 is reported for comparison (dotted line)

Table 1: Abundance of some Interstellar Molecules, expressed in number of atoms that they include, versus total number of H nuclei, in regions where they are observed.

Molecule	n_{at}/nH	
H_2	1	
CO	2×10^{-4}	(radio)
PAH_s	5×10^{-5}	(IR)
H_2CO	5×10^{-8}	(radio)
NH_3	5×10^{-8}	(-)
HCN	2×10^{-8}	(-)
HC_9N	10^{-9}	(-)

At the present level of our knowledge, it is a family of molecules which is identified rather than precise individual species, as opposed to the identification made by radioastronomy. The expected rotational spectrum of such molecules is so complex that there is very little hope to identify one of them in radio wavelength spectra. An advantage of IR spectroscopy is that it is specific to atomic groups (e.g., CH, CC) in a given environment, allowing the identification of these aromatic species that would have been undetected if the radio astronomy was the only tool to identify molecules, although they are extremely abundant. An example of the variety of species is given by the analysis of a natural mixture of PAHs: extract of carbon black (Peaden et al., 1980, Fig. 3).

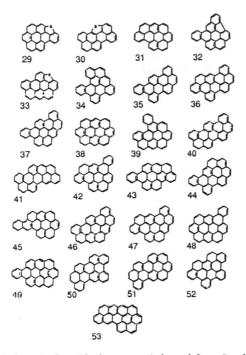

Figure 3. Some PAHs in a Carbon Black extract (adapted from Peaden et al., 1980).

However, this identification has some real specificity. For instance, one can assert that the interstellar PAHs are compact rather than noncompact (with prominent aromatic cycles) because the latter have a strong band at 6.7 μm which is not observed. In the same direction, the very attractive molecule C_{60} (buckminsterfullerene) which has been evidenced by Kroto et al. (1985), and recently produced in large quantities (Krätschmer et al., 1990) cannot be a major constituent of the Interstellar Small Grains because its IR spectrum has been measured in the laboratory and has lines [7.0, 8.45, 17.3, and 18.9 μm] which are clearly not the major ones in the astronomical spectra. This is even true for aromatic species that have five member rings and are curled because they have a band at 7.0 μm (Barth and Lawton, 1966) which is not observed in space.

4.   IMPLICATIONS

The abundant presence of these molecules in the Interstellar Medium has many implications, and predictions can be derived from models of the interstellar matter when they are included. Several of these predictions have been verified (Léger et al., 1989), the last one has been the observation of a band at 5.25 μm which is seen in all the laboratory spectra of PAHs and therefore was expected in the astronomical data. It has been observed in 1990 (Allamandola et al.)

If Interstellar material is incorporated with limited processing in comets and carbonaceous chondrites, *PAHs are a major source for further organic chemistry*. As an example, Schock and Schulte (1990) have proposed that they were at the origin of the amino acids observed in carbonaceous meteorites as illustrated in Fig. 4.

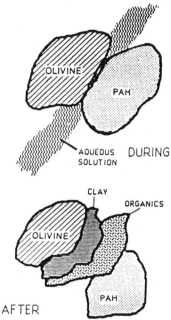

Figure 4.   Schematic diagram explaining the possible simultaneous formation of clay and organic species as amino acids in carbonaceous condrites (Shock and Schutte, 1990).

References

- Allamandola, L.J., 1984: Galactic and Extragalactic IR Spectroscopy, eds. Kessler, M.F., and Phillips, T.P., Reidel, Holland.
- Allamandola, L.J., 1990: Ap. J. Let.
- Andriesse, C.D., 1978: Aston. Astroph. 66, 169.
- Barth, W.E., and Lawton, R.G., 1966: J.A.C.S. 88, 380.
- Krätschmer, W., Lamb, L.D., Fostiropoulos, K., and Huffman, D.R., 1990: Nature 347, 354.
- Kroto, H.W., Heath, J.R., O'Brien, S.C., Curl, R.F., and Smalley, R.E., 1985: Nature 318, 162.
- Léger, A. and Puget, T.L., 1989: Astron. Astroph. Let. 137, L5.
- Léger, A., d'Hendecourt, L., and Defourneau, D., 1989: Astron. Astroph. 216, 148.
- Peaden, P.A., Lee, M.L., Hirata, Y., and Novotny, M., 1980: Anal. Chem. 52, 2268.
- Puget, J.L., and Léger, A., 1989: Ann. Rev. Astron. Astroph. 27, 161.
- Schok, E.L., and Schulte, M.D., 1990: Nature 363, 728.
- Sellgren, K., 1984: Ap. J. 227, 623.

PRIMITIVE EVOLUTION: EARLY INFORMATION TRANSFER AND CATALYSIS BY PURINES

Marie-Christine Maurel

Institut Jacques Monod, Tour 43, 75251 Paris, France

1. INTRODUCTION

The biochemistry of primitive molecular evolution addresses two basic questions: the first concerns the origin of information transfer, and, in particular, the nature of the first self-replicating molecules. The second concerns primitive catalysis and the nature of the first metabolic pathways.

Contemporary nucleic acids (DNA or RNA) are the replicated molecules that carry the genetic information. In the field of primitive replication, most current studies are directed toward the search for self-replicating polymers structurally related to the standard nucleic acids (Joyce et al., 1987), but with differences involving the chemistry of ribose-phosphate backbone, nucleosides, and base pairing.

Some modifications are necessary because of the difficulty in synthesizing standard "biological" DNA or RNA under prebiotic conditions.

In the chemical synthesis of sugar, a complex and unstable mixture was obtained in which ribose (the sugar of RNA) was not the majority (Buttlerow, 1861; Miyoshi et al., 1983). Attempts to synthesise nucleic acid bases yielded substantial purine but little pyrimidine (<0,1%), and attempts to connect a base to a ribose led to structural analogs of RNA nucleosides, with non-standard base-sugar linkages (Orgel and Lohrmann, 1974; Ferris, 1987).

Acknowledging these difficulties, the modern trend is to replace the biological sugar-phosphate backbone of DNA and RNA by a simpler acyclic chain (see Fig. 1 from Joyce et al., 1987).

Small peptides or even minerals such as clays have often been nominated as the most primitive prebiotic catalysts. But currently, the emphasis is on primitive RNA catalysts, following the discovery of RNA catalysis (Zaug et al., 1983; Guerrier-Takada et al., 1983; and Schvedova et al., 1987). We have been able to demonstrate (Maurel and Ninio, 1987) that a ribosylated derivative of adenine, N6-ribosyl adenine (Fig. 2), synthesised in presumably prebiotic conditions (Fuller et al., 1972), is of particular interest: its imidazole group is free for a type of catalysis normally achieved only by proteins. Our purpose here is to also demonstrate the potential importance of purines in early metabolic pathways.

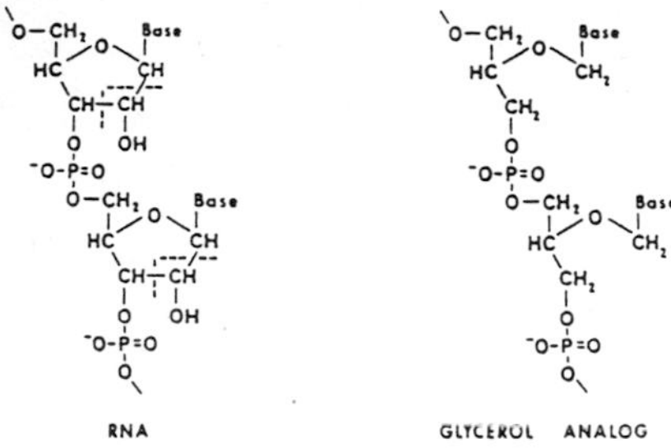

Figure 1: Natural nucleic acid (left) and analog (right).

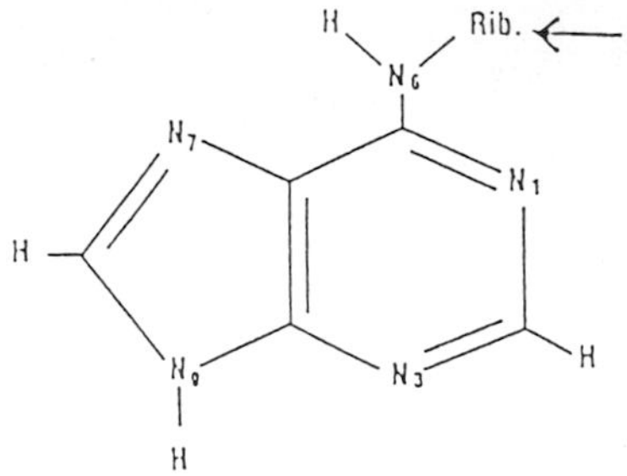

Figure 2: Structural formula of N6-ribosyladenine.

2. RESULTS AND DISCUSSION

Among the modified nucleotides (as reviewed by Adams et al., 1976), are found all the reactive groups of amino acids in proteins except for the imidazole group. But imidazole is a constituant of histidine, an amino acid which is often present at the active site of proteins, and of purines which are basic building blocks of nucleic acids (Fig. 3).

Thus, we demonstrated that N6-ribosyl adenine acts like an analog of histidine in the model reaction of hydrolysis of paranitrophenylacetate (PNPA) (Table 1).

In a second study we confirmed by mass spectroscopy and proton NMR (Maurel and Convert, 1990), that during the condensation of adenine with ribose, the sugar is preferentially linked to the N6 of adenine. Four isomers are formed, of which the α and β isomers of ribopyranosyl adenine are the two predominant forms. The minority is composed of furanose forms, which argues in favour of early replicating molecules somewhat different from the biological nucleic acids (Schwartz and Orgel, 1985).

Finally, we studied the effects of pH and substrate concentration on catalysis by N6-ribosyl adenine (Fig. 4).

Catalysis as a function of pH. No measurable activity is observed at a pH below 6.8. Above pH 6.8., the catalytic activity of both isomers rises steadily, and there is a difference between the isomers beyond pH 8.5.

Catalysis as a function of substrate concentration. At 10^{-4} M of catalyst and for substrate concentrations ranging from 10^{-4} to 8.10^{-3} M, Michaelis kinetics are observed (Fig. 5).

Most importantly, we showed that N6-ribosyl adenine is an authentic but slow catalyst. It hydrolyzes two molecules of substrate in five hours.

Table 1: Catalytic efficiencies of imidazole and related compounds.

Compound	Catalytic Efficiency
Histidine	1
Imidazole	5
AMP, UMP, GMP, CMP	<0,1
Adenosine, Uridine, Cytidine, Inosine	<0,1
Adenine	0,45
"Prebiotic Adenosines"	
Compound 1	1
Compound 2	1

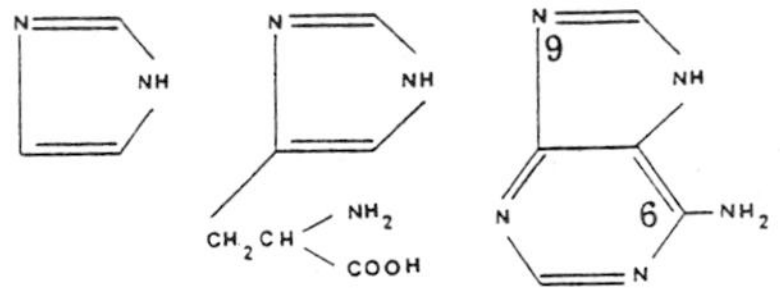

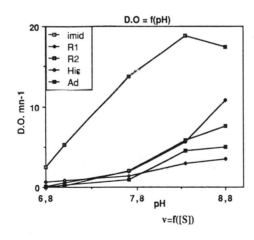

Im	His	Ade

Figure 3: Structural formulas of imidazole (lm), histidine (His), and adenine (Ade).

D.O = f(pH)

- imid
- R1
- R2
- His
- Ad

D.O. mn-1

pH

v=f([S])

Figure 4. PNPA hydrolysis assays: Dry PNPA, purchased from Sigma, was dissolved just before use in methanol (Merck) 30 mg/50 ml) and kept in ice. The assays were in 0.8 ml containing, unless otherwise specified, 10^{-4} M PNPA, 4×10^{-4} M of catalyst and 0.2 M of $NaH_2PO_4/Na_2H\,PO_4$ buffer at pH 6.8 to 8.8. The reaction was initiated by the addition of PNPA, and the appearance of paranitrophenate ion was followed in a Perkin-Elmer spectrophotometer at 400 nm, assuming a molar extinction coefficient of 1.8×10^{-4} M^{-1} cm^{-1} for the paranitrophenate anion at pH 8.5 (Bender et al., 1967).

Michaelis-Menten parameters for each isomer are shown in Table 2.

Table 2: Michaelis-Menten parameters for the hydrolysis of PNPA by R1 and R2 at ph 7.7.

Isomers	Km, M	Vmax, M. mn^{-1}	kcat/Km,M^{-1} mn^{-1}
R1	3,62. 10^{-3}	3,19. 10^{-2}	8,82
R2	3,42. 10^{-3}	2,2. 10^{-2}	6,54

As adenine is one of the most abundant heterocyclic compounds in the biochemical world, it is among the easiest to synthesize in primitive Earth-simulation experiments, and purines in modern metabolism are both constituents of nucleic acids and important components of many coenzymes. These factors support the hypothesis of early catalysis by purines. Our results strengthen this hypothesis.

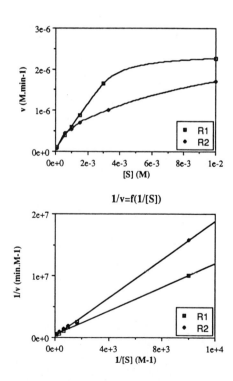

Figure 5. PNPA hydrolysis assays as a function of substrate concentration with same conditions as those in Fig. 4. Reactions were in 0.8 ml containing 10^{-4} M of catalyst; substrate concentration varied from 10^{-4} to 10^{-2} M in the same buffer at pH 7.7.

More generally, White suggested that modern coenzymes are fossils of an earlier stage of biological evolution that involved polynucleotide catalysts (White, 1976). We assume that purines and nucleoside-like molecules performed quasigenetic and catalytic functions, and played a key role in setting up primeval metabolism and its evolution. The fact that histidine is the only amino acid in modern metabolism which is synthesized from a nucleotide primer strengthens White's suggestion that some amino acids are an outgrowth of coenzyme evolution and the hypothesis of an "all-purine precursor" ancestor of nucleic acids (Wächtershäuser, 1988).

Our aim is now to incorporate the pseudohistidine, namely, N6-ribosyl adenine, into short oligomers to enhance their catalytic activity.

We did obtain products (following the procedure of Shimizu et al., 1984) having a far higher catalytic activity than the starting material in the model reaction of PNPA hydrolysis. However, this gain seems to be due to a chemical alteration (under study) of the monomer rather than oligomerization per se. We are also studying in collaboration with J.L Decout (L.E.D.S.S. Grenoble), the activity of purine oligomers linked via their six positions by a polyethylene chain.

Such knowledge will help us in designing further prebiotically plausible primitive catalysts.

3. Acknowledgment

I thank Jacques Ninio for many valuable suggestions and for his constant interest in this work.

4. Abbreviations

im : imidazole   R1 : N6 β ribopyranosyl adenine
Ade : adenine   R2 : N6 α ribopyranosyl adenine
His : histidine   PNPA : paranitrophenylacetate

References

Adams, R.L.P., Burdon, R.H., Campbell, A.M., and Smellie, R.M.S. 1976, The Biochemistry of Nucleic Acids, 8th edition (Chapman and Hall: London).

Bender, M.L., Kezdy, F.J., and Wedler, F.C. 1967, *J. Chem.* Ed., **44**, 84.

Butlerow, A. 1861, *Justis Liebigs Ann. Chem.*, **120**, 295.

Ferris, J.P. 1987, *Cold. Spring Harb. Symp. Quant. Biol.*, **52**, 19.

Fuller, W.D., Sanchez, R.A., and Orgel, L.E. 1972, *J. Mol. Biol.*, **67**, 25.

Guerrier-Takada, C., Gardiner, K., Marsh, T., Pace, N., and Altman, S. 1983, *Cell*, **35**, 849.

Joyce G.F., Schwartz A.W. Miller S.L., and Orgel L.E. 1987, *Proc. Natl. Acad. Sci.*, **84**, 4398.

Maurel, M-C. and Ninio, J. 1987, *Biochimie*, **69**, 551.

Maurel, M-C. and Convert, O. 1990, *Origins of Life and Evolution of the Biosphere*, **20**, 43.

Miyoshi, E., Kobayashi, A., and Yanagisawa, and S., Shirai, T. 1983, *The Chemical Society of Japan*, **10**, 1449.

Orgel, L.E. and Lohrmann, R. 1974, *Accounts of Chem. Res.*, **7**, 368.

Schvedova, T.A., Korneeva, G.A., Ostrochchenko, V.A., and Venkstern, T.V. 1987, *Nucleic Acids Res.*, **15**, 1745.

Schwartz, A.W. and Orgel, L.E. 1985, *Science*, **228**, 585.

Shimizu, T., Yamana, K., Kanda, N., and Maikuna, S. 1984, *Nucleic Acids Res.*, **12**, 3257.

Wächtershäuser, G. 1988, *Proc. Natl. Acad. Sci. USA*, **85**, 1134.

White, H.B. 1976, *J. Mol. Evol.*, **7**, 101.

Zaug, A.J., Grabowski, P.J., and Cech, T.R. 1983, *Nature*, **301**, 578.

Discussion

D. BRIN: This is fascinating. Earlier, Clifford Matthews showed how easily protein-like polymers arise out of HCN in a wide variety of circumstances. Now you demonstrate how catalysts arise out of purines such as adenine. For years there has been argument over which must have come first, protein or coded RNA-like molecules, but your paper and Matthews seem to say that both come about quickly, easily, and probably simultaneously. Can we conclude that there is no dichotomy after all?

And finally, given all we have seen about the ubiquity of adenine in early life processes, would it be fair to call this the most important simple molecule, likely to be used by life forms in other locales?

M-C. MAUREL: I think that the question, "Which have come first?" is a wrong question. Proteins and RNA are today high technological molecules and it's quite implausible that they were in ancestral competition between themselves, since they were not present!

The question is, "Which was able to do what because of the origin of information transfer and of the primitive catalysis?" We have noticed that, for instance, in the modified bases of RNA, we have essentially the same range of functional groups as proteins. So we demonstrate that both free puric bases and a prebiotic nucleoside containing imidazole group have ability to replace histidine in the hydrolysis of PNPA. Here perhaps is the link between RNA catalysis and protein catalysis (primitive evolution could have proceeded by such "bridges"). For the second question, I think that purines (like adenine) linked together, pairings, etc....performed quasigenetic and catalytic functions and provided some kind of scaffold, and played a key role in setting up primeval metabolism.

THE SEARCH FOR H_2O IN EXTRATERRESTRIAL ENVIRONMENTS

Harold P. Larson

Department of Planetary Sciences
University of Arizona
Tucson, Arizona 85721

INTRODUCTION

The H_2O molecule combines two biogenic elements considered necessary for living systems, it participates in the formation of both organic and inorganic compounds, and it provides a medium that promotes the concentration of trace constituents and facilitates the synthesis of complex molecules. The availability of H_2O in extraterrestrial environments is therefore an important consideration for assessing their potential to support biological evolution, assuming that the terrestrial analog is relevant. Water, especially as a liquid, is essential to terrestrial biology since it is the vital solvent in living organisms where the chemical processes of life take place. Even without its biological associations, however, the water molecule is an important probe of physical and chemical conditions in extraterrestrial environments. Parameters that may be retrieved include its abundance, phase, distribution, temperature, and velocity. These measurements relate to such diverse topics as the characterization of primitive matter in comets, aqueous alteration of minerals in asteroids, grain mantle evolution and kinematics in dense molecular clouds, and the origin and evolution of planetary atmospheres. Another reason for studying H_2O (or at least the sources of H and O from which it could be formed) in other objects is for its use in life support systems and in the manufacture of propellants.

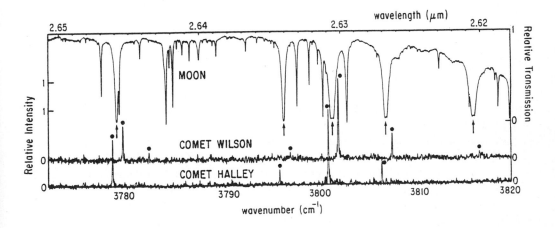

Figure 1. Emission from gaseous H_2O in comets. Spectrum recorded at the KAO with resolving power $\approx$ 10^5. (From Larson et al., 1989.)

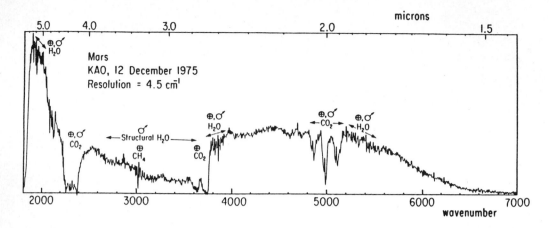

Figure 2. Absorption by hydrated minerals ($\approx$2.7-3.5 μm) in Mars. Spectrum recorded at the KAO with resolving power $\approx$ 700. (Larson, unpublished data.)

Remote studies of H_2O depend upon a variety of spectroscopic instruments in order to distinguish it from other molecules and to provide the type of information (i.e., multiple lines, resolved band shape) required for spectrochemical analyses. However, the choice of techniques is influenced most by two fundamental problems. One is Earth's atmosphere which itself contains large amounts of H_2O that block detection of extraterrestrial H_2O features. The only recourse is to conduct observations from the highest mountaintop sites, use stratospheric platforms such as aircraft, or go into space. The second complication is that ultraviolet (UV) photons photodissociate H_2O, thus limiting most remote studies of molecular H_2O to its long wavelength rotation and vibration-rotation spectra. The net result of these factors is that astronomical studies of H_2O became possible only in the last two decades as infrared (IR), sub-mm, and radio spectroscopic techniques were developed and applied at high altitude observing sites. This progress complements that achieved in remote spectral studies of other volatiles such as CO, CO_2, CH_4, and H_2CO. Collectively, this work constitutes important input to studies of the chemical evolution of primitive matter within and beyond our Solar System.

REMOTE SPECTRAL SIGNATURES OF H_2O

No other cosmochemically abundant volatile displays such a broad range of observable spectral features with such high potential for interpretation as does H_2O. The spectra in Figs. 1-6 were chosen to illustrate this diversity with actual astronomical data. Some of the obvious differences are due to the physical state in which H_2O is observed. The solid state spectra in Figs. 2, 4, and 6 display broad, relatively featureless bands whereas the gas phase H_2O spectra in Figs. 1, 3, and 5 contain narrow, isolated lines. Other major differences in appearance are due to the H_2O excitation mechanism. The interstellar H_2O emission line in Fig. 3 may be masing, the cometary H_2O emission lines in Fig. 1 are due to fluorescent excitation by solar IR photons, and the H_2O ice absorption bands in Fig. 6 are formed in reflected sunlight. Thermal emission from the object may carry the spectral signatures of H_2O. Thus, the H_2O absorption lines in Fig. 5 result from radiative transfer of thermal flux in Jupiter's atmosphere and the H_2O ice emission features in Fig. 4 are formed in thermal emission from the circumstellar ice grains themselves. The detailed features contained in these H_2O spectra support many types of analyses.

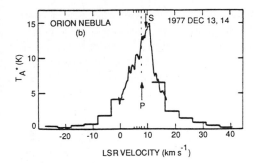

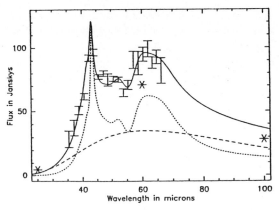

Figure 3. Emission from gaseous H_2O in OMC. Spectrum recorded at the KAO with resolving power $= 6 \times 10^4$. (From Waters et al., 1990.)

Figure 4. Emission from H_2O ice in IRAS 09371. Spectrum recorded at the KAO with resolving power ≈ 30. (From Omont et al., 1990.)

The H_2O line strengths, for example, lead to molecular and elemental abundance ratios, excitation temperatures, ortho/para ratios (OPR), and the characterization of excitation conditions. The H_2O line positions convey kinematic information (molecular velocities, outflow patterns) and the H_2O line widths may be proportional to the ambient pressure. The H_2O line shape is influenced in important ways by mineralogy (e.g., bound H_2O and OH mineral groups), mixtures with other ices (e.g., NH_3, CO), and phase (amorphous and crystalline H_2O ice).

The examples in Figures 1-6 demonstrate the most important instrumental requirements for remote spectral studies of H_2O. Information content is usually proportional to the achieved spectral resolution. For solids, resolving power $\lambda/\Delta\lambda$ between 10^2-10^3 is necessary to resolve subtle band shape differences (e.g., hydrate and hydroxyl mineral groups) while $\lambda/\Delta\lambda \geq 10^5$ is required for some gas phase studies

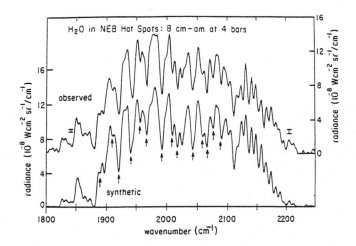

Figure 5. Absorption by gaseous H_2O in Jupiter. Spectrum recorded on Voyager with resolving power ≈ 650. (From Bjoraker et al., 1986.)

(e.g., kinematics in cometary comae). The spectral bandwidth in which observations of H_2O are conducted is also important. The ability to record multiple gas phase lines, for example, is essential to many analyses (e.g., OPR and kinetic temperatures). The examples in Figs. 1-6 also demonstrate the broad spectral interval in which opportunities exist to observe H_2O: from <2 μm in Fig. 6 to 1.64 mm in Fig. 3. Finally, remote studies of H_2O depend critically on specialized observing facilities. Only one of the examples included here represents ground-based work (Fig. 6); the others used aircraft (Figs. 1-4) and spacecraft (Fig. 5).

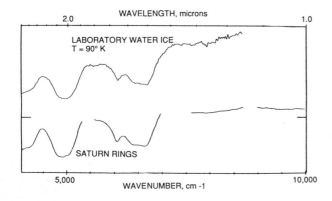

Figure 6. Absorption by H_2O ice in Saturn's rings. Spectrum recorded at the Catalina 1.5 m telescope with resolving power $\approx$ 100. (From Fink et al., 1976.)

EXAMPLES OF RECENT WORK

Several areas of current research are briefly discussed in the sections below. The choices represent three locations in our galaxy where H_2O is an essential, but poorly understood, participant in their chemical evolution: the interstellar medium, where H_2O is created; comets, where H_2O is stored; and Mars, where H_2O once flowed on the surface.

1. <u>The Interstellar Medium</u>. The H_2O molecule is believed to be created in ion-molecule gas phase reactions in molecular clouds. Conditions favorable to the stability of H_2O and other interstellar molecules occur only in dense molecular clouds where more than 90 species including many exotic trace constituents have been detected, primarily at radio wavelengths. Many of the most abundant predicted interstellar molecules have not yet been detected, however. As a result we are not really sure where and in what form the most abundant elements C, N, O, and S occur. One key molecule to understanding interstellar chemistry is H_2O, both as a reservoir for O and because of its importance to gas and solid phase chemistry. Important research goals therefore include observing both gas and solid H_2O in quiescent regions of dense molecular clouds, confirming the H_2O production mechanism, and establishing H_2O gas-solid interactions. Interstellar H_2O ice has been observed in molecular clouds. In OMC, for example, spectral analyses demonstrate that the H_2O ice is amorphous and mixed with NH_3 ice (Knacke et al., 1982). Laboratory simulations suggest that these icy grains may be important sites for the synthesis of complex molecules (see e.g., Greenberg et al., 1983). The characterization of gas phase H_2O in quiescent interstellar material has been much more difficult. Although there are many detections of gaseous H_2O in molecular clouds, most at radio wavelengths, the observable transitions are associated with hot, dense cloud cores and some are masing. These observations are very useful for many aspects of cloud physics and chemistry, but they are not directly relevant to the quiescent material.

An alternate approach, near-IR absorption spectroscopy, is being used to search for interstellar molecules that are difficult (e.g., H_2O) or impossible (e.g., CH_4) to observe at radio wavelengths. The method involves observing a background stellar source through quiescent foreground material. Interstellar features would then appear as absorption lines in the continuum spectrum of the stellar source, as successfully demonstrated in near-IR detections of CO (see e.g., Black and Willner, 1984). Knacke et al. (1988) reported the detection of an interstellar H_2O line in airborne observations at 2.63 μm of the BN object in the Orion molecular cloud (OMC). Follow-up work (Knacke and Larson, 1991) confirmed this detection and improved the evidence for additional interstellar H_2O lines. The results of their analysis include $[H_2O]/[H_2] = 10^6\text{-}10^5$, $H_2O_{gas}/H_2O_{ice} < 0.05$, and ortho-$H_2O$/para-$H_2O$ = 1.0 $\pm$ 0.5 from which $T_{equil} \approx 20$ K. Knacke and Larson concluded that grain processes dominate the physics of H_2O in OMC: most of the H_2O is frozen out on grains and less than 1% of the O is in gas phase H_2O. Moreover, it appears from the observed OPR that the gaseous H_2O may have "recently" sublimed from icy grains. The low equilibrium temperature deduced from the OPR could have been set when H_2O condensed onto interstellar grains, whose temperature in dense clouds is < 20 K. Since the OPR is thought to be a cosmogonic invariant that is unaffected by subsequent phase change, thermal cycling, or collisions (see e.g., Mumma et al., 1987), T_{equil} is actually a "fossil temperature" characteristic of a previous physical environment of the observed H_2O. There must therefore be significant cycling of mass between the gas and solid phases of interstellar matter in OMC for the observed OPR to be so low.

2. Comets. Since comets are chemical fossil remnants of the primitive solar nebula, knowledge of their compositions could resolve fundamental questions concerning thermochemical evolution in the presolar nebula and the interstellar medium. However, it is first necessary to understand the physics and chemistry of mass loss processes in cometary comae in order to relate their highly altered compositions to that of the parent molecular material in the nucleus. Observations of H_2O, the most abundant volatile in the coma, are essential for this work. Although H_2O was anticipated in comets for decades, definitive observational proof was lacking until comet Halley's last apparition when Mumma et al. (1986) detected multiple IR transitions in gaseous H_2O at 2.63 μm from the KAO. The high information content of these spectra created new opportunities to reconstruct conditions in the coma and relate some of them to properties of the nucleus. Cometary outbursts, for example, are frequently observed but poorly understood in terms of the structure and composition of the nucleus. An outburst of H_2O in comet Halley was fortuitously recorded on UT 1986 March 20.7; its H_2O brightness increased by a factor of $\approx$2 in less than 10 min (Larson et al., 1990). The event was attributed to a nuclear explosion that dispersed $\approx$1.0 x 10^{-3} km^{-3} of matter into the coma as gas and icy grains. The most intriguing aspect of this event is the condition in the nucleus that powered it. Possibilities include pockets of compressed gas (e.g., CO, CO_2), thermal stress, chemical explosions (e.g., NH), and phase change in H_2O ice. The actual process cannot be uniquely identified with remote data, but exothermal phase change from amorphous to crystalline H_2O ice triggered by solar heating of interior ice zones is most consistent with the scale of the outburst (it could have left a crater of width $\approx$100 m), thermal and chemical models of cometary nuclei, and the observed OPR ($T_{equil} \approx 25$ K). If Halley's interior contains amorphous H_2O ice that originally condensed near 25 K, its formation region is restricted to the distant outer region of the solar nebula, including in situ formation in the Oort cloud. The composition of cometary matter may therefore be similar to minimally processed interstellar material rather than to chemically evolved remnants left over from planetary formation in the solar nebula.

An important objective of remote sensing studies is to deduce kinematic properties of source regions. Larson et al. (1991) used velocity-resolved spectral line profiles of H_2O in comets to characterize the spatial distribution and expansion velocity of the neutral gas. The analysis utilized an outflow model

(Hu et al., 1991) to synthesize molecular line profiles for comparison with observations. The results demonstrate that considerable anisotropy exists in the neutral gas outflow, most likely due to surface structure. They also retrieved expansion velocity profiles from the shapes of cometary spectral lines. The velocity is a manifestation of energy balance in the coma and its measurement is very important for guiding the development of outflow models that must treat multiple heating and cooling processes in the expanding coma. The empirical velocity curves show promising agreement with recent models that incorporate such coma conditions as spatial anisotropy, H_2O recondensation, and optical trapping. These kinematic measurements represent initial attempts to extract maximum information from velocity-resolved line profiles of parent molecules. Although in situ measurements might be the most desirable approach, opportunities are very limited and spacecraft experiments have their own sensitivity and performance limitations.

3. <u>Mars</u>. The evidence for abundant H_2O in Mars takes many forms: hydrated minerals in the surface, gaseous H_2O in the atmosphere, H_2O ice in the polar caps, and morphological features due to liquid H_2O. In spite of these superb data, however, compelling questions remain associated with the evolutionary history of H_2O in Mars, including its role in biochemical evolution and characteristics of the present crustal reservoir of H_2O in Mars. New input is being obtained from studies of deuterated water (HDO) in Mars' atmosphere. The basic premise is that the D/H ratio observed in martian water vapor is influenced by fractionation processes acting over geologic time. Photodestruction of atmospheric water vapor would lead to the preferential escape of H over D, thus concentrating HDO in the atmosphere, but replenishment of atmospheric H_2O from the crustal reservoir would dilute this enhancement. A geochemical model incorporating these and other processes could therefore relate an atmospheric D/H measurement to details of Mars' geochemical evolution. Two problems to exploiting this connection, lack of a suitable geochemical model and lack of observations of martian HDO, were recently eliminated. The first achievement was the detection of martian HDO in ground-based observations (Owen et al., 1988). The accompanying data for martian H_2O was poor, however, so the derived D/H ratio was quite uncertain although it still indicated an appreciable enhancement (factor of 6 ± 3) over the terrestrial value. Prompted by this new observational result, Yung et al. (1988) developed a geochemical model that related the enhancement in martian D/H to the amount of crustal water that exchanged with the atmosphere. The next input came from airborne observations of Mars at 2.63 μm. Dozens of lines of martian H_2O and HDO were observed simultaneously, thus, allowing a definitive measurement of the D/H enhancement (5.2 ± 0.1; Bjoraker et al., 1991). According to Yung et al.'s geochemical model, only ≈ 35 cm of crustal H_2O could have exchanged with the atmosphere, thus implying that the bulk of martian H_2O has been immobilized over geologic time. This conclusion contrasts dramatically with the geologic record which requires ≈ 500 m of groundwater to carve surface channels, support polar cap deposits, and form hydrated minerals. The spectroscopic and geologic data are themselves excellent; only the interpretations are so far irreconcilable. This situation should provoke new thinking about surface-atmosphere interactions on Mars that will eventually advance our understanding of this planet.

FUTURE PROSPECTS

The ability to study H_2O in extraterrestrial environments from ground-based, sub-orbital, Earth-orbiting, and deep space facilities will improve dramatically during the next decade. New large-aperture (up to 8 m) ground-based telescopes with spectrometers incorporating IR array detectors will allow unprecedented combinations of spatial and spectral resolution. SOFIA, the next generation will allow unprecedented combinations of spatial and spectral resolution. SOFIA, the next generation airborne telescope, will continue NASA's airborne astronomy program for at least two decades. SOFIA will be particularly important for high resolution spectroscopic work because it readily accommodates large instruments and its larger aperture (2.5 m compared with 0.9 m on the KAO) will significantly improve remote sensing capabilities in the sub-mm wavelength region. Finally, cryogenic Earth-orbiting telescopes (SIRTF and ISO) promise remarkably high sensitivity at IR wavelengths while deep space missions such as CRAF, Galileo and Cassini permit compositional studies using both remote and direct sampling techniques.

References

Bjoraker, G.L., Larson, H.P., Kunde, V.G.: 1986, Astrophys. J. 311, 1058

Bjoraker, G.L., Mumma, M.J., Larson, H.P.: 1991, Astrophys. J., submitted

Black, J.H., Willner, S.P.: 1984, Astrophys. J. 279, 673

Fink, U., Larson, H.P., Gautier, T.N., Treffers, R.R.: 1976, Astrophys. J. Letters 207, L63

Greenberg, J.M., van de Bult, C.E.P.M., Allamandola, L.J.: 1983, J. Phys. Chem. 87, 4243

Hu, H-Y., Larson, H.P., Hsieh, K.C.: 1991, Icarus, in press.

Knacke, R.F., McCorkle, S., Puetter, R.C., Erickson, E.F., Krätschmer, W.: 1982, Astrophys. J. 260, 141

Knacke, R.F., Larson, H.P., Noll, K.S.: 1988, Astrophys. J. Letters 335, L27

Knacke, R.F., Larson, H.P.: 1991, Astrophys. J., 367, 162.

Larson, H.P., Weaver, H.A., Mumma, M.J., Drapatz, S.: 1989, Astrophys. J. 338, 1106

Larson, H.P., Hu, H-Y., Mumma, M.J., Weaver, H.A.: 1990, Icarus, 86, 129.

Larson, H.P., Hu, H-Y., Hsieh, K.C., Weaver, H.A., Mumma, M.J.: 1991, Icarus, in press.

Mumma, M.J., Weaver, H.A., Larson, H.P., Davis, D.S., Williams, M.: 1986, Science 232, 1523

Mumma, M.J., Weaver, H.A., Larson, H.P.: 1987, Astron. Astrophys. 187, 419

Omont, A., Moseley, S.H., Forveille, T., Glaccum, W.J., Harvey, P.M., Likkel, L., Loewenstein, R.F., Lisse, C.M.: 1990, Astrophys. J. Letters 355, L27

Owen, T., Maillard, J.P., de Bergh, C., Lutz, B.: 1988, Science 240, 1767

Waters, J.W., Gustincic, J.J., Kakar, R.K., Kuiper, T.B.H., Roscoe, H.K., Swanson, P.N.: 1980, Astrophys. J. 235, 57

Yung, Y.L., Wen, J-S., Pinto, J.P., Allen, M., Pierce, K.K., Paulson, S.: 1988, Icarus 76, 146

Discussion

W.M. IRVINE: Your results for the ortho/para ratio for water toward BN in Orion are very interesting. What are the uncertainties in the ratio due to uncertainties in the excitation, both for Orion and for Comet Halley?

I might point out that Y. Minh (Ph.D. thesis, University of Massachusetts, 1990) has measured the ortho/para ratio for thioformaldehyde in the Orion "compact ridge" cloud. To our surprise, it indicates equilibrium at about 20K, although the kinetic temperature is at least 100K. This may well indicate that H_2CS, like H_2O, has sublimed from icy grain mantles.

H.P. LARSON: The uncertainties in the OPR's in Orion and comet Halley are due primarily to uncertainties in the measured relative intensities of individual H_2O lines. Uncertainties due to excitation conditions are more difficult to quantify at this time. The excitation of H_2O in comets, for example, involves multiple processes (fluorescence, collisions, etc.) whose effects on spectral line intensities are still being modeled. For Orion, uncertainties associated with excitation include inhomogeneous cloud conditions along the line of sight to BN and the low signal-to-noise ratios in the observed line intensities. The values of the H_2O OPR in these objects must therefore be critically reassessed as better data and more realistic excitation models become available.

T.L. WILSON: The formation temperature deduced from your H_2O ortho-to-para ratio is consistent with temperatures deduced from the D/H ratio (from NH_2D/NH_3) in the Orion "hot core," which is ~15" south of BN. An emission line measurement of H_2O in this source (diameter 10") would be very worthwhile, since maps in the radio lines show large differences in gas phase material between the direction of the "hot core" and BN.

H.P. LARSON: The diameter of the field of view in the near-IR observations of interstellar H_2O in Orion was 20". The spectrum would therefore not contain any possible emission from H_2O in the "hot core" region.

L.V. KSANFOMALITY: The estimated amount of water on Mars including permafrost is ranging from some units to hundreds of meters. Why the estimation is critical from the point of view of your presentation?

H.P. LARSON: Available data that relate to the evolutionary history of the H_2O on Mars give incomplete perspective on the interactions between the sources, sinks, and loss mechanisms for martian H_2O. The atmospheric D/H ratio is a measure of the amount of H_2O that exchanged between the crustal reservoir and the atmosphere, which need not be the same as the amount of H_2O on or below the surface that shaped Mars' terrain. The large difference between the spectroscopic and geologic estimates of the amount of H_2O on Mars must still be accounted for with a comprehensive geochemical model; hence the need to reassess current interpretations whenever a new type of measurement becomes available.

C.F CHYBA: I would like to comment on the last question about the Martian D/H ratio. Michael Carr has a paper in press at Icarus addressing this question; the key issue is how much of the Martian water inventory actually exchanges with the atmosphere, which is not well known. Moreover, a single, typical cometary impact on Mars will completely reset the atmosphere D/H ratio.

In addition, I would like to comment on your remark that the Halley 25 Kelvin ortho/para ratio pushes Halley's formation outside the planetary region. Triton, admittedly a very bright object, has a temperature of only about 38°K. Doing the simple heliocentric distance-temperature scaling will then give you 25°K at only 50 AU or so, which is hardly outside the planet-forming region. Certainly it does not require formation to have taken place out in the ISM.

H.P. LARSON: It is important to distinguish between the physical temperature of an object and the equilibrium temperature associated with an observed ortho/para ratio. The latter applies to the time and location where cometary H_2O last equilibrated. If that event was low temperature condensation in the primordial solar nebula prior to the comet's formation and the H_2O ice incorporated into the comet was not subsequently reprocessed in a way that would reset the OPR, then the original condensation temperature may be retrieved from the observed OPR in gaseous H_2O independent of the thermal history of the comet. This "fossil temperature" may then be compared with temperature gradients in models of the solar nebula in order to locate the formation region of the comet.

SELF-ASSEMBLY PROPERTIES OF PRIMITIVE ORGANIC COMPOUNDS

D. W. Deamer

Department of Zoology, Univ. of California, Davis, CA 95616

Summary

A central event in the origin of life was the self-assembly of amphiphilic compounds into closed microenvironments. If a primitive macromolecular replicating system could be encapsulated within a vesicular membrane, the components of the system would share the same microenvironment, representing a step toward indivuality and true cellular function.

What molecules might have been available on the early Earth to participate in the formation of such boundary structures? We have investigated primitive organic mixtures present in carbonaceous meteorites such as the Murchison meteorite, which typically contain 1-2 percent of their mass in the form of organic compounds. It is likely that such compounds contributed to the inventory of organic carbon on the prebiotic earth and were available to participate in chemical evolution leading to the emergence of the first cellular life forms. We found that Murchison components extracted into non-polar solvent systems are surface-active, a clear indication of amphiphilic character (Deamer and Pashley, Origins of Life and Evolution of the Biosphere 19 (1989) 21-33). One acidic fraction self-assembles into vesicular membranes that provide permeability barriers to polar solutes. Other evidence indicates that the membranes are bimolecular layers similar to those formed by contemporary membrane lipids. We conclude that bilayer membrane formation by primitive amphiphiles on the early Earth is feasible. However, only a minor fraction of acidic amphiphiles assembled into bilayers, and the resulting membranes required narrowly defined conditions of pH and ionic composition to be stable. It seems unlikely, therefore, that meteoritic infall was a direct source of membrane amphiphiles. Instead, the hydrocarbon components and their derivatives would provide an organic stock for chemical evolution in which membranogenic amphiphiles were generated. One possible reaction is photochemical oxidation of hyrdrocarbons. We found that pyrene, a major polycyclic aromatic hydrocarbon of carbonaceious meteorites, acts as a photosensitizer in the hydroxylation of long chain hydrocarbons. This reaction is significant, in that a non-polar species (the hydrocarbon) becomes a surface-active molecule which would be available to partake in self-assembly of early membranes.

Discussion

H. DAVIES: Were you showing budding of a mother cell into daughter cells in your slide of liposomes?

D. DEAMER: When lipid or other membrane-forming amphiphilic molecules are hydrated after being dried, the membranous vesicles that appear are produced by a swelling process as water penetrates the main mass of material. This can give rise to the apparent "budding" that you noticed in the micrograph.

N. EVANS: Can you describe how you have ruled out terrestrial contamination?

D. DEAMER: Several lines of evidence weight against significant terrestrial contamination in the total extract, and in one sub-fraction as well.

1. Common contaminants that would occur from previous handling of the specimens, such as oils and triglycerides, do not have fluorescent components resembling those observed in the meteoritic extracts. Furthermore, only interior samples were used in our research, eliminating gross surface contaminants.

2. The amounts extracted were substantial, 1-2 mg per gram meteorite.

3. The organic acid fraction that assembles into membrane vesicles has undergone two independent estimates of δ D/H ratios. Both gave corrected values of +200, well outside expected terrestrial ranges.

COMETARY STUDIES: BIOASTRONOMICAL PERSPECTIVES

A. Chantal Levasseur-Regourd
Université Paris VI/Service d'Aéronomie
BP 3, 91371 - Verrières le Buisson, France

ABSTRACT

The main points of interest for bioastronomers are presented, with special emphasis on recent results (obtained either from in-situ probes or remote observations), on physico-chemical properties of the nucleus, on deuterium enrichment, and on cometary dust structure and composition. The question of the possible supply, during the period of heavy bombardment of the young Earth, with water and complex organic molecules (from cometary or interplanetary origin) is addressed. Finally, and since many questions still remain unanswered, the relevance of future space missions to comets (GEM, CRAF, and Rosetta), and specially of samples analysis is shown.

INTRODUCTION

Suggestions concerning the development of prebiotic material in comets and their contribution to terrestrial oceans and organic compounds have noticeably preceded the return of Halley's comet in 1985/1986 (Ponnamperuma, 1981). The bioastronomical interest in comets was indeed strengthened once it was generally accepted that:

- Small grains with organic mantles are likely to exist in interstellar molecular clouds, and complex organic material is slowly processed on these grains by cosmic radiation;

- Comparable chemical compounds are detected in molecular clouds and in cometary comae;

- Complex organic compounds are found in carbonaceous chondrites, that is, primitive meteorites;

- Life appeared on Earth at a time where the influx of cometary and meteoritic material was larger than now.

These topics have been addressed in various papers, for instance Cronin and Pizzarello (1986); Delsemme (1982); Greenberg (1982); and Irvine et al., (1985). They allow us to draw a sketch going from molecular clouds and the protosolar nebula, to cometary nuclei (which remained "frozen" for gigayears on the fringes of the solar system), to comets (with eventually some asteroids which may be defunct cometary nuclei), to zodiacal dust, and finally to impact on Earth, (Figure 1).

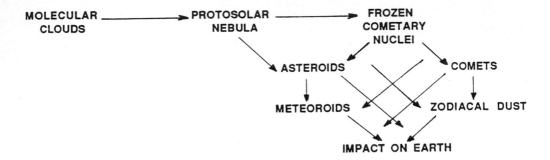

Figure 1. From molecular clouds to delivery to the Earth

I. PHYSICO-CHEMICAL PROPERTIES OF HALLEY'S NUCLEUS

The most spectacular results obtained from the comet Halley flybys in March 1986 are related to its nucleus morphology. It is found, from detailed analysis of Vega and Giotto images, that its size was about 16 x 8.2 x 7.5 km, its surface topography was similar to chains of hills, mountains and ridges, and active regions covered about 10% (Keller et al., 1987).

The central depression region, together with the so-called "duck tail" mountain seen on the night side on Giotto images, suggests that the building blocks making up the nucleus of comet Halley when it was formed could have been in the 1 to 5 km range (Keller, 1987).

The rotation of the nucleus is still poorly understood. Observations could be fit with a 2.2 days rotation period (along an axis orthogonal to the long axis), with a longer period of precession or nutation motion, as well as with a 14.6 days rotation period with a 2.1 days precession period, or even with a more complex rotation and spin motion (Belton, 1991).

However, of greater interest for the understanding of the physical and chemical properties of comets are the results relevant to the surface of the comet Halley nucleus. At the time of the flyby, it was generally warm ($\approx$ 350 K), with a geometric albedo of a few percent and a reddish black colour (Rickman, 1991). The polarization of scattered light was found to be negative (direction of polarization in the scattering plane) at small phase angles, both for the nuclear region and for the coma. It can be concluded from these results that the surface of the nucleus is made of fluffy aggregates of porous and insulating material (Levasseur-Regourd, 1989).

Of the many parent or daughter molecules detected in Halley's comet coma, H_2O was by far the most abundant ($\approx$ 80% of the volatiles); also found were CO (10 to 20%), CO_2 (2 to 3%), HCN, CN, C_2, C_3 in smaller proportions, and CH_4 and NH_3 with low relative abundances (Jessberger et al., 1988; and Kawara et al., 1988). The dust to gas ratio could be as high as 2 (Mc Donnell et al., 1991; and Jessberger and Kissel, 1991).

The elemental abundances derived from these results fit broadly with those of the Sun, apart from the expected deficiency in H and He. Since the CH_4 and NH_3 abundances are low, it can be inferred that Halley was formed further away from the Sun than the giant planets. If the nucleus is indeed a pristine aggregate of interstellar matter, its composition has to be similar to that of the interstellar medium; most of the hydrogen is trapped in water and hydrocarbons, while an important fraction of carbon is trapped in dust.

II. THE D/H RATIO IN COMETS

Almost the entire amount of deuterium has been synthesized in the early universe. The D/H ratio, which is of the order of 10^{-5} in the interstellar medium (Vidal-Madjar, 1983), is therefore a critical parameter for the estimation of the baryonic density, that is, for the test of cosmological models. However, the observed D abundance may be affected by ion-neutral reactions which, in dense molecular clouds, take place at very low temperatures, and concentrate D at the expense of H_2. If comets formed in a dense environment at low temperature, then a substantial enhancement in D/H is to be observed (Vanysek and Vanysek, 1985).

Although H, as explained previously, is an important species in comets, the D/H ratio is not obtained directly by spectrography (Levasseur-Regourd, 1988): the Ly α deuterium line, on the wing of the strong Ly α hydrogen line, is hardly detectable since UV spectrography lacks of very high resolution. The D/H ratio has to be derived from deuterated molecules measurements, either by spectrography (remote sensing) or by mass spectrometry (in-situ analysis).

International Ultraviolet Explorer satellite (IUE) observations of comets give an upper limit of OD/OH of the order of 4 x 10^{-4} (A'Hearn et al., 1985; and Schleicher and A'Hearn, 1986). Giotto neutral gas spectrometer observations during the March 1986 flyby indicate that the H_2DO^+/H_3O^+ ratio is between 0.6 x 10^{-4} and 4.8 x 10^{-4} (Eberhardt et al., 1987). The cometary volatile material is therefore enriched in deuterium by a factor of about 10 relative to the protoplanetary nebula.

As can be seen in Figure 2, primordial abundances are observed in the solar system in the atmospheres of Jupiter and Saturn, while an enrichment in D is observed in comet Halley, and also in Uranus, Neptune, and Titan (Vanysek, 1991). As proposed by Owen et al. (1986), two distinct reservoirs of deuterium could have existed in the solar system; the enrichment of one of these reservoirs probably occurred before the solar system and the cometesimals were formed. It may be of interest to notice that D/H ratios are comparable in Standard Mean Ocean Water (SMOW) and in comet Halley, and that even greater enrichments have been measured in some pristine meteorites.

It should be added that, among other isotopic ratios, $^{12}C/^{13}C$ has also been tentatively determined from mass spectrometry measurements of cometary grains (Jessberger and Kissel, 1991). The large dispersion of the data, together with its departure from the solar system value, is still disputable. It could provide some evidence for the presence of presolar grains in cometary nuclei.

III. THE COMETARY DUST

The composition of the dust released from the surface of the comet Halley nucleus has been measured by the dust spectrometers on board Vega 1, Vega 2, and Giotto (Kissel et al., 1986). Due to the high velocity flyby (about 70 km s^{-1}), only the elemental composition is obtained by mass spectrometry. Light elements (H,C,N,O) are frequently detected in the spectra: the dust particles are not only built up of inorganic, but also of organic "CHON" material (Kissel and Krüger, 1987).

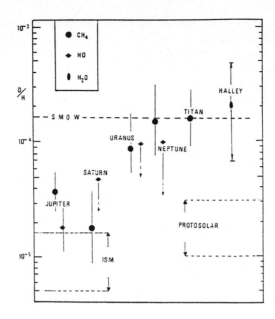

Figure 2. D/H ratios in the solar system (from Vanysek, 1991). Note that D/H measurements from HD molecule for Saturn, Uranus, and Neptune are upper limits.

The size distribution of dust has been measured by the impact system on board Giotto. Large grains, up to 100 μm, are detected. An unexpected large amount of submicron size grains is also found, especially in the outer cometary coma (Vaisberg et al., 1986; Simpson et al., 1986). Since CO and CN are detected in some cometary jets, both by remote spectral analysis and by in-situ optical probing (A'Hearn et al. 1986; and Levasseur-Regourd et al., 1986), it is likely that the dust grains are built of very small particles, which are released as the organic yellow-brown material binding them is slowly sublimated (Greenberg et al., 1989).

The masses of ions found inside Halley's coma have been recently derived from Giotto positive ion cluster analyzer, since they can be assumed to be proportional to the measured energy/charge ratio. Peaks at 35, 37, and 39 atomic mass units are attributed to H_2S^+, C_3H^+, and $C_3H_3^+$. The progenitors of these ions are computed to be CHON (S) dust particles, rather than gas (Korth et al., 1989; and Marconi et al., 1989, 1990).

The $C_3H_3^+$ ion could be a dissociation product of hydrocarbon chain molecules. The existence of complex molecules of a prebiotic type, which could have been produced by cosmic irradiation and thermal processing on ices or dust icy coating, is indeed strongly suggested by the detection of an ordered series of mass peaks at 31, 45, 60, 75, 91, and 105 amu. The regular spacing of about 15 amu could be attributed to dissociation products of polyoxymethylene, "POM," a linear chain of polyformaldehyde (Huebner, 1987).

The H_2CO molecule has recently been identified in IR and radio spectra (Mumma and Reuter, 1989; and Snyder et al., 1989). However, signals which would correspond to broken POM chains have not yet been found in the Vega or Giotto data. Besides, the regular spacing of 15 amu seems to be a general characteristic of molecules with CHON (Mitchell et al., 1989)

The existence of biological material in cometary dust is obviously even more speculative than that of POM. Infrared spectra of comet Halley, and more recently of Wilson, Bradfield, and Austin, have shown a strong emission feature in the 3.2 to 3.6 μm region (Combes et al., 1986; and Wickramasinghe and Allen, 1986). The width and shape of the band suggest that it is indicative of solid grains of organic material with a C-H bond. Good fits have been obtained with bacterial grains (Wallis et al., 1989); but fits can also be obtained with irradiated methane clathrates (Chyba and Sagan, 1987).

To summarize, the cometary dust appears to be made of aggregates of submicron-sized particles, which could be imbedded in a carbon-rich matrix or glued by various polymers. It may be of interest to notice that, as the cometary dust injected in the interplanetary dust cloud spirals towards the Sun under the Poynting-Robertson effect, its albedo increases and its polarization degree decreases (Levasseur-Regourd and Dumont, 1990); the sublimation of organic material by heating, together with the sputtering by solar wind particles would indeed produce a decrease in roughness, porosity, and size with decreasing solar distance (Levasseur-Regourd et al., 1990).

IV. COMETARY IMPACTS

The properties of cometary deuterium and dust grains, as presented above, have led to the question of knowing whether comets (and interplanetary dust) have contributed to the Earth's oceans and organic molecules formation or not (Delsemme, 1989). It is difficult to accurately estimate the cratering flux from carbonaceous asteroids, comets, and micrometeoroids in Earth crossing orbits, because of uncertainties with regard to mass distribution, total populations, and possible temporal variations in the flux (Weissman, 1990). The debate on impacts from comets showers as a cause of biological extinction events is far from being closed.

However, recent studies of the lunar impact record allow a fair estimate of the lunar cometary impacts during the period of heavy bombardment, about 4.5 to 3.8 Gyr ago (Chyba, 1987). Taking into account the probability of collision with the Earth, and the competition between impact delivery of new volatiles and impact erosion of those already present, a net amount corresponding to approximately 0.2 to 0.7 Earth's oceanic mass is obtained (Chyba, 1990).

It is indeed possible (but not demonstrated, even with comparable D/H ratios) that comets have supplied the terrestrial planets with oceans of water. At an earlier epoch where the zodiacal cloud was much denser than now, comets and carbon-rich interplanetary dust particles (Maurette et al., 1987) could also have supplied terrestrial planets with organic material processed in the interstellar medium or on cometary surfaces.

V. FUTURE COMETARY MISSIONS

All these subjects are still extremely speculative, and the origin and evolution of the solar system, including our planet Earth, are far from being understood. Future cometary missions should hopefully allow us to obtain a better understanding of the isotopic, elemental, molecular, and structural properties of comets.

The next mission to take place should be the Giotto Extended Mission (GEM). The Giotto spacecraft was successfully reactivated on February 19, 1990 after four years of hibernation. More than half of the experiments on board have been found to be fully operative. Following an Earth gravity assist on July 2, 1990, Giotto should fly by Grigg-Skjellerup (P≈5.1 yrs, i≈21°, q≈1 au) on July 10, 1992 and probe its inner coma. Later, the Comet Rendezvous Asteroid Flyby (CRAF), should orbit for several years a cometary nucleus and deploy a penetrator into the comet's crust (Neugebauer and Weissman, 1989). The proposed target is periodic comet Tempel 2.

The most ambitious project, to take place early in the next century, is Rosetta, a Comet Nucleus Sample Return mission. After rendezvous with a comet, surface mapping, landing site selection, descent, and landing, the samples collection (approx. 15 kg), return to Earth, reentry in the Earth's atmosphere, and analysis in laboratories would take place (Atzei et al., 1989). Only then would some of the most pristine material in the solar system be extensively studied, allowing us to establish the missing links between interstellar matter, comets, interplanetary dust, and primitive meteorites, and to know what role comets have played in the biological evolution on the early Earth.

Acknowledgements are made to P. Weissman and T. Yamamoto for their relevant comments.

References

A'Hearn, M.F., Schleicher, D.G., West R.A.: 1985, Astrophys. J., **297**, 826.

A'Hearn, M.F., Hoban, S., Birch, P.V., Bowers, C., Martin, R., Klinglesmith III, D.A.: 1986, Nature, **324**, 649.

Atzei, A., Schwehm, G., Coradini, M., Hechler, M., de Lafontaine J., Eiden M.: 1989, ESA Bulletin, **59**, 19.

Belton M.J.S.: 1991, in Comets in the post-Halley era, ed. R. Newburn et al., Kluwer Ac. Pub., 691.

Chyba, C. F.: 1987, Nature, **330**, 632.

Chyba, C. F.: 1990, Nature, **343**, 129.

Chyba, C. F., Sagan C.: 1987, Nature, **330**, 350.

Combes M., Moroz V.I., Crifo J.F., Lamarre J.M., Charra J., Sanko N.F., Soufflot A., Bibring J.P., Cazes S., Coron N., Crovisier J., Emerich C., Encrenaz T., Gispert R., Grigoriev A.V., Guyot G., Kranopolsky V., Nicolsky Y.V., Rocard F.: 1986, Nature, **321**, 266.

Cronin J.R., Pizzarello, S.: 1986, Geochim. Cosmochim. acta, **50**, 2419.

Delsemme A.H.: 1982, in Comets, ed. L.L. Wilkening, U. of Arizona Press, 85.

Delsemme A.H.: 1989, Adv. Space Res., **9**, 6, 25.

Eberhardt P., Dolder U., Schulte W., Krankowsky D., Lämmerzahl P., Hoffman J.H., Hodges R.R., Berthelier J.J., Illiano J.M.: 1987, Astron. Astrophys., **187**, 435.

Greenberg J.M.: 1982, in Comets, ed. L.L. Wilkening, U. of Arizona Press, 131.

Greenberg J.M.: 1989, Ann. Phys. Fr., **14**, 103.

Huebner W.F.: 1987, Science, **237**, 628.

Irvine W.M., Schloerb P., Hjalmarson A., Herbst E.: 1985, in Protostars and Planets II, D.C. Black and M.S. Matthews eds., U. of Arizona Press, 579.

Jessberger E.K., Christoforidis A., Kissel J.: 1988, Nature, **332**, 691.

Jessberger E.K., Kissel J.: 1991, in Comets in the post-Halley era, ed. R. Newburn et al., Kluwer Ac. Pub., 1075.

Kawara K., Gregory B., Yamamoto T., Shibai H.: 1988, Astron. Astrophys., **207**, 174.

Keller H.U.: 1987, in *Symposium on the Diversity and Similarity of Comets*, ESA SP278, 447; in *Comets in the post-Halley era*, ed. R. Newburn, Kluwer Ac. Pub., in press.

Keller H.U., Delamere W.A., Huebner W.F., Reitsema H.J., Schmidt H.U., Whipple F.L., Wilhelm K., Curdt W., Kramm R., Thomas N., Arpigny C., Barbieri C., Bonnet R.M., Cazes S., Coradini M., Cosmovici C.B., Hughes D.W., Jamar C., Malaise D., Schmidt K., Schmidt W.K.H., Seige P.: 1987, *Astron. Astrophys.*, **187**, 807.

Keller H.U., Kramm R., Thomas N.: 1988, *Nature*, **331**, 227.

Kissel J., Sagdeev R.Z., Bertaux J.L., Angarov V.N., Audouze J., Blamont J.E., Büchler K., Evlanov E.N., Fechtig H., Fomenkova M.N., von Hoerner H, Inogamov N.A., Khromov V.N., Knabe, W., Krüger F.R., Langevin Y., Leonas V.B., Levasseur-Regourd A.C., Managadze G.G., Podkolzin S.N., Shapiro V.D., Tabaldyev S.R., Zubkov B.V.: 1986, *Nature*, **321**, 280.

Kissel J., Krüger F.R.: 1987, *Nature*, **326**, 755.

Korth A., Marconi M.L, Mendis D.A., Krüger F.R., Ritcher A.K., Lin R.P., Mitchell D.L., Anderson K.A., Carlson C.W., Rème H., Sauvaud J.A., d'Uston C.: 1989, *Nature*, **337**, 53.

Levasseur-Regourd A.C., Bertaux J.L., Dumont R., Eichhorn G., Festou M., Giese R.H., Giovane F., Lamy P., Le Blanc J.M., Llebaria A., Weinberg J.L.: 1986, *BAAS*, **18**, 3, 790.

Levasseur-Regourd A.C.: 1988, *J. Phys.*, **49**, 3, C1, 25.

Levasseur-Regourd A.C.: 1989, in *Workshop on physics and mechanics of cometary materials*, ESA SP **302**, 209.

Levasseur-Regourd A.C., Dumont R.: 1990, *Adv. Space Res.*, **10**, 3, 163.

Levasseur-Regourd A.C., Dumont R., Renard J.B. : 1990, *Icarus*, **86**, 264.

Marconi M.L., Korth A., Mendis D.A., Lin R.P., Mitchell D.L., Rème H., d'Uston C. : 1989, *Astrophys. J.*, **343**, L77.

Marconi M.L., Mendis D.A., Korth A., Lin R.P., Mitchell D.L., Rème H.: 1990, *Astrophys. J.*, **352**, L17.

Maurette M., Jehanno C., Robin E., Hammer C.: 1987, *Nature*, **328**, 699.

McDonnell J.A.M., Lamy P., Pankievicz G.S.: 1991, in *Comets in the post-Halley era*, ed. R. Newburn et al., Kluwer Ac. Pub., 1043.

Mitchell D.L., Lin R.P., Anderson K.A., Carlson C.W., Curtis D.W., Korth A., Rème H., Sauvaud J.Λ., d'Uston C., Mendis D.A.: 1989, *Adv. Space Res.*, **9**, 2, 35.

Mumma M.J., Reuter D.C.: 1989, *Astrophys. J.*, **344**, 940.

Neugebauer M., Weissman P.: 1989, *Eos*, **70**, 633.

Owen T., Lutz B.L., de Bergh C.: 1986, *Nature*, **320**, 244.

Ponnamperuma C. (ed.): 1981, *Comets and the origin of life*, D. Reidel

Rickman H. : 1991, in *Comets in the post Halley era*, ed. R. Newburn et al., Kluwer Ac. Pub., 733.

Schleicher D.G., A'Hearn M.F.: 1986, in *New insights in astrophysics*, ESA SP **263**, 31.

Simpson J.A., Sagdeev R.Z., Tuzzolino A.J., Perkins M.A., Ksanfomality L.V., Rabinowitz D., Lentz G.A., Afonin Y.V., Erö J., Keppler E., Kosorokov J., Petrova E., Szabo L., Umlault G.: 1986, *Nature*, **321**, 278.

Snyder L.E., Palmer P., de Pater I.: 1989, *Astron. J.*, **97**, 246.

Vaisberg O.L., Smirnov V.N. Gorn L.S., Iovlev M.V., Balikchin M.A., Klimov S.I., Savin S.P., Shapiro V.D., Shevchenko V.I.: 1986, *Nature*, **321**, 274.

Vanysek V., Vanysek P.: 1985, *Icarus*, **61**, 57.

Vanysek V.: 1991, in *Comets in the post-Halley era*, ed. R. Newburn et al., Kluwer Ac. Pub. 879.

Vidal-Madjar A.: 1983, in *Diffuse matter in galaxies*, NATO ASI series 110, R. Reidel, 57.

Wallis M.K., Wickramasinghe N.C., Hoyle F., Rabilizirov R.: 1989, *Adv. Space Res.*, **9**, 2, 55.

Weissman P.: 1990 in *Conf. on global catastrophes in Earth history*, in press.

Wickramasinghe N.C., Allen D.A.: 1986, *Nature*, **323**, 44.

Discussion

C. MATTHEWS: The reddish-black color of the nucleus of Halley could be further evidence for the presence of black HCN polymers, which have been shown by infrared reflective studies (Cruikshank et al.) to be reddish in color. The 'missing' methane and ammonia could also be locked up in these HCN polymers, since these compounds together are readily converted to hydrogen cyanide, which has been detected in the coma.

C. LEVASSEUR-REGOURD: Comment rather than question = no answer.

C. MCKAY: Since most models of the interstellar medium include significant carbon in the form of graphite (or amorphous carbon), is there any indication from the Halley results for grains composed purely of carbon?

C. LEVASSEUR-REGOURD: Due to the high velocity flyby of Halley, only the <u>elemental</u> composition is obtained. However, the organic grains, as seen on the mass spectrometer spectra, seem to be "CHON" molecules rather than pure "C."

D. BRIN: You point out that dust grains evolve after leaving the cometary nucleus, breaking up into submicron particles as the "glue" holding them together sublimates. Is the size gradient smooth? Or are there different slopes to the distributions of grain sizes emitted, and those finally spread through the solar system? Are the tiny grains held together by a different glue than that binding the larger ones in the nucleus?

C. LEVASSEUR-REGOURD: All these points are still poorly known. It is indeed most important to accurately estimate the various slopes in the size distribution.

W. IRVINE: I believe it is important to verify the H_2CO identification in P/Halley by seeking H_2CO in other comets. Are you familiar with the recent search for H_2CO in C/Austin by the Meudon comet group using the IRAM 30 meter telescope?

C. LEVASSEUR-REGOURD: Hopefully, this will be answered within a few weeks.*

D. WHITMIRE: You said comet Halley formed beyond the giant planets. Can you give a minimum distance or maximum temperature? Also, is it known if any long-period comets formed beyond the giant planets?

C. LEVASSEUR-REGOURD: The distance could be greater than 20 to 30 AU (T smaller than 50 to 100K) or may be even greater than 100 AU (from S_2 molecule observations).

*Note added in proof:
Positive results are published by Bockelée-Morvan D., Colom P., Crovisier J., Despois D., and Paubert G. in <u>Nature</u>, 1991, **350**, 318.

ORIGIN OF THE BIOSPHERE OF THE EARTH

A. H. Delsemme
Department of Physics & Astronomy
The University of Toledo
Toledo, Ohio 43606, U.S.A.

ABSTRACT

The paradigm that has emerged to describe the origin of the solar system excludes the presence of water and of carbon in the planetesimals that agglomerated to form the proto-Earth. An unlikely but possible primary atmosphere of solar composition was transient enough not to play any significant role in the retention of water or carbon. However, the latter evolution of the planetesimals formed in the zone of the Jovian planets, brings a large number of objects made at cooler temperatures into the zone of the terrestrial planets. These objects are mainly the comets, that are going to bring to the Earth more water than needed to explain our oceans, and more carbon than needed to explain the carbonates and the biosphere. This general mechanism seems to work in the late evolution of numerous accretion disks around young stars, and promises to bring enough water and volatile compounds on rocky planets that would have otherwise remained barren.

ACCRETION DISKS

Observational evidence has recently established the abundance of accretion disks around very young stars. The Infrared Astronomical Satellite IRAS has found around many stars an infrared excess which is the signature of interstellar dust. Many of these stars are very young T Tauri stars; some, like FU Orionis, are still probably accreting mass. Finally, optical pictures in the visual have resolved dusty disks of size 500 to 1000 AU, like in Beta Pictoris (Smith and Terrile 1984). The existence of accretion disks has therefore been substantiated suddenly and has become the accepted explanation on the way Nature succeeds in making single stars: namely by shedding the angular momentum in excess to the expanding margin of an accretion disk during the buildup of the stellar mass. The first consequence of this explanation is that many single stars are likely to make a planetary system out of their accretion disk. The basic theory of the viscous accretion has been given by Lynden-Bell and Pringle (1974). Numerical models have been developed by Cameron (1985), Lin and Papaloisou (1985); Morfill et al. (1985), Wood and Morfill (1988), and Morfill (1988), Morfill and Wood (1989). Although momentum transfer by gravitational torques has also been considered, we will use only the viscous accretion disk model for simplicity.

ORIGIN OF THE SOLAR SYSTEM

To describe the origin of the Solar System by an accretion disk model, we need the collapse rate as well as the viscous friction. These unknown parameters can be combined into a single variable that can be adjusted to empirical data, for instance to the condensation temperatures of the planets established by Lewis (1974). This adjustment is possible because of the smooth temperature gradient established radially in about r^{-1} (Lewis 1974) versus the temperature gradient found in the models, like in $r^{-0.9}$ (Morfill 1988) in the mid-plane of the accretion disk. Figure 1 shows the type of adjustment reached in this case. Lewis' temperatures are based on the condensation of solids from a cooling gas of solar composition at thermochemical equilibrium. They specify the temperatures of the dust grains, at the epoch when they were removed from this thermochemical equilibrium, that is when they sedimented to the mid-plane of the disk and started accreting into larger solid bodies that effectively isolated the dust from the gas.

Lewis' original model used a cooling sequence starting at high temperature from a totally vaporized nebula, but early models of the accretion disk have shown that this is not what happened: interstellar grains, made far out from chemical equilibrium, were not completely vaporized, except at very close distances to the protosun. However, there are at least three independent telltales of an early accretion phase where the disk was heated to higher temperatures, at least in the asteroid belt. Two of the clues come from the chondrites only: the pervasive chondrules were partially molten at temperatures beyond 1700 K; the calcium-aluminum inclusions (CAI) were fractionated near 1800 K. Both imply high rates of cooling, suggesting transient heat episodes only. The third clue is shared by the chondrites, the Earth and the dust of comet Halley: it indicates that a variable fractionation of total Fe relative to the other elements has occurred among these three groups of objects. In particular the Fe/Si ratio goes from 75% to 25% solar, using the new solar values of Anders and Grevesse (1989). Iron condenses earlier than silicates in the temperature range beyond 1500 K, suggesting a longer and variable heating beyond 1500 K in the accretion disk at the distances of the Earth and of the asteroids, and for a fraction of the dust later incorporated into comets of the Halley type.

Figure 1. Mid-plane temperature of the accretion disk as a function of radius. The crosses are Lewis (1974) aggregation temperatures of planets and satellites. The solid line is the adjustment of the disk model (Morfill 1988). The two dotted lines correspond to accretion rates 10 times smaller or larger.

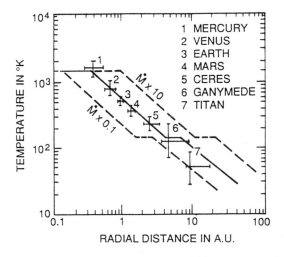

The scenario of the iron fractionation remains somewhat uncertain and the temperature history of the accretion disk is probably more complicated than that suggested by oversimplified models. For instance, most models assume that the accretion rate was constant; it was probably very variable, because the collapsing cloud fragment was not a sphere of homogeneous density, but an irregular nodule with density fluctuations in its structure. The collapse was probably initiated by that substructure that reached the highest density, hence the accretion rate was larger to begin with and subsided with large fluctuations.

If the collapse started at a rate ten times as large as the average rate, Fig. 1 shows that the early temperature distribution in the disk was about 2-1/2 times the late temperature distribution. Fluctuations in the accretion rate with early large peaks could explain the variable iron fractionation, the CAI and the chondrules at the proper distances. The cooler temperature distribution coming later would be consistent with the existence of a low-temperature component in chondrites; a very small fraction of individual interstellar grains could also be preserved in the accretion disk during the final subsidence of the collapse rate. Since the details of the actual fluctuations are not known, significant numerical models cannot be computed, but the steady-state models (Fig. 1) remain useful to give limits and orders of magnitude.

The essential for the following argument is that Lewis' temperatures were probably not the highest temperatures reached during accretion. Therefore Lewis' temperatures were assumedly reached by cooling, hence the kinetics had ample time to reach chemical equilibrium before the separation of dust from gas, which took place by grain sedimentation to the mid-plane. This chemical kinetics has been thoroughly discussed by Lewis et al (1979) and by Lewis and Prinn (1980); they find that near 1000 K, the significant time constants are near one century, a time much shorter than the coagulation time of dust into meter-size bodies (Weidenschilling 1988). We conclude that chemical equilibrium was easily reached before the solid grains were removed and lost contact with the gas phase, even in circumstances that do not imply a previous hot spike.

THE EXOGENOUS ORIGIN OF CARBON AND WATER ON THE EARTH

The complex organic molecules discovered in the carbonaceous chondrites have been explained by Anders (1986) by a Fischer-Tropsch type (FTT) catalysis on Fe-Mg-silicate grains between 360 and 430 K. This catalysis reproduces most observed features: branching peculiarities as well as isotopic anomalies. Since carbonaceous chondrites are believed to come from beyond 2.6 AU in the solar system, this interpretation implies that, when dust was separated from gas in the accretion disk, the temperature had cooled down to 430 K at 2.6 AU. The temperature gradient of the disk (Fig. 1) and, more appropriately, the adiabats of Fig. 2, show then that 800-900 K was reached in the accretion zone of the Earth (1 AU).

This corresponds to the adiabat CD of Fig. 2 (interpolated from Cameron 1985). One sees that for such an adiabat, all carbon is in gaseous CO in the accretion zone of the Earth. Therefore no carbon could condense on the Earth's grains of dust at that time. Later on, the grains were removed from contact with the gas phase, by sedimentation to the mid-plane and accumulation into larger and larger planetesimals. On Fig. 2, models C, CD and D can be interpreted as a cooling sequence, and the organic molecules present in the chondrites coming from beyond 2.6 AU can be interpreted as the telltale of the fact that dust grains were removed from the chemical equilibrium with the nebular gas along adiabat

CD, not sooner and not later. This also implies that the Earth's dust was removed from this equilibrium at a temperature of 850 K, in agreement with Lewis' assessment based on totally different considerations.

The iron grains of the Earth's zone had been fractionated from silicates earlier, at or beyond 1500 K, but the disk was cooling down because assumedly the accretion rate had diminished by a factor of 10 to 20. At 1043 K (Curie Temperature) a fraction of the iron grains were transported elsewhere, possibly by the magnetic field of the disk (ionized corona in rotation). At 850 K, the residual grains had sedimented to the mid plane with other silicate grains. When they were removed from the gaseous equilibrium, let's note that the iron grains were still completely reduced (near 850 K) by CO in presence of a large excess of H_2, a prerequisite for them to sink eventually into the Earth's core (neither iron oxides nor silicates can either sink to the Earth's core because of their smaller density, or be reduced when imprisoned in the Earth's silicate mantle). This sequence of events also explains why, if it were homogenized, the Fe/Si ratio of the Earth is at most 45 to 50% of solar: half of its iron grains have been lost just before the accretion process took place. The fate of the lost grains remains uncertain (see Delsemme 1990).

DID A PRIMARY ATMOSPHERE EXIST?

At 850 K, all carbon was in CO and all hydrogen in H_2. Iron and silicate grains had been outgassed and dehydrated before accretion, hence the bulk of carbon and the oceans of the Earth must have an exogenous origin. It could be argued that a primary atmosphere of solar composition could have been captured in the nebular gas by the growing gravity of the protoplanet. However, the isotopic composition of the heaviest inert gases (Kr, Xe) still present in our atmosphere is far from solar, indicating that if such an atmosphere ever existed, it was unimportant or has dissipated early.

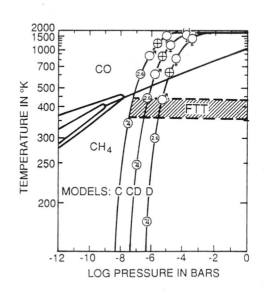

Figure 2. - Thermochemical equilibrium for carbon in a gas of solar composition. The quasi vertical curves are Cameron's 1985 adiabats. These adiabats can be interpreted as a cooling sequence. Adiabat CD brings the distance 2.6 AU at the high end of the FTT temperature zone, implying that dust sedimentation took place at that time. The same adiabat shows that, in the Earth's zone (symbol ⊕) all carbon was then in gaseous CO.

Hayashi et al (1985) argue that the accretion of the protoplanet takes place in 10^6-10^7 years, but its primary atmosphere would dissipate later in 10^6 years. If the bulk of our atmosphere is not primary, we must now elucidate its origin.

ORBITAL DIFFUSION OF PLANETESIMALS

After the accretion of planetesimals, larger planetary embryos grow from numberless low-velocity collisions. As soon as there are larger bodies whose gravitational attraction cannot be neglected, close encounters enlarge the inclinations and the eccentricities of the smaller objects' orbits. The mean relative velocities grow, at steady state, in proportion to the escape velocity of the largest body, hence the zones swept by the minor bodies grow with the size of the protoplanet nearby. The final growth of the giant planets brings more volatile material (namely comets) into the zone of the terrestrial planets, at a time when the growth of the latter is virtually finished.

Computing the contribution of the different zones that are going to bring a veneer of more volatile material on the protoearth, I find that the contribution of the zones of Uranus and Neptune is negligible; the chondrites of the asteroids' zone bring a veneer of 1 km of siliceous material on the Earth, but only 10 m of water and 5 m of organics. The comets of the zones of Jupiter and Saturn provide the major contribution: a veneer of 2.5 km of siliceous material, but 5 km of water (the double of our present oceans), 1.2 km of organic material and 240 bars of atmospheric gases. This veneer of volatile materials is brought 10^7 to 10^8 years after the nebular gas has been blown away by the T Tauri wind of the Sun, hence much later than the dissipation of the possible primitive atmosphere (Delsemme 1981, 1984, 1990a & b).

Recent evaluations of the amount of water go beyond my results. Matsui and Abe (1986) found four times the mass of our oceans, Ip and Fernandez (1988) ten times, consistent with Chyba's (1987) data. Chyba uses the visible craters on the Moon to estimate the early bombardment on the Earth. He finds that, if only 10% of the extant lunar craters are due to comets, the total mass of our oceans is explained. This suggests that a large fraction of this water and even a larger fraction of the atmosphere have been eroded by those giant impacts that are implied by the late stages of the theory of the accumulation of the protoplanets. Hence the early crust of the Earth, including the labile elements could have been mostly produced by exogenous silicates coming from 30% of chondrites and 70% of comets. Of course this veneer material was subjected to melting and magma formation due to the gardening of the impacts, but this is another story. It is however easy to imagine that, as soon as an atmosphere was strongly established on the primitive Earth, the conditions that have been able to bring carbonaceous chondrites down to Earth recently without destroying their organic chemistry, were duplicated. Even if a tiny fraction of the early organic veneer was preserved, it would contain countless tons of amino-acids (as in the Murchinson meteorites) and of purines and pyrimidines (as in Comet Halley).

The process by which the terrestrial planets were first deprived of volatiles to make them rocky, then sowed with prebiotic material and provided with water is likely to be a general feature of the accretion disks that explain the formation of single stars. At the proper distance from each single star, the existence of at least one planet with liquid oceans containing prebiotic molecules is more likely to be the norm than the exception. Has life-as-we-know-it appeared on all these planets? This is the step we cannot yet clear, in our ignorance of how life has emerged out of prebiotic molecules.

More extended portions of this work are being published elsewhere (Delsemme 1990 a&b). This work was supported by NASA (CRAF mission).

References

Anders, E. 1986, p. 31-39 in *Comets Nucleus Sample Return*, ESA SP 249.

Anders, E. and Grevesse, N. 1989 Geochim. Cosmochim. Acta, 53, 197-214.

Cameron, A.G.W. 1985, p. 1073-1099 in *Protostars and Planets II*, Black and Matthews (eds), Univ. of Arizona Press, Tucson.

Chyba, C. F. 1987, Nature 330, 632-635.

Delsemme, A. H. 1981, p. 141-159 in *Comets and the Origins of Life*, Ponnamperuma (ed.), Reidel Publ., Dordrecht.

Delsemme, A. H. 1984, Origins of Life 14, 51-60.

Delsemme, A. H. 1990a, in *Comets in the Post-Halley Era*, Newburn et al, (eds.) Kluwer Acad. Press, Dordrecht (in press).

Delsemme, A. H. 1990b, in Proc. COSPAR XXVIII Plenary Meeting, The Hague (in press).

Hayashi, C., Nakazawa, K., and Nakagawa, Y. 1985, p. 1100-1153 in *Protostars and Planets II*, Black and Matthews (eds.), Univ. of Arizona Press, Tucson.

Ip, W. H., and Fernandez, J. A. 1988, Icarus 74, 47-62.

Lewis, J. S. 1974, Science 186, 440-443.

Lewis, J. S., Barshay, S. S., Noyes, B. 1979, Icarus 37, 190-206.

Lewis, J. S. and Prinn, R. B. 1980, Astrophys. J. 238, 357-364.

Lin, D.N.C., Papaloisou, J. 1985, p. 981-1072 in *Protostars and Planets II*, Black and Matthews (eds.), Univ. of Arizona Press, Tucson.

Lynden-Bell, D., and Pringle, J. E. 1974, M.N. Roy. Astron. Soc. London 168, 603-637.

Matsui, T., and Abe, Y. 1986, Nature 322, 526-528.

Morfill, G. E. 1988, Icarus 75, 371-379.

Morfill, G. E., Tscharnuter, W., Völk, H. J. 1985, p. 493-533 in *Protostars and Planets II*, Black and Matthews (eds.), Univ. of Arizona Press, Tucson.

Morfill, G. E., and Wood, J. A. 1989, Icarus 82, 225-243.

Smith, B. A. and Terrile, R. J. 1984, Science 226, 1421-1424.

Weidenschilling, S. J. 1988, p. 348-371 in *Meteorites and the Early Solar System*, J. F. Kerridge and M. S. Matthews, eds., Univ. of Arizona Press, Tucson.

Wood, J. A., and Morfill, G. E. 1988, p. 329-347 in *Meteorites and the Early Solar System*, J. F. Kerridge and M. S. Matthews, eds., Univ. of Arizona Press, Tucson.

Discussion

M. HARRIS: Your proposed accretion disk differs from the more familiar accretion disks (around x-ray sources, for example) because it contains a large proportion of solid objects. One would intuitively expect, therefore, that the disk viscosity α would be very low. It is not clear that there are self-consistent hot accretion disks with very low α.

A. DELSEMME: The accretion disk describing the origin and evolution of the Solar Nebula does not contain more solid objects than any interstellar cloud (98-99% gas versus 1-2% dust). The disk viscosity may be low, but compensated by the accretion rate that is also unknown. These two parameters can be combined into a single variable that can be adjusted to empirical data (see my Fig. 1).

The accretion disks described in the literature are self-consistent, even if their viscosity coefficient is unknown or if the source of their viscosity has not been clarified (see Morfill 1988, Cameron 1985 and others in my reference list).

A. LEGER: If we are correct saying that 10% of cosmic carbon is in aromatic compounds in IS matter, where is it in the objects bombarding the Earth?

A. DELSEMME: The poly-aromatic-hydrocarbons (PAH)) that you have studied in interstellar space have been recognized in the thick mantle of organic refractory material covering the cores of many silicate grains of comet Halley (Kissel and Krueger 1987). They have been incorporated in my model (see Delsemme 1990 Table XIII).

G. MARX: If the Moon is formed by an impact on Earth, how do you explain that its orbit is in plane of Ecliptic (like those of other moons)? The parallel angular moments seem to indicate that most moons were formed simultaneously with the planets themselves, due to angular momentum conservation.

A. DELSEMME: The orbit of the Moon is <u>not</u> in the plane of the ecliptic, but about 5° from it; (otherwise, there would be solar and lunar eclipses every month). Since there is transfer of energy and momentum from the Earth to the Moon because of the tides, the Moon goes away from us at a secular rate that is well documented. When the Moon was much closer, in the distant past, the inclination of its orbit was correspondingly much larger, perhaps 50° or 60° from the ecliptic, which implies a random process like a collision with a photoplanet of the size of Mars. See Cameron (1988).

E. MARTIN: Is the typical timescale of planetesimal agglomeration (formation of planets) of the same order of magnitude than the duration of viscous disk accretion onto a young solar-type star?

A. DELSEMME: The evolution of the hot viscous accretion disk typically takes a time of the order of 10^5 years, but the order of magnitude can be changed by changing the infall rate of the interstellar nodule that collapses and feeds the disk; typically, this infall rate is $10^{-5 \pm 1}$ solar masses per year. When dust sediments to the mid-plane, it coagulates into meter-size bodies in a few thousand years, into asteroid-size bodies in 10^5 years. Then the process slows down and it takes 10 to 100 million years to bring 99% of the mass available in a single zone into one planet like the Earth. The final orbital diffusion and bombardment of comets will take some 500 million years, in a gas-free environment, because the gale due to the T Tauri phase of the Sun has blown away the gaseous remnants of the nebula, in a time scale of 1 to 10 million years after the end of the infall that brought the Sun to its final mass.

S. ISOBE: The comet Austin discovered last December did not show its brightning depending on its distance from the Sun, as expected from previously observed comets. This suggests that the comet had thin layers of material with low sublimation temperature surrounding a solid surface of the comet, and that this comet was formed at different conditions since this comet had highly eccentric or parabolic orbit.

A. DELSEMME: For each individual comet, we are never sure whether it is "new" or not. A "new" comet is a comet that comes from the Oort cloud for the very first time. Unfortunately, the planetary perturbations due to a previous passage through the inner solar system, may send a comet back on an orbit that is undistinguishable from a "new" comet's orbit. There is about one chance over twenty for this to happen. The fact that comet Austin did not brighten up to the expectations for a "new" comet, may only mean that its more volatile materials have been lost during a previous perihelion package close to the Sun. It happened also for comet Kohoutek.

CARBON-RICH MICROMETEORITES AND PREBIOTIC SYNTHESIS

M. Maurette,[1] Ph. Bonny,[1,2] A. Brack,[3] C. Jouret,[4]
M. Pourchet,[5] P. Siry[6]

[1]C.S.N.S.M., Batiment 108, 91406 Orsay
[2]Département de Thermophysique, ONERA, B.P. 72, 92322 Chatillon
[3]Centre de Biophysique Moléculaire, CNRS, 45045 Orleans Cedex
[4]Laboratoire d'Optique Electronique, BP4347, 31055 Toulouse
[5]Laboratoire de Glaciologie, BP 96, 38402 St. Martin d'Heres
[6]Laboratoire de Biophysique, Faculté de Medecine, 94010 Créteil

Paper Presented by M. Maurette

ABSTRACT

About 5000 unmelted and well preserved "giant" chondritic micrometeorites with sizes ~50-200 μm have been extracted from ~100 tons of antarctica blue ice. They have been unexpectedly well shielded against both terrestrial weathering and frictional heating in the atmosphere. Mineralogical studies indicate that they are all related to primitive "unequilibrated" meteorites (mostly carbonaceous chondrites). About 50% of them are made of friable and porous aggregates of submicron-sized grains, that represent a highly desequilibrated assemblage of minerals, including hydrous silicates and anhydrous clasts, metal oxides and sulfides, and some carbonaceous material related to the broad family of "hydrogenated refractory carbon." Each carbon-rich micrometeorite might have behaved as a "minicenter" of prebiotic synthesis on the early Earth, through the "in-situ" catalyzed hydrolysis of this carbonaceous material.

1. INTRODUCTION

According to the classical scenario (1,2), life on the Earth might have emerged about 4 billion years ago from a large scale production of organic molecules in a reducing atmosphere, leading to a soup of amino acids, simple sugars and heterocycle bases, that constitute the chemical building blocks of cells. "Spontaneous self organization" (3) might have played a major role in the linkage of such building blocks into the complex proteins and nucleic acids of modern cells.

Because of the difficulties in reconstructing an early reducing atmosphere, geochemists favor now an atmosphere dominated by CO_2, in which the production of nitrogen containing compounds appears difficult (4). Moreover the Earth probably accreted from some differentiated material strongly depleted in carbonaceous material. Thus the building blocks and/or their precursor material should have originated from other sources. A promising one is the accretion of carbonaceous extraterrestrial materials on the Earth (5,6). Until 1988 only the contribution of large extraterrestrial bodies was considered (meteorites up to km-sized planetesimals).

Since 1988 both our group (7,8,9) and Anders (10) suggested that the much smaller micrometeorites (sizes ≤ 200 μm) might be the dominant source of precursor C-rich material on Earth. The major purpose of this paper is to outline recent analyses of micrometeorites extracted from the ultra-clean ice of the Antarctica ice sheet. These analyses suggest that they could have individually functioned as "minicenters" for the synthesis of prebiotic molecules, when they left the hostile interplanetary medium, to end up in a variety of favorable sites, such as ponds of hot and highly mineralized waters, on the heavily cratered regolith of the early Earth.

2. THE CAP-PRUDHOMME ANTARCTICA MICROMETEORITES

2.1 Collection and Presentation of Micrometeorites.

About 5000 well preserved <u>unmelted</u> cosmic dust grains and ~5000 cosmic spherules (i.e., melted) with sizes ≥ 50 μm, all showing a "chondritic" composition (see below), have been found in ~10 g of glacial sediments with size ≥ 50 μm, extracted from ~100 tons of Antarctic blue ice, near Cap-Prudhomme at ~6 km from the French station of Dumont d'Urville (8,11). About 2/3 of the unmelted grains are found in ~200 mg of the 50-100 μm size fraction, which is the purest "mine" of unmelted micrometeorites found on Earth, yet. It contains ~10% of extraterrestrial grains, as well as an unexpectedly high ratio (~80%) of unmelted to melted chondritic grains.

Extraterrestrial grains in the sediments are preselected with an optical microscope, relying on simple criteria such as being dark and showing angular shapes. Their bulk chemical composition is next analyzed with the energy dispersive X-Ray spectrometer (EDX analysis) of a scanning electron microscope (SEM). Only grains showing either a "chondritic" composition (see section 2.2) or an Fe/Ni composition are selected as micrometeorites (the proportion of Fe/Ni grains is about 1%).

Indeed the measurements of cosmogenic nuclides ([10]Be and [26]Al; see ref. 12 and 13) and/or neon isotopes (14,15,16) in these chondritic grains clearly show that they are extraterrestrial, and that their parent bodies in space were interplanetary dust particles. They have not been released from much larger meteorites impacting the Earth.

2.2 Chemical and Mineralogical Compositions.

About 300 unmelted chondritic micrometeorites from the 50-100 μm size fraction of the Cap-Prudhomme sediments have been analyzed by M. Christophe Michel-Levy and G. Kurat, relying on accurate electron microprobe analyses of clasts, minerals and matrix material (16).

These analyses reveal a chondritic composition in Mg, Al, Si and Fe, which is depleted in Ca, Ni and S with regard to the "bulk" composition of C1 carbonaceous chondrites (17). With the exception of ~6 grains that might be chunks of "ordinary" chondrites, all 300 grains only bear similarities with the <u>fine grained matrix</u> of the rare "primitive/unequilibrated" meteorites (about 4% of all meteorite falls), that mostly include carbonaceous chondrites only considered yet in exobiology.

The grains appear as porous aggregates of submicron-sized grains when ultramicrotomed sections are examined at high magnification with a high resolution electron microscope (fig. 1). About 80% of them contain a few "large" (size ~5-20 μm) crystals of olivine and pyroxene. Their mineral assemblage, which is highly disequilibrated, frequently shows the association of hydrous silicates and anhydrous clasts. They also contain fine-grained metallic oxides and sulfides, and about 50% of them show ultrathin coatings of Fe-rich material on both their external surface and some micropores. They comprise an amazingly high diversity of types, and only two "similar" grains have been found in this set of 300 chondritic grains.

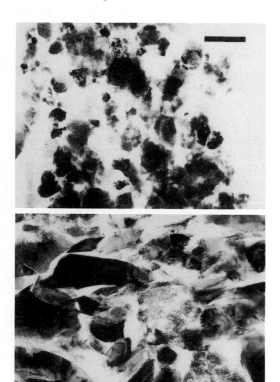

Figure 1. High voltage electron micrographs of ultramicrotomed sections of G21 (top) and G20 (bottom), showing their texture of porous "composite-nanophase" material (scale bar ~0.2 μm) (Courtesy P. Veyssières).

2.3 <u>Negative Test For Biogenic Weathering</u>.

We have developed an "etch canal" method to assess the type and the degree of weathering (either chemical or biogenic) of any micrometeorite collection (18). This method relies on SEM observations of "barred" chondritic cosmic spherules (fig. 2), that are composed of olivine and interstitial glass bars (light and dark phases, respectively, with the bright inclusions of magnetic being embedded into glass).

In deep sea spherules the glass bars are etched out preferentially, in accordance with a process of chemical etching in water. But in spherules extracted from Greeland sediments made of cocoons of siderobacteria ("cryoconite") olivine is now etched out preferentially through "biogenic" weathering. In Antarctic spherules ≥90% of the spheres are unetched. Occasionally a sphere shows very shallow etch canals up to depths of a few μm, that clearly follow the glass bars (fig. 2). This demonstrates that

Antarctic micrometeorites were never exposed to any biogenic activity (i.e., a potential source of severe contamination in organic material).

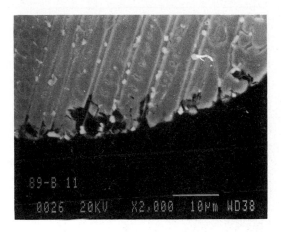

Figure 2. SEM micrograph of etch canals in a chondritic barred spherules from the Cap-Prudhomme collection (scale bar: 10 μm).

2.4. High Voltage Electron Spectroscopy.

We microprobed the carbon content of micron-size crushed fragments (fig. 3A) from a peculiar subset of 12 chondritic antarctica micrometeorites with size ~200 μm, that all showed a high porosity. The frictional heating of such "giant" micrometeorites in the atmosphere was certainly more severe than that of ~50-100 μm-size grains, and consequently more defavorable to the retention of their carbonaceous component.

For this analysis with the electron energy loss spectrometer of a 1 MV electron microscope (19) we had first to successfully tackle the severe problem of carbon contamination. A reliable search for carbon excludes the use of ultramicrotomed sections impregnated with organic resin, and that of grain dispersion on ordinary and/or holley carbon membranes. We crushed directly chunks of micrometeorites on gold and platinum grids (without membrane) held between two glass slides. This technique works beautifully: crushed grains from a single crystal of olivine did not show any measurable carbon peak

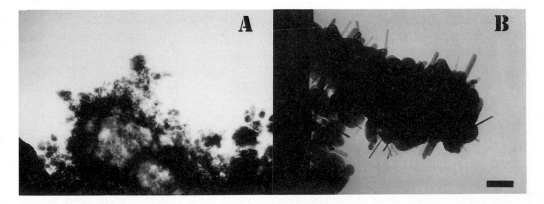

Figure 3. High voltage electron micrographs of crushed fragments from micrometeorite G20, before (A) and after (B) pyrolysis at 1000°C.

after one hour of exposure in the electron beam of the microscope (fig. 4, top) and, the grains can be heated up to 1000°C without being lost while suffering drastic textural changes (fig. 3B).

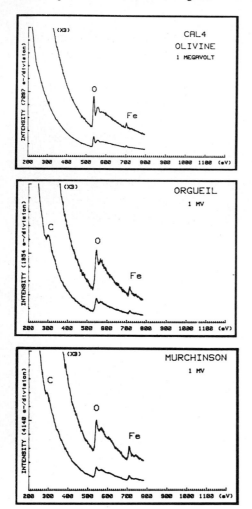

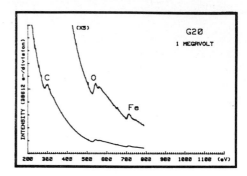

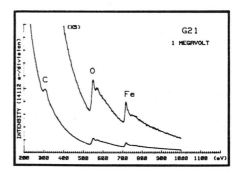

Figure 4. Electron energy loss spectra of micron-sized crushed fragments. Left: a single olivine crystal; two carbonaceous meteorites (Orgueil, Murchison). Above: two Antarctic micrometeorites (G20, G21).

For each micrometeorite, and for chunks of the Orgueil and Murchinson meteorites (used as the calibration standard), we selected ten micron-size grains (for each object) showing a favorable zone: where oxygen has a double peak structure (see the magnified upper spectra in each plot). This indicates that the sample is thin enough (≤0.5 μm) as to allow a good determination of the C/O atomic ratio with an appropriate reduction procedure (19).

In fig. 4 we reported a few spectra corresponding to the highest C/O ratios measured in the grains. For Orgueil and Murchinson, 9 zones out of 10 show a detectable peak with C/O ratios either ranging from 0.4 to 0.2 (Orgueil) or clustering around 0.1 in Murchinson. In two micrometeorites, including G20, all 10 zones yield a high carbon peak with variable C/O ratios ranging from 1.1 to 0.8. G21 belongs to a major group of 5 micrometeorites, where the C content is much lower but still detectable, as two of the 10 zones show a C peak corresponding to C/O ratios ~0.3. No carbon peak was observed in the remaining micrometeorites. Nitrogen in the spectra of both micrometeorites and the matrix of carbonaceous chondrites was below our detection limit of 1%. HVEM micrographs indicate that the carbon-rich material is associated with an amorphous component appearing as a light phase in fig. 3A.

Grains with high C contents were run in the reaction chamber of a 3MV HVEM, where their dynamical transformation upon prolysis up to 1000°C (at 10^{-7} torr) was monitored with a video camera. In Antarctic micrometeorites textural transformations of the amorphous component both start at high temperatures (700-800°C), and lead to diverse evolutions around 900-1000°C, such as a "whisker" type recrystallization (fig. 3B), a gradual thinning, a softening that induced rounded habits and so forth. In contrast the amorphous component of a much smaller (~10 µm) micrometeorite collected in the stratosphere suffered a drastic shrinking at 300°C, but no further transformation was observed up to our highest temperature of 1000°C.

The high thermal stability of the carbon-rich component of Antarctic micrometeorites suggests its filiation with a broad class of "refractory hydrogenated carbon" (including polycyclic aromatic hydrocarbons that might be among the most abundant interstellar molecules, see ref. 20), that has still not reached the stage of graphitization.

The matrix of ~20% of unmelted Antarctic micrometeorites with sizes ~200 µm show carbon contents larger than those observed in that of carbonaceous meteorites. Both micrometeorites and carbonaceous chondrites are depleted in nitrogen with regard to the so-called universal abundances. Grains released by comet Halley, that impacted the time-of-flight mass spectrometer on board the Giotto spacecraft (21), also show a 5-fold depletion in nitrogen.

About 90% of Antarctic micrometeorites are found in the 50-200 µm size fraction, that corresponds to the peak in the mass distribution of the cosmic mass influx to Earth (22). About 10,000-20,000 tons of this material is accreted each year by the Earth, while the contribution of all meteorites with mass ≥100 g amounts to ≤100 tons. Moreover the ~300 micrometeorites investigated as yet are all related to carbonaceous chondrites. In contrast only ~4% of all meteorite falls belong to this group of meteorites, which is the only one to be seriously considered in a discussion of prebiotic synthesis. As ~20% of the micrometeorites contain carbon concentrations that exceed those measured in the most C-rich meteorites, their accretion by the Earth looks like a dominant source of C-rich material on the Earth.

3. HYPERVELOCITY CAPTURE OF EXTRATERRESTRIAL MATERIAL BY THE EARTH: THE DOMINANT ROLE OF MICROMETEORITES

Micrometeorites survive amazingly well upon their hypervelocity impact with the atmosphere. Indeed ~80% of those found in the 50-100 µm size fraction are unmelted, and ~40% of them contain very high concentrations of extraterrestrial neon (14,15). This high survival probability has to be contrasted with previous predictions inferred from classical "ablation" models (23), stating that "all" particles with sizes ~100 µm should be completely destroyed (i.e., at least completely melted) upon frictional heating in the atmosphere. However, even in the framework of these classical models, the recent "statistical" computations of Bonny (9) predict an astonishingly good preservation of the C-rich component of the grains, which is mostly related to the very short duration of their "flash" heating in the atmosphere.

This high relative abundance of micrometeorites was probably even further enhanced ~4by ago, near the end of the accretionary tail, when the inner solar system was swept up by both leftover planetesimals, comets and asteroids deflected by the giant planets, and interplanetary dust. At this time the flux of extraterrestrial debris infalling on the Earth might have been $\geq 10^3$ times higher than by contemporary flux (see ref. 10). It might also be that the extended and denser terrestrial atmosphere was even more favorable to the survival of micrometeorites, as aerodynamical braking started at higher elevation.

For comparison, the C-rich constituent of most "giant" extraterrestrial bodies with masses $\geq 10^3$ tons would be likely destroyed during their cataclysmic explosive impact with the Earth, because aerodynamical braking in the atmosphere becomes negligible for such bodies. The very detailed modeling presented by Chyba et al (24) predicts that a small fraction of carbonaceous asteroids and comets would have survived upon impact on early seas, bringing "intact" extraterrestrial organic molecules at a rate of 10^3 to 10^4 tons/yr, which is about ≥ 10 times smaller than that deduced for "early" micrometeorites. All factors again favor a dominant contribution of micrometeorites as a source of carbonaceous material on the early Earth, without worrying too much about their exact origin (cometary and/or asteroidal).

But other important characteristics of micrometeorites, that were previously overlooked, suggest that they might have even functioned as "minicenters" of prebiotic synthesis on the early Earth.

4. "MINICENTERS" OF PREBIOTIC MOLECULES ON THE EARLY EARTH

Each C-rich micrometeorite is made of millions of tiny mineral grains. It concentrates within a microscopic "environment" of $\sim 10^{-6}$ cm^3, all favorable ingredients postulated in previous scenarios of prebiotic synthesis, namely: a high concentration of precursor organic material and a dominant component of fine grained hydrous silicates, that might present some catalytic activity.

Micrometeorites show additional characteristics that are favorable to "in-situ" synthesis of complex organic molecules during hydrolysis of their organic component, and such as: the high specific area of their micropores/vesicles with dimensions extending down to molecular sizes, where high local concentrations of reactive species could build up through some type of internal "exudation" and/or weathering; a probable depth variation in the composition of the carbonaceous precursor material, that would reflect different temperature histories during aerodynamical braking in the atmosphere; a rich mixture of minerals, ultrathin coatings of Fe-rich materials that were likely produced during atmospheric entry, as well as varieties of fine-grained metallic oxides and sulfides, that might further catalyze some important chemical reactions; a large diversity of micrometeorite types, that greatly complicates their classification, but that would be favorable to prebiotic synthesis, as it warrants a concomitant diversity of carbonaceous material and host mineral phases to be captured by the Earth.

During the period of "early intense bombardment" we estimate that about one trillion micrometeorites were isotropically captured each year by the Earth. Each thousand years, each m^2 of the Earth surface was thus "hitted" by $\sim 10^6$ micrometeorites showing a huge diversity of types, while roughly only one carbonaceous meteorite (mass ≥ 100 g) was expected over an area of $\sim 10,000$ m^2. Meteorites could thus miss favorable prebiotic "spots" already populated by a large number of very diverse micrometeorites.

This scenario is supported by related observations concerning the synthesis of organic material in carbonaceous chondrites during their hydrolysis on a "damp" extraterrestrial regolith. McSween and collaborators (25,26) first showed that carbonaceous chondrites are not the "most primitive" solar system objects, as their precursor material was subjected to aqueous alteration in some "damp" extraterrestrial regolith. Chang and Bunch (27) suggested that amino acids and other C-rich compounds found in these primitive meteorites resulted from such an "hydrolysis," as catalyzed by their fine grained hydrous silicates. Recently Shock and Schulte (28) have used thermodynamic computations to show that hydrolysis of polycyclic aromatic hydrocarbons (PAHs) in meteorites might yield a rich mixture of organic compounds.

5. FUTURE PROSPECTS

Micrometeorites might have individually behaved as "minicenters" of prebiotic synthesis in the "damp" and hot regolith of the early Earth. The initial carbonaceous components of micrometeorites would belong to a continuum of materials ranging from hydrogenated refractory carbon to PAHs. The partial pyrolysis of these compounds during their "flash" heating in the atmosphere possibly generated a "kerogen" type material.

There are several problems still attached to this microscopic scenario of prebiotic synthesis, such as the depletion of nitrogen compounds in micrometeorites, which might be a general property of cometary matter, as grains from comet Halley are also depleted in this element. It would be important to find a chemical reaction network allowing atmospheric nitrogen to react with the organic component of micrometeorites. We have also to definitively eliminate the possibility that the carbonaceous component of micrometeorites does not result from some exotic terrestrial contamination.

Some aspects of this scenario are presently investigated at "Centre de Biophysique Moléculaire du CNRS," relying on amino-acid condensation already described by Brack (29). This procedure allows us to check whether or not mixtures of PAHs and carbonaceous chondrites, and/or an aliquot of a few hundred Antarctica micrometeorites, have any detectable catalytic activity for the synthesis of peptides and oligonucleotides from the corresponding building blocks.

References

(1)   Oparin, A. I. in "L'origine de la vie sur Terre," Masson (1965).
(2)   Stribling, S., Miller, S. L., Origins of Life 17, 261 (1987).
(3)   See "Artificial Life II," February 1990, Sante Fe.
(4)   Schlesinger, G. and Miller, S. L. J., Molec. Evol. 19, 376-382 (1983).
(5)   Oro, J., Nature 190, 389 (1961).
(6)   Greenberg, J. M., Adv. Space Res. 9, 15 (1989).
(7)   Maurette, M., Pourchet, M.,, Bonny, Ph., Jouret, C., Siry, P. in "Comptes Rendus des Journées de l'ATP de Planétologie," p. 359, Observatoire de Besançon, 24-26 Mai 1988.
(8)   Maurette, M., Hammer, C., Pourchet, M. in "From Mantle to Meteorites," Indian Academy of Sciences, p. 87 (1990).
9)   Bonny, Ph., L'Astronomie, 103, 498 (1989).
(10)   Anders, E., Nature 342, 255 (1989).

(11)   Maurette, M., Pourchet, M., Bonny, Ph., de Angelis, M., Lunar Plan. Sci. <u>20</u>, 644 (1989).

(12)   Raisbeck, G., Yiou, F., Bourles, D., Maurette, M., Meteoritics <u>21</u>, 487 (1986).

(13)   Raisbeck, G., Yiou, F., Meteoritics, <u>22</u>, 485 (1987).

(14)   Maurette, M., Olinger, C., Walker, R.M., Hohenberg, C., Lunar Plan. Sci. <u>20</u>, 640 (1989).

(15)   Olinger, C., Maurette, M., Walker, R. M., Hohenberg, C., Earth Plan. Sci. Lett., accepted for
         publication (1990).

(16)   Maurette, M., Olinger, C., Christophe-Michel Levy, M., Kurat, G., Pourchet, M., Brandstater,
         F., Bourrot-Denisse, M., Nature (1991) in press.

(17)   Anders, E., Ebihara, M., Geochim. Cosmochim. Acta <u>46</u>, 2363 (1982).

(18)   Callot, G., Maurette, M., Pottier, L., Dubois, A., Nature <u>328</u>, 147 (1987).

(19)   Kihn, Y., "Thèse de Doctorat es Sciences," n°1211, Universite Paul Sabatier (Toulouse, 1985).

(20)   Leger, A., Puget, J. L., Astron. Astrophys. <u>137</u>, L5 (1984).

(21)   See "Proc. Workshop on Analysis of Returned Comet Nucleus Samples," in press (1990).

(22)   Grun, E., Zook, H. A., Fechtig, H., Gleen, R. H., Icarus <u>62</u>, 244 (1985).

(23)   Brownlee, D. E., Ann. Rev. Earth Plan. Sci. <u>13</u>, 147 (1985).

(24)   Chyba, C. F., Thomas, P. J., Brookshaw, L., Sagan, C., Science <u>249</u>, 366 (1990).

(25)   McSween, H. Y., Geochim. Cosmochim. Acta <u>51</u>, 2469 (1987).

(26)   Tomeoka, K. T., McSween, H. Y., Buseck, P. R., Proc. NIPR Symp. Antarct. Meteorites <u>2</u>, 221
         (1989).

(27)   Chang, S., Bunch, T. E. in "Clay Minerals and the Origin of Life," (ed. Cairns-Smith, A. G.),
         p. 116 (Cambridge University Press, 1986).

(28)   Shock, E. V., Shulte, M. D., Nature <u>343</u>, 728 (1990).

(29)   Brack, A., Origins of Life <u>17</u>, 367 (1987).

TITAN'S ATMOSPHERE PROBED BY STELLAR OCCULTATION

A. Brahic*†, B. Sicardy*†, C. Ferrari†, and D. Gautier†

†Observatoire de Paris, France
*Universite Paris VII, France

Paper Presented by A. Brahic

Summary

Saturn's giant moon Titan shares with the Earth and Triton a predominantly nitrogen atmosphere. Most of current information about Titan's atmosphere was obtained by Voyager I fly-by on November 12, 1980. This satellite will remain a mysterious object until the ESA-NASA Cassini-Huygens mission. Titan's atmosphere may be conveniently divided into three regions: (i) the troposphere (with pressure $> \sim 50$ mbar, or an altitude $< \sim 50$ km), dominated by convection, (ii) the middle atmosphere, or stratosphere (from ~ 50 mbar to ~ 5 nbar, i.e. between ~ 50 and 800 km), dominated by radiative equilibrium, and (iii) the thermosphere and the exosphere (above the ~ 5 nbar level), dominated by conduction. The Voyager I radio occultation experiment probed the atmosphere of Titan's ground until an altitude of 200 km, as did the infrared spectrometer. On the other hand, the Voyager UV spectrometer explored higher levels in the thermosphere, between 1000 and 1300 km altitude. Many physical processes (heating, cooling, haze variations, vibrational relaxation of molecules, ionization phenomena, molecular absorption, organic condensation,...) are involved in the region between 1 mbar and 1 nbar, where the temperature profile is uncertain. This is why a direct measurement is important.

On July 3, 1989, the bright star 28 Sagitarii, or SAO 187255 (V ~ 5.5), passed behind the satellite (V ~ 8.3). The event was visible from Europe, North America and Middle-East. This observation probed Titan's atmosphere in the ~ 250-500 km altitude range, or the ~ 250-1μ bar pressure range, where up to now there was an "information gap" between infrared and UV Voyager observations. This was the first stellar occultation by Titan ever observed. B. Sicardy, C. Ferrari, J. Lecacheux, F. Roques, J. Arlot, F. Colas, W. Thuillot, F. Sevres, J.L. Vidal, C. Blanco, S. Cristaldi, C. Buil, A. Kletz and E. Thouvenot observed this occultation at Paris Observatory and Catania Observatory.

These observations provide information on Titan's stratosphere between the 250 and 500 km altitude levels. We derive a mean-scale-height of 48+/-3 km at 450 km altitude (3μbar pressure level). This constrains the mean temperature to be in the range 149 $<$ T $<$ 178 K at that level. Titan's shadow center passed within a few tens of kilometers of Paris, and the central flash observed there is an unique opportunity to constrain the apparent oblateness of Titan's atmosphere at the 0.25 mbar level (250 km altitude), giving a value which may be as high as 0.014. This atmospheric oblateness could be due to a super rotation of the atmosphere or a differential thermal effect. If entirely due to rotation, the oblateness would require a minimum 26-hour atmospheric rotation period which is much smaller than the solid body rotation period (assumed to rotate with the synchronous period of about 16 days). Then Titan's large scale stratospheric motions may be of the same class as those of Venus.

STUDY OF TRANSMITTED LIGHT THROUGH TITAN'S ATMOSPHERE

D. Toublanc, J.P. Parisot, J. Brillet
Obs. Bordeaux, BP 89 33270 Floirac, France

Paper Presented by D. Toublanc

Poster Paper

Search for chiral molecules on extraterrestrial objects is of great interest for the studies of the origin of life. Titan's unique atmosphere is particularly conducive to chemical evolution: CH4 photolysis in the mesosphere and stratosphere may produce chiral derivatives of the carbon atom substituted by hydrocarbon or nitrogen radicals. Thus it is necessary to know the intensity of the available light at each level of the atmosphere and particularly at the surface of the satellite.

To describe the behavior of the atmosphere we use the first order development of the Redlich-Kwong equation modified by Wilson: $P = nkT(1+\epsilon)$ and the hydrostatic law: $dp = $ -nmg dz. Because of the saturation law of methane, its mixing ratio doesn't exceed 0.033 between 30 and 800 km (turbopause). A Monte Carlo model has been developed which simulates the multiple-scattering in the atmosphere. The physical effects taken into account in the model are Rayleigh and Mie scattering, pure absorption by atmospheric gases and aerosols (McKay & al Icarus 80, Khare & al Icarus 60), and ground albedo. The model output consists of solar flux as a function of altitude and direction of photons.

A photon is initially emitted from the sun with an elevation angle θ. It is then advanced a distance τ (in units of optical depth) in the direction θ to the point of its first collision. Since the probability of penetration obeys the exponential absorption law $p(\tau) \, d\tau = e^{-\tau} \, \delta\tau$, we generate τ with a random number r; $\tau = -\ln(r)$. It is necessary to first decide whether the scattering is Rayleigh or Mie. A random number r_i is generated, and if $r_i \leq f_r$ where $f_r = \tau$Rayleigh/(τRayleigh + τMie) a Rayleigh scattering occurs, otherwise, it is a Mie scattering. In both cases an exact calculation is performed for the new scattering direction. The simulation process keeps going until the photon escapes from the atmosphere or is absorbed by the ground.

Figure 1 shows the penetration of light through the atmosphere at different wavelengths,

UV: 250 nm, visible: 470, 520, 650 nm and IR: 950 nm.

The maximum of photodissociation rates of methane (Figure 2) by the solar flux in the atmosphere and thus the place where most hydrocarbons molecules are formed occurs above 600 km.

It is then shown that the solar light is going through the atmosphere and reaches the ground. The amount of light is the same as a foggy day on earth.

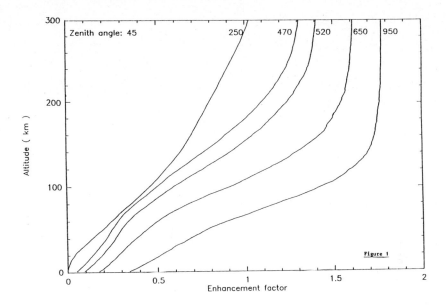

Figure 1: Penetration of Light Through the Atmosphere at Different Wavelengths

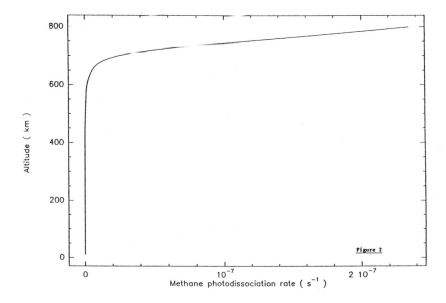

Figure 2: Maximum Photodissociation Rates of Methane

UV RADIATION FIELDS IN DENSE CLOUDS: PHOTOCHEMICAL EFFECTS

S. Aiello, S. Tine'
Dept. of Astronomy and Space Science
University of Florence

C. Cecchi-Pestellini
Dept. of Astronomy and Space Science
University of Florence and Harvard-Smithsonian Center
for Astrophysics

Paper Presented by S. Aiello

Poster Paper

In diffuse clouds and in the outer parts of dense clouds, photodissociation by interstellar UV photons is the dominant destruction mechanism of the neutral molecules, while in deeper regions inside the clouds, UV radiation is generally assumed to play a negligible role because of screening by dust particles. However, how efficiently and how deeply the UV photons penetrate inside the clouds depends upon the local extinction and scattering properties of dust particles.

Inside dense clouds and in region of recent star formation, dust grains are likely to have different properties. In particular, their size distribution appear to be biased towards larger radii, and this results in an extinction lower than the mean interstellar extinction by a factor of 2 to 3.

Moreover the excitation to the Lyman and Werner systems of the hydrogen molecule by cosmic-ray particles (Prasad-Tarafdar mechanism)[1] may originate a significant flux of UV photons inside dense clouds. The chemical importance of such photons has been investigated and corroborated in a number of recent papers.[2,3,4]

In this work we investigate the effects of dust properties on the transfer of UV radiation of interstellar origin, as well as of cosmic-ray generated photons. As a first approximation, the total radiation field has been obtained simply by adding together the two photon fluxes. The transfer equation has been solved by adopting the spherical harmonic method[5]. In computing the production of the cosmic-ray-induced photons, we took into account the contribution of both primary protons and secondary electrons. The interstellar proton spectrum adopted is the one measured by Webber and Yushak[6].

We present the results for a homogeneous spherical cloud with gas density $n(H_2) = 10^4$ cm^{-3} and visual optical depth (edge to center) of 10, for two possible values of R, the ratio of the total to selective extinction: 3.2, characteristic of the diffuse medium, and 4.2, found in the inner part of the ρ Oph cloud.[7] The relative extinction laws have been computed following Cardelli et. al[8]. For the sake of simplicity we adopt $\omega = g = 0.5$. The interstellar UV radiation used is taken from Van Dishoeck[9].

Finally we computed the photodissociation rates for a number of molecules observed in dense clouds. Some of them, H_2O, NH_3, CH_4, HCN, CH_2O_2, CH_2NH, and CH_3NH_2 are of particular interest for the prebiotic evolution. The cross-sections adopted were kindly provided by Prof. E. Herbst. Figs 1 and 2 show some of the computed photodissociation rates as a function of the distance from the cloud center, for R = 4.2, without and with cosmic ray induced photons respectively. It appears that the UV photons generated through the Prasad-Tarafdar mechanism can affect significantly molecular lifetimes in deeper parts of dense clouds.

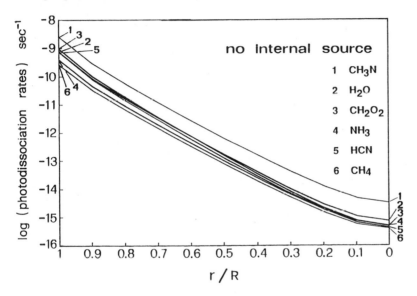

Figure 1. Computed Photodissociation Rates

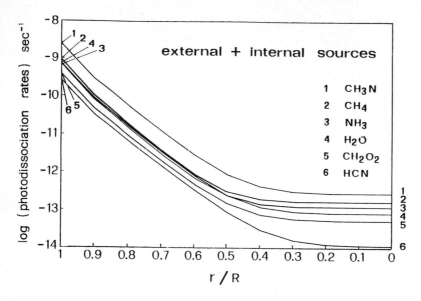

Figure 2. Computed Photodissociation Rates

References

1. Prasad, S.S. and Tarafdar, S.P.: 1983, Ap.J. _267_, 603
2. Sternberg, A., Dalgarno, A., Lepp, S.: 1987, Ap.J. _320_, 676
3. Gredel, R., Lepp, S., Dalgarno, A., Herbst, E.: 1989, Ap.J. _347_, 289
4. Hartquist, T.W. and Williams D.A.: 1990, MNRAS, _247_, 343
5. Flannery, D.P., Roberge, W., Bybicki, G.D.: 1980, Ap.J. _236_, 598
6. Webber, W.R. and Yushak S.M.: 1983, Ap.J. _275_, 391
7. Wu, C.C., Girla, D .P., Van Duimen, R .T.: 1980, Ap.J. _241_, 173
8. Cardelli, J.A., Clayton, G.C., Mathis, J.S.: 1989, _345_, 245
9. Van Dishoeck, W.M.: 1987, in Astrochemistry, M.S. Vardya and S.P. Tarafdar eds., Reidel, pp.51-63

III. PRIMITIVE EVOLUTION

PREBIOTIC CHEMISTRY IN PLANETARY ENVIRONMENTS

François Raulin
Université Paris Val De Marne, Créteil, France

ABSTRACT

The formation of reactive organic compounds, such as HCN or HCHO, followed by their evolution in solution is one of the earliest steps in chemical evolution which might have led to the emergence of life on the Earth. Such organics are key ingredients of the prebiotic chemistry, since, in the presence of liquid water, they can give rise to the building blocks of living systems. Similar processes are going on in present planetary environments, especially on Titan, but in the absence of liquid water. With a dense reduced atmosphere mainly composed of N_2 and CH_4, rich in organic compounds in the gas and aerosol phases, and with the likely presence of an ocean of liquid methane and ethane, this moon appears as a natural laboratory for studying prebiotic organic chemistry at a planetary scale.

1. INTRODUCTION

Within almost 40 years of experimental studies in the field of prebiotic chemistry, detailed scenarios have been elaborated to explain the origin of life on the Earth. They all agree on the general idea of a chemical evolution from simple compounds towards high molecular weight organics, including the biomacromolecules, with the appearance of the complex chemical machinery of a primitive metabolism. In fact, these scenarios are based on a subtle mixture of observations and speculation. On one side, they use the results of laboratory works studying the chemistry of a specific compound - carbon - in a specific solvent - water. On the other side, they assume that these results can be extrapolated to the environment of the primitive Earth, although we have no direct information on this environment, and in spite of the limitations due to the simplicity, size and very short (geologically speaking) duration of the "simulations".

Space exploration has opened the logical approach of exobiology (=bioastronomy), providing a way to search directly for extra-terrestrial life on extra-terrestrial planets. But it has also provided a way to test, in real planetary environments, our theories of chemical evolution. Exobiology includes the study of the origin of Life on the Earth, and of its first step, the synthesis and evolution of organics in planetary environments. It is clear now that organic chemistry is widely distributed in the universe. In particular, it is present on many planetary bodies of the outer solar system, such as Jupiter, Saturn and especially Titan. With a dense N_2 atmosphere rich in organics, the largest satellite of Saturn offers a place to observe and study prebiotic chemistry evolving on a real planetary scale.

This paper presents a (very short) comparative review of the experimental data related to prebiotic chemistry on the Earth, and the data from observation and modeling related to Titan's organic chemistry.

2. PREBIOTIC CHEMISTRY ON EARTH

Since the pioneer work of Stanley Miller (1953), which demonstrated the possible abiotic formation of amino acids from the chemical evolution of a model atmosphere, the "prebiotic chemist" has checked that most of the biomonomers can be obtained from simple organics. These are small molecules with multiple bonds in their structure, such as nitriles and aldehydes, which may have been formed in the atmosphere of the primitive Earth.

The evolution of HCN in aqueous solution can give rise to a tetramer. After reaction with a fifth molecule of HCN, this tetramer can produce adenine, one of the purine bases, which, like the pyrimidine bases are the constituents of the nucleotides, building blocks of the nucleic acids. HCN polymerization processes can also produce complex oligomers which, after hydrolysis, release purines (adenine and guanine), pyrimidine bases and amino-acids (Fig. 1). Similarly, the chemistry of HC_3N in aqueous solution can produce pyrimidine bases (cytosine and uracil) and that of formaldehyde can produce the biological sugars (for a review, see Raulin, 1990). In addition, C_2N_2 can act as a chemical agent, allowing the condensation of the monomers to form the biopolymers. Some experimental data (Matthews and Moser, 1966) suggest that the chemistry of HCN could even directly provide polypeptides, fragments of proteins. Thus from only a small number of different organic compounds as starting material, the "prebiotic chemist" can produce most of the building blocks of the biomacromolecules.

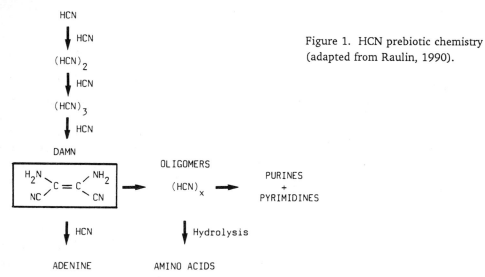

Figure 1. HCN prebiotic chemistry (adapted from Raulin, 1990).

However, it must be emphasized that all of these prebiotic syntheses require very narrow physical-chemical conditions, (of pH for instance), corresponding to very constrained environmental conditions. For instance, this is the case of the prebiotic synthesis of sugars from HCHO. In addition, this reaction provides very low yields of pentoses with a wide chemical dispersion. In fact, so far, one should admit that there is no really convincing prebiological pathway for the biopentoses, and one starts searching for possible ancestors of the present biological sugars, of simpler structure.

Generally, to be prebiotic, this chemistry also requires large quantities of starting ingredients, especially if the primitive oceans were formed rapidly, producing large liquid bodies and dilution of the reactive atmospheric precursors. Those compounds, at low concentration, cannot play their prebiotic role. For instance, at a concentration lower than about 0.01 M, HCN and HCHO are irreversibly hydrolyzed into formic acid. By extrapolation of their yields of synthesis (observed during experimental simulation or theoretical modeling) one can estimate their atmospheric fluxes, given in Table 1, for different atmospheric compositions. Now, if we take into account the possible flux of water on the primitive Earth and assume that this water readily condenses, it is possible to derive the concentration of the starting organics. Table 1 shows that the range of concentration needed by the "prebiotic chemist" may have been reached from atmospheric syntheses if the atmosphere was reducing. It is not established that such a condition was followed. However, some of the starting materials may have also been brought by meteoritic or cometary impacts, or may even have been formed together with biomonomers in the primitive submarine hot springs. But such contributions will need to be quantified.

Table 1: Plausible atmospheric fluxes of some starting ingredients for prebiotic synthesis on the primitive Earth and their upper limit of mean concentration in the primitive aqueous environment (assuming a lower limit of water flux of 5×10^{11} liter/yr).

Starting Ingredient	Atmospheric Flux mol/yr Model Atmosphere			Upper limit of Concentration mole/l
	CO_2	CO	CH_4	
HCHO	$5\text{-}8 \times 10^9$	$2\text{-}3 \times 10^{11}$	$10^{10}\text{-}10^{11}$	0.2
HCN	5×10^9	5×10^{10}	$2\text{-}5 \times 10^{11}$	1
HC_3N	-	-	5×10^{10}	0.1

How complex were the organics synthesized in the atmosphere of the primitive Earth? What were the relative roles of the different energy sources available in this environment, especially that of UV light and electric discharges? Were atmospheric organic syntheses efficient enough to provide the starting material necessary for chemical evolution toward living systems? What are the respective roles of the processes in the gas phase and in solution, and what is the influence of temperature? How far can prebiotic chemistry evolve in the absence of liquid water? More generally, to what extent can the results of laboratory experiments be extrapolated to acquire understanding of real planetary environments? The observation of Titan may provide some answers to these questions.

3. PREBIOTIC CHEMISTRY ON TITAN

Several of the conditions necessary for prebiotic chemistry to evolve toward complex organic systems are present on Titan:

A dense (1.5 bar surface pressure), mid-reducing atmosphere.

Titan's atmosphere is mainly composed of nitrogen with noticeable mole fraction of methane and very low mole fraction of hydrogen (Table 2). Simulation experiments in the laboratory suggest that such an atmosphere is one of the most favourable to the synthesis of organics (for a review, see Raulin and Frère, 1989). With the exception of C_4N_2, all of the organics already detected in Titan's atmosphere were expected to be present from the results of experiments simulating the evolution of a model of Titan's atmosphere submitted to energy deposition. They include six hydrocarbons and four nitriles, including those which play a crucial role in prebiotic chemistry. The good agreement between

Table 2. Chemical composition of Titan's stratosphere (adapted from Raulin & Frère, 1989; Coustenis, 1989; Coustenis et al, 1989)

Constituents	Mixing Ratio	Remarks
Main constituents and hydrocarbons		
Nitrogen N_2	0.757 - 0.99	inferred indirectly
Argon Ar	0 - 0.21	id°
Methane CH_4	0.005 - 0.037	
Hydrogen H_2	0.002 - 0.006	
Ethane C_2H_6	1.3×10^{-5}	at the equator
Acetylene C_2H_2	2.2×10^{-6}	id°
Propane C_3H_8	7.0×10^{-7}	id°
Ethylene C_2H_4	9.0×10^{-8}	id°
Diacetylene C_4H_2	1.4×10^{-9}	id°
	2.0×10^{-8}	at the north pole
N-Organics and O-Compounds		
Hydrogen cyanide HCN	1.9×10^{-7}	at the equator
Cyanoacetylene HC_3N	7.0×10^{-8}	at the north pole
Cyanogen C_2N_2	6.0×10^{-9}	id°
Dicyanoacetylene C_4N_2	solid phase	id°
Carbon dioxide CO_2	1.4×10^{-8}	at the equator
	1.0×10^{-10}	at the north pole
Carbon monoxide CO	6.0×10^{-5}	in the troposphere
	$< 4 \times 10^{-6}$	

the detection of minor constituents in Titan and data from simulation experiments suggests that one can extrapolate the results of such experiments to the real atmosphere. This approach has been used as a guide to the search for IR (Cerceau et al, 1985 ; Raulin et al, 1990) and UV (Bruston et al, 1989), signatures of additional nitriles and hydrocarbons in Titan, and to estimate the upper limit of their mean abundances.

The presence of several sources of energy in the atmosphere.

UV light and energetic electrons from Saturn's magnetosphere allow an efficient transformation of the main atmospheric constituents into more complex compounds. Photochemical models developed in particular by Yung, et al (1984), show that there is an important coupling between the high atmosphere low pressure and low atmosphere high pressure processes. For instance, HCN formation, involving the dissociation of N_2 by energetic electrons, would mainly occur at very high altitudes, while the other nitriles would be produced in the stratosphere, from HCN photolysis.

The presence of a low temperature tropopause, coupled to the existence of atmospheric hazes.

Hazes made of submicron particles are present in the high atmosphere of Titan. They are probably composed of oligomers produced by ethylene, acetylene or HCN photopolymerization. By diffusion down to the stratosphere, they may act as condensation nuclei and induce the condensation of organic compounds of smaller molecular weight. The resulting particles precipitate down to the troposphere, increasing in dimension as altitude decreases (Frère et al, 1990; Frère and Raulin, 1991). Their layered structure would be formed of compounds more and more volatile from the core to the outside (Fig. 2). Haze particles in the stratosphere become hail in the high troposphere and rain in the low troposphere. These particles irreversibly carry most of the organics of very low vapour pressure from the high atmosphere down to the surface. Such processes can protect the organics from further destruction in the gas phase.

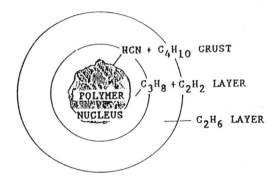

Figure 2. Modelled structure of the low stratospheric aerosol (adapted from Frère, 1989; Frère and Raulin, 1991)

The likely presence of liquid ocean at the surface.

We have no direct evidence of the nature of the surface of Titan. However, the analysis of physical chemical processes occurring in the atmosphere strongly suggests that the surface is covered by a liquid ocean (Lunine et al., 1983; Lellouch et al, 1989). Although Titan's ocean must be very different from the terrestrial oceans, since it is supposed to be mainly composed of liquid ethane and methane, it may play an important role in Titan's chemical evolution. By dissolution of atmospheric organics in this liquid milieu, the oceanic medium provides an additional way to protect the organics formed in the gas phase from destruction in the atmosphere. In addition, high energy cosmic rays reaching the surface of the ocean can induce new organic processes involving the main constituents. These can give rise to organic oligomers, but also to exotic organics of low stability at room temperature, such as simple azides CH_3N_3 or $C_2H_5N_3$. From this step, chemical evolution on Titan may have followed a very different path compared to the Earth.

To check such hypotheses and try to answer the many associated questions, it is necessary to study in detail each of the components that compose "Titan's geofluid": atmosphere, aerosols and surface. This is one of the main objectives of the Cassini-Huygens mission and of most of the instruments on board the Huygens probe, which is scheduled to sound Titan's atmosphere in 2004. In particular, the Gas Chromatograph-Mass Spectrometer will provide vertical concentration profiles of the main atmospheric constituents and of minor species, already detected and new organics, and the

Table 3. Ocean properties in nominal hypotheses at 92.5 and 101 K (from Dubouloz et al, 1989)

	92.5 K		101 K	
Surface temperature	92.5 K		101 K	
Atmospheric argon	$Y_{Ar} = 0$		$Y_{Ar} = 0.17$	
Atmospheric methane	$Y_{CH4} = 0.0155$		$Y_{CH4} = 0.211$	
Main composition (ocean mole fractions)				
C_2H_6 (+ C_3H_8)	90.9 %		5 %	
CH_4	7.3 %		83.4 %	
N_2	1.8 %		6 %	
Ar	-		5.6 %	
Depth	695 m		9.4 km	
Solutes*	X	$H_D(m)$	X	$H_D(m)$
C_2H_2	4.1×10^{-4}	107	3.7×10^{-4}	123
HCN	8.5×10^{-6}	18.0	2.3×10^{-6}	18.8
CH_3CN	6.3×10^{-5}	0.20	1.5×10^{-5}	4.1×10^{-2}
C_2H_3CN	2.2×10^{-5}	0.28	5.0×10^{-6}	0.22
C_2HCN	7.2×10^{-6}	2.2	2.0×10^{-6}	2.3
C_2N_2	1.4×10^{-6}	0.83	1.5×10^{-7}	0.84

*X is the real ocean mole fraction, H_D is the height of solid deposit at the bottom of the ocean.
-- = no saturation.

Aerosol-Collector Pyrolyser will provide the chemical composition of the aerosols. If such instruments provide evidence of the presence of organic compounds of biochemical interest on Titan, it will have strong implication in Exobiology.

4. CONCLUSIONS

Depending on the planetary environment, chemical evolution may evolve differently. On the primitive Earth, in the presence of liquid water, prebiotic organic chemistry starting from simple reactive molecules allowed the emergence of life. On Titan, with very low temperature conditions, chemical evolution is still going on, but in the absence of liquid water. Studies of the prebiotic organic chemistry on this planetary body may provide information about the atmospheric processes which occurred on the primitive Earth. But such studies could also indirectly furnish important information on the role of water in chemical evolution, since prebiotic chemistry on Titan has evolved in the absence of this universal solvent. However, Titan's ocean may have evolved in the presence of noticeable traces of dissolved ammonia, playing the role of water. If this was the case, there may be a pseudo-biochemistry going on in Titan's ocean, where ammonia substitutes for water. There may be purine and pyrimidine bases present on Titan, and pseudo-polypeptides included in Titan's organic oligomers.

In the next decades, many space missions will have implications for the study of extra-terrestrial organic chemistry and exobiology. One can expect that the most important will be Mars missions, cometary missions, and the Cassini-Huygens mission to Titan. The Titan mission offers the possibility of studying ongoing organic chemistry in a natural planetary environment, and over a very long period of time.

References

Bruston P., H. Poncet, F. Raulin, C. Cossart-Magos and R. Courtin, 1989, Icarus 78, 38-53.

Cerceau F., F. Raulin, R. Courtin and D. Gautier, 1985, Icarus 62, 207-220.

Coustenis A., 1989, Thèse de Doctorat, Université Paris 7.

Coustenis A., B. Bézard and D. Gautier, 1989, Icarus 80, 54-76.

Dubouloz N., F. Raulin, E. Lellouch and D. Gautier, 1989, Icarus 82, 81-96.

Frère C., 1989, Thèse de Doctorat, Université Paris 12.

Frère C., F. Raulin, G. Israël and M. Cabane, 1990, Adv. Space Res. 10 (1), 159-163

Frère C. and F. Raulin, 1991, submitted.

Lellouch E., A. Coustenis, F. Raulin, N. Dubouloz and C. Frère, 1989, Icarus 79, 328-349.

Lunine J.I., D.J. Stevenson and Y.L. Yung, 1983, Science 222, 1229-1230.

Matthews C. and R. Moser, 1966, Proc. Nat. Acad.Sci. (USA), 56, 1087-1094.

Miller S.L., 1953, Science 117, 528-529.

Raulin F. and C. Frère, 1989, J. British Interplanetary Soc. 42, 411-422.

Raulin F., 1990, J. British Interplanetary Soc. 43, 39-45.

Raulin F., B. Accaoui, A. Razaghi, M. Dang-Nhu, A. Coustenis and D. Gautier, 1990, Spectrochim. Acta, 46A (5), 671-683.

Yung Y.L., M. Allen and J.P. Pinto, 1984, Astrophys. J. Suppl. Ser. 55, 465-506.

Discussion

R. BROWN: I think It is important to look for new ideas about the production of primitive biopolymers rather than the constituent building blocks (amino acids, purines, pyramidines, ribose, phosphoric acid). Many years of effort have still not produced convincing evidence that condensation to biopolymers can effectively occur, even under highly selected, favorable conditions in the laboratory. Prospects on the surface of the Earth look much less favorable.

We should therefore seek primitive templates that will absorb crucial species (such as ribose) from complex mixtures and locate constituents in favorable relative positions for condensation. Or perhaps we need some cosmic polymers that will be converted, under primitive terrestrial conditions, into crude biopolymers resembling RNA. C. Mathews' work on HCN polymers possibly shows the way in which primeval small polypeptides may have formed, avoiding the problem of condensating amino acids on the young Earth.

F. RAULIN: Prebiotic condensation of amino acids into polypeptides is not a problem. It has been obtained by several authors. In particular, recently André Brack was able to get polypeptides from Leuch's anhydride of L amino acids, without chemical dispersion (chiralty is preserved, no branched condensation, only αamino acids are condensed etc.). Although Cliff Matthews' idea is great, it has not been, so far, clearly demonstrated that "his" HCN "polymer" has the right chemical structure. However, your remark is important, and I agree: we do need to look for new ideas in the field of prebiotic chemistry, especially in relation to the prebiotic synthesis of polynucleotides.

D. DEAMER: In your paper, you listed several conceptional difficulties now facing research on prebiotic evolution. For instance, the concentrations of important monomers were probably quite dilute. I would add phosphate to the list you showed. Phosphate is in the micromolar range in today's ocean, essentially a saturated calcium phosphate solution at pH 8. One might invoke phosphate minerals as a source of phosphate. However, the amount present in known mineral deposits is negligible. Furthermore, in a high CO_2 primitive atmosphere, the resulting lower pH ranges (pH 5-6) would markedly reduce the ability of calcium phosphate to precipitate as apatite or brushite. Micromolar phosphate, then, may be all we have to work with, at least in solution chemistry. Of course, concentrating mechanisms such as drying-wetting cycles (tide pools) could obviate these concerns.

F. RAULIN: Indeed, the problem of concentration, in particular for phosphate, has already been pointed out by many authors. And indeed, cycles of evaporation-rehydrating processes can solve the problem. This has been tested experimentally (D. White, in particular) with success.

C. MATTHEWS: As Ron Brown just pointed out, there is an impasse in today's studies of the origin of life. Monomers are readily formed, but how did they join together to form polymers? The dilute soup model is clearly in trouble. As you know, Francois, the cyanide model proposes instead that the primitive Earth (compare Halley's Comet today) was covered with hydrogen cyanide polymers. The polyamidine portion of this polymeric material would be converted by water to polypeptides (primitive proteins). Moreover, it may be that this same material was the original dehydrating agent on Earth that joined together nitrogen bases, sugars and phosphates to form nucleosides, nucleotides and nucleic acids. You rightly point out this polyamidine structure has yet to be fully established (though we do have good NMR results). It is certain, though, that the HCN polymer yields amino acids after hydrolysis. I think you agree the model should be thoroughly investigated--it explains so much with one simple idea!

F. RAULIN: I agree that HCN polymers do release amino acids and are an easy way to get those compounds. They may be a very interesting way of directly getting polypeptides if they have the proposed simple structure of polyamidine (although much simpler than that of a polypeptide and without any chiral selection--another big problem).

However: 1) You use pure liquid HCN to get these polymers--such conditions are difficult to place in an astrophysical or planetological context, where a lot of other organics may be simultaneously present.

2) The simple structure you proposed is not fully established, as shown by the controversy between your group and Jim Ferris' group (Science, 203, p.1135, 1979), for instance.

3) The main problem in prebiotic synthesis of biomacromolecules concerns polynucleotides, not polypeptides. It has not been shown, so far, that "your" polymer can be used as a dehydrating agent for nucleotides and polynucleotides synthesis.

But it is a good idea to be checked!

TERRESTRIAL ACCRETION OF PREBIOTIC VOLATILES AND ORGANIC MOLECULES DURING THE HEAVY BOMBARDMENT

C. F. Chyba and C. Sagan
Laboratory for Planetary Studies, Cornell University, Ithaca, NY 14853, USA

L. Brookshaw
Lawrence Livermore National Laboratory, P.O. Box 808, Livermore, CA 94550, USA

P. J. Thomas
Department of Physics and Astronomy, University of Wisconsin, Eau Claire, WI 54702, USA

Paper Presented by C. F. Chyba

The possible prebiotic importance of impacts of carbonaceous chondrites and comets has a speculative history extending back at least to the beginning of this century (Chamberlin and Chamberlin, 1908). Such speculations have received support from modern nonhomogeneous terrestrial accretion models (Wasson, 1971; Turekian and Clark, 1975; Anders and Owen, 1977), as well as dynamical models for outer planet formation (Fernandez and Ip, 1983; Shoemaker and Wolfe, 1984), in which Earth receives the bulk of its surface volatiles as a late-accreting impactor veneer. However, independently of solar nebula chemistry or planetary formation models, we may ask what the observed lunar cratering record tells us about terrestrial accretion of key prebiotic volatiles (e.g., carbon, nitrogen, and water) and organics during the period of heavy bombardment 4.5 to 3.5 billion years ago. This approach has the advantage of minimizing the conclusions' model-dependence, by basing the calculations as much as possible on the available extant data.

In the following discussion, we summarize the reasoning that leads from the cratering record on the Moon to the conclusion that a substantial fraction of the Earth's oceans (Chyba, 1990a) and surface carbon (Chyba, 1990b) inventories were indeed accreted during the heavy bombardment. These calculations are then extended to a consideration of the terrestrial nitrogen inventory, and an assessment is made of the role of atmospheric erosion of carbon and nitrogen by large impacts. Finally, we summarize recent results (Chyba et al., 1990) for the much more difficult problem of quantifying the impact delivery of intact organic molecules.

Fig. 1 shows cumulative lunar crater density as a function of surface age (Basaltic Volcanism Study Project, 1981), and an analtyical fit to these data (Chyba, 1990a). The data, for craters bigger than a diameter D, are well modeled by the equation

$$N(t,D) = \alpha[t+\beta(e^{t/\tau} -1)](D/4 \text{ km})^{-1.8} \text{ km}^{-2}, \tag{1}$$

where α and β are determined by two-dimensional χ^2 minimization. [Eq.(1) is just the mathematical statement of the observation that cratering has been roughly constant for the past 3.5 Gyr, while increasing exponentially into the past prior to that time. Since Fig. 1 is a cumulative crater plot, its data are fit by the integral (over time t) of this constant plus exponential, giving a sum of a linear and exponential term.] In the following discussion, we take the decay constant τ of the cratering flux to be 144 Myr, i.e. equivalent to a 100 Myr half-life. (Choices for τ ranging from 70 to 220 Myr have been made in the literature; we discuss the consequences of such different choices below.) Crater diameter D is related to impactor mass m via an equation of the form (Schmidt and Housen, 1987):

$$m = \gamma D^{3.8} v^{-1.67}, \tag{2}$$

where γ is a constant that depends on surface gravity and impactor and target densities, and v is incident impactor velocity. Eqs.(1) and (2) may then be combined to obtain the number of objects with mass $> m$ that have impacted the Moon as a function of time:

$$n(>m,t) = \alpha[t+\beta(e^{t/\tau} -1)][m/m(4 \text{ km})]^{-0.47} \text{ km}^{-2}, \tag{3}$$

where $m(4 \text{ km})$ is given by Eq.(2) with $D = 4$ km.

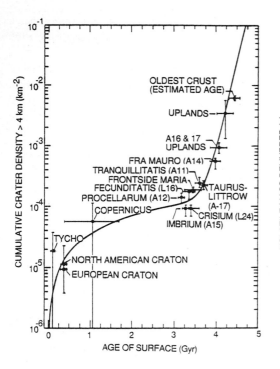

Fig. 1. Analytical fit to cumulative lunar crater density as a function of surface age.

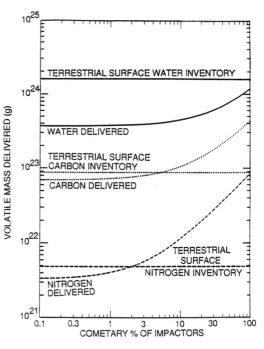

Fig. 2. Volatile mass delivered to Earth during the heavy bombardment, as a function of the cometary % of the impactors. Horizontal lines indicate the current terrestrial surface (atmosphere, ocean, and sedimentary column) inventories of water, carbon, and nitrogen.

The total mass, $M(t)$, incident in impactors with masses in the range m_1 to m_2 on a lunar surface of age t is then:

$$M(t) = \int_{m_2}^{m_1} m[\partial n(>m,t)/\partial m]dm \sim D_{max}^{2.0} \, v^{-1.67}, \tag{4}$$

where we have explicitly displayed the dependence of $M(t)$ upon the largest crater diameter D_{max} used in the integration, and upon our choice of typical impactor velocity v. Note that $M(t)$ depends much more weakly on D_{max} than one might have anticipated from the Schmidt-Housen scaling, Eq.(2). Nevertheless, uncertainties in D_{max} and v mean that our final results are uncertain to at least a factor of several.

Fits to the data of Fig. 1 with $\tau = 144$ Myr (Chyba, 1990a) or $\tau = 220$ Myr (Melosh and Vickery, 1989) lead to values of $M(t=4.5$ Gyr) which differ by a factor of about 3, comparable to other uncertainties in the problem. However, a fit to a different lunar cratering data set, using $\tau = 70$ Myr (Maher and Stevenson, 1988), gives a value for $M(t=4.5$ Gyr) that is more than two orders of magnitude higher (Chyba, 1990a). This latter fit can almost certainly be excluded, on the grounds that it delivers to the Moon $\sim$40 percent of the present lunar mass subsequent to 4.5 Gyr ago (i.e., subsequent to the time of lunar formation!), as well as tens of oceans of water to the Earth. The fit using $\tau = 144$ Myr, on the other hand, is in good agreement with geochemical data for lunar accretion subsequent to the solidification of the lunar crust $\sim$4.4 Gyr ago (Chyba, 1990a), as well as with an estimate of the lunar cratering chronology based on a crustal viscosity model (Oberbeck and Fogleman, 1989).

$M(t=4.5$ Gyr) from Eq.(4) is then gravitationally scaled to determine the total mass incident on Earth during the heavy bombardment. The impactors comprising this bombardment may have been $\sim$50 percent (Chyba, 1990a), and possibly >80 percent (Hartmann, 1987), carbonaceous asteroids. It is unclear what fraction of this bombardment was cometary, although it seems unlikely to have been more than $\sim$10 percent (Chyba, 1987). Using known volatile compositions of carbonaceous asteroids and comets, we calculate the mass of carbon, nitrogen, and water delivered to Earth by this bombardment (Fig. 2). The results show approximate agreement between these results and the known terrestrial surface inventories of these volatiles.

However, these results have not yet incorporated the effect of atmospheric erosion by energetic impacts. The conditions for such erosion have been described by Melosh and Vickery (1989); the key result for our discussion is that impacts by sufficiently fast and massive comets and asteroids lead to an expanding vapour plume which carries off the entire atmospheric cap above the tangent plane to the point of impact. In this case, the impactor's volatiles will escape from Earth, and previously-accreted terrestrial volatiles in the atmosphere and oceans will also be removed.

The net effect of these erosive impacts may be calculated from integrals similar to Eq.(4), of the form:

$$M(t) = \int_{m_2}^{m^*} f(m)[\partial n(>m,t)/\partial m]dm, \tag{5}$$

where m_* is the minimum impactor mass sufficient to cause erosion. For atmospheric erosion, we put $f(m) = m_{cap}$, the mass of the eroded atmospheric cap; for erosion of condensed oceans, we have $f(m) = \pi r_p^2 d_{ocean} \rho_{ocean}$, where r_p is the radius of the impacting projectile, d_{ocean} is the depth of the terrestrial ocean, and ρ_{ocean} is the ocean density (Chyba, 1990a). We find that condensed volatiles are protected against atmospheric erosion, whereas volatiles in the atmosphere may be substantially eroded if present at sufficient concentrations. In particular, it is possible to erode "excess" carbon and nitrogen delivered in quantities greater than those currently found in the terrestrial surface inventory, provided these excess quantitites reside primarily in the atmosphere (see Fig. 3).

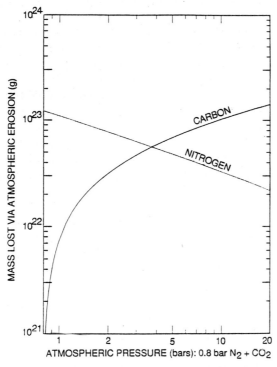

Fig. 3. Atmospheric erosion of carbon (as CO_2) and nitrogen (as N_2), as a function of the CO_2 pressure of atmosphere. This simple model assumes N_2 is maintained at a constant 0.8 bar pressure, while CO_2 is held at a pressure between 0 and 20 bar. Under these conditions "excess" carbon and nitrogen are easily eroded from the atmosphere.

The bulk of carbon in carbonaceous asteroids and comets is organic [see, for example, Chyba et al. (1990), and references therein]. Over the last decade, this potential extraterrestrial source of organic molecules for the primitive Earth has taken on new importance, as an emerging consensus in planetary science has replaced previous models of a primordial reducing (CH_4/NH_3-rich) terrestrial atmosphere with that of a neutral (CO_2/N_2) one. Synthesis of key prebiotic molecules in CO_2 atmospheres has been shown experimentally to be much more difficult than in reducing ones, with production efficiencies dropping precipitously by many orders of magnitude as the ratio H_2/CO_2 falls below unity (Schlesinger and Miller, 1983a,b). However, while these results are suggestive, there is certainly at present no requirement to invoke exogenous organics to account for the evolution of life on Earth. Many other speculative mechanisms for terrestrial prebiotic synthesis have been suggested. Our goal has been to quantify as best as possible prebiotic organic delivery by comets, so that the importance of cometary organics may be weighed against other possible sources. We are currently extending this work to carbonaceous asteroids.

A comprehensive treatment of comet/asteroid interaction with the atmosphere, ensuing surface impact, and resulting organic pyrolysis is required to determine whether more than a negligible fraction of the organics in incident comets and asteroids actually survived collision with Earth. Results of such an investigation (Chyba et al., 1990), using a smoothed particle hydrodynamic simulation of cometary impacts into both oceans and rock, demonstrate that organics will not survive impacts at velocities >10 km s^{-1}, and that even comets and asteroids as small as 100 m in radius cannot be aerobraked to below this velocity in 1 bar atmospheres. However, for plausible dense ~10 bar CO_2) early atmospheres, there will be sufficient aerobraking during atmospheric passage for some organics to survive the ensuing impact. Combining these results with Eq.(4) above shows that 4.5 Gyr ago Earth was accreting at least 10^9-10^{10} g yr^{-1} of intact cometary organics, a flux which thereafter declined with a ~100 Myr half-life. These results may be placed in context by comparison with it *in situ* organic production from a variety of terrestrial energy sources, as well as organic delivery by interplanetary dust. Which source dominated the early terrestrial prebiotic inventory appears to depend on the nature of the early terrestrial atmosphere. [A preliminary quantification of these effects may be found in Chyba and Sagan (1990).] However, there is an intriguing symmetry: It is exactly those dense CO_2 atmospheres, where *in situ* atmospheric production of organic molecules should be the most difficult, in which intact cometary organics would be delivered in large amounts.

This work was supported in part by grants from the National Aeronautics and Space Administration, and the Kenneth T. and Eileen L. Norris Foundation.

References

Anders, E., and Owen, T., 1977, Science 198, 453-465.

Basaltic Volcanism Study Project 1981, Basaltic Volcanism on the Terrestrial Planets (Pergamon, New York).

Chamberlin, T.C., and Chamberlin, R.T., 1908, Science 28, 897-911.

Chyba, C.F., 1987, Nature 330, 632-635.

Chyba, C.F., 1990a, Nature 343, 129-133.

Chyba, C.F., 1990b, Eos 71, 644.

Chyba, C., and Sagan, C., 1990, Bull. Amer. Astron. Soc. 22, 1097.

Chyba, C.F., Thomas, P.J., Brookshaw, L., and Sagan, C., 1990, Science 249, 366.

Fernandez, J.A., and Ip, W.-H., 1983, Icarus 54, 377-387.

Hartmann, W.K., 1987, Icarus 71, 57-68.

Maher, K.A., and Stevenson, D., 1988, Nature 331, 612-614.

Melosh, H.J., and Vickery, A.M., 1989, Nature 338, 487-489.

Oberbeck, V.R., and Fogleman, G., 1989, Origins of Life 19, 549-560.

Schlesinger, G., and Miller, S.L., 1983a, J. Mol. Evol. 19, 376-382.

Schlesinger, G., and Miller, S.L., 1983b, J. Mol. Evol. 19, 383-390.

Schmidt, R.M., and Housen, K.R., 1987, Int. J. Impact Engng. 5, 543-560.

Shoemaker, E.M., and Wolfe, R.F., 1984, Proc. Lunar Planet. Sci. Conf. 15, 780-781.

Turekian, K.K., and Clark, S.P., 1975, J. Atmos. Sci. 32, 1257-1261.

Wasson, J.T., 1971, Earth Planet. Sci. Lett. 11, 219-225.

Discussion

W. IRVINE: Given the low density, composite nature of cometary nuclei as discussed by Dr. Levasseur-Regourd, would such nuclei not break up in the atmosphere, thus increasing the flux of surviving organic material?

C. CHYBA: There are some estimates of cometary tensile strengths, derived from observations of comets tidally disrupting as they make close passes to the Sun and Jupiter. However, by no means are all comets making such close passes tidally disrupted. Hence, such estimates provide information only for the weakest comets. But in these cases, at least, you're absolutely right; such comets, with radii less than about 1 kilometer, will probably disrupt during atmospheric passage. Probably this will substantially increase the delivery of intact organics, as such an airburst will greatly enhance the surface area/volume ratio of the impactor, leading to much greater atmospheric deceleration. However, note that bodies much bigger than 1 kilometer or so, will not airburst; in such bodies, the atmosphere-induced shock wave will not have time to traverse the impactor before the impactor collides with the Earth.

GEOPHYSIOLOGY AND HABITABLE ZONES AROUND SUN-LIKE STARS

D.W. Schwartzman
Dept. Geology & Geography, Howard University, Washington, D.C. 20059

T. Volk
Dept. Applied Science, New York University, New York, N.Y. 10003

Paper Presented by D. W. Schwartzman

Discussions of the subject of habitable zones around stars date back to the early years of the SETI research program (e.g., Dole, 1964; Shklovskii and Sagan, 1966). Hart (1978, 1979) introduced the notion of the continuously habitable zone (CHZ), the zone around a star in which an Earth-like planet could maintain habitable conditions for the evolution of complex life. He computed an inner boundary at 0.95 AU for the sun, from a model of climatic evolution; this limit is set by a runaway greenhouse at this distance from the sun (D). Kasting (1988) argued that water would be rapidly lost from Earth-like planets by photodissociation at $D \leq 0.95$ AU, in coincidental agreement with Hart, since Kasting computed runaway greenhouse conditions at $D \leq 0.85$ AU. Hart computed an outer boundary to the CHZ of 1.01 AU, limited by the onset of global glaciation. However, negative feedback controls on climate involving the cycling of carbon could significantly widen the CHZ (Schwartzman, 1981; Tang, 1982; Kasting et al., 1988; Schwartzman and Rickard, 1988a, 1988b).

If atmospheric composition some 2 billion years ago were the same as today's, the Earth's oceans would have been frozen because of lower solar luminosity. This is known as the "faint young sun paradox" (Sagan and Mullen, 1972). Since the Earth has had liquid oceans for at least the last 3.5 billion years, since marine sedimentary rocks date back that far, a regulatory mechanism for surface temperature is needed, probably involving the greenhouse gas carbon dioxide. The carbonate-silicate cycle provides just such a geochemical mechanism for the stabilization of climate since the early Archean (Walker et al., 1981). This mechanism entails the chemical weathering of CaMg silicates on land (reaction with carbonic acid producing bicarbonate) and the deposition of calcium carbonate on the ocean floor (a net "sink" for atmospheric carbon dioxide), with a steady-state atmospheric carbon dioxide level being achieved by balancing the sink with the volcanic source. The atmospheric carbon dioxide level along with solar luminosity determines temperature. The heart of this climatic stabilizer is the temperature dependence of the weathering rate. This rate increases with increasing temperature; thus negative feedback is obtained, regulating global temperature.

An important question is whether the carbonate-silicate cycle would keep the Earth's surface habitable, without the presence of the biota, as Kasting et al. (1988) have argued. Precipitation of calcium carbonate would surely occur on an abiotic Earth, but what would be the weathering rate in the absence of life? Lovelock and Watson (1982) proposed that the atmospheric carbon dioxide level (and therefore surface temperature) has been regulated by biological acceleration of weathering. We have argued that the Earth would be likely uninhabitable to all but thermophilic microbes, i.e., with temperatures $\geq$ 50°C, were the Earth abiotic, because of the absence of the biotic amplification of weathering rates (Schwartzman and Volk, 1989, 1990). The biotic enhancement of weathering is apparently a factor of at least 100 times, plausibly 500 times or more, ie., the weathering sink for CO2 is $\geq$ 100 times the rate for an abiotic Earth surface at the same atmospheric CO2 level and temperature. This geophysiological (Lovelock, 1987) rather than purely geochemical regulator includes the biota as an integral part of the exogenic system. It apparently arose with the origin of life near 100°C and the initial microbial colonization of land in the early Archean.

Biotic acceleration of chemical weathering arises mainly out of soil stabilization insuring a high surface area/land area for reaction of silicates with carbonic acid. Other biotic effects contributing to higher weathering rates include carbon dioxide elevation in soils and organic acid production. Before higher plants, microbes alone apparently stabilized soil (a modern analogue is cryptogamic soil, now largely restricted to desert areas where algae and lichens face little competition from higher plants). In spite of some 2 billion years of evolution, apparently, eucaryotes cannot survive at temperatures higher than 50-60°C, because of the thermolability of their proteins and cellular membranes. Thus, we take 2 billion years and 50°C to be the minimum time and maximum temperature, respectively, required for habitability of Earth-like planets around sun-like stars leading to human-like intelligence, with another 2 billion years required to evolve eucaryotes from a procaryotic ancestor.

Following the criteria argued above, we calculate the inner boundary of the habitable zone with the potential for the emergence of human-like intelligence for an abiotic Earth surface , given the probable biotic enhancements of weathering and geologic limits on degassing rates of carbon dioxide and continental growth (land area) as functions of time. These calculations incorporate the inferred variation of solar luminosity over geologic time (Gough, 1981). We do not consider effects of variation of planetary mass, composition, eccentricity etc. (see Dole, 1964); this is a strictly Earth model with variable D, the distance of Earth to the sun (in AU).

(1) $Te = 255/ \{(D^{0.5}) (1 + 0.087 t)^{0.25}\}$,

where Te is the effective radiating temperature of the Earth (°K) (no greenhouse effect) and t is time in billion of years before present.

Equation (1) was derived from expressions for the variation of solar luminosity with time and Te as a function of L (see Kasting, 1987):

$L(t)/Lo = 1/\{1 + 0.4 (t/4.6)\}$,

where Lo is the present solar luminosity.

$Te^4 = (S/4 \sigma) (1 - A)$,

where A is the global albedo (assumed = 0.3), σ = Stefan-Boltzman constant and S is the amount of sunlight reaching Earth.

$S \propto L$, $S \propto 1/D^2$, hence $Te \propto 1/D^{0.5}$.

Using the greenhouse function, slightly modified, from Walker et al (1981):

(2) $Ts = 2\,Te + 4.6\,(P)^{0.364} - 226.4,$

where Ts is surface temperature of Earth, P is the ratio of atmospheric carbon dioxide pressure at time t ("pCO_2") to pCO_2now.

On present Earth the chemical weathering rate (carbon sink) Bo, biotically-enhanced, equals the hypothetical abiotic rate, Babiotic, multiplied by factors which constitute the biotic enhancement factor (in { }):

$Bo = Babiotic\ \{P^{\alpha}\ e^{(\beta\Delta T)}\ e^{(\gamma\Delta T)}\},$

where α, β and γ are factors expressing the dependence of silicate weathering rate on pCO_2, temperature and runoff respectively, with $\alpha = 0.4$, $\beta = 0.056$ and $\gamma = 0.017$; ΔT is the temperature elevation required for an abiotic condition over the present global mean, taken as 288°K ($\Delta T = Ts - Tnow$).

The elevation of atmospheric pCO_2 and resultant temperature increase under abiotic conditions is simply the result of requiring atmospheric pCO_2 to "do all the work" in generating a weathering rate which balances the volcanic emission rate (Schwartzman and Volk, 1989). We define B as the biotic enhancement factor:

$B = Bo/Babiotic$

For an abiotic Earth surface of age t:

$Bo\ (V/Vo) = Babiotic\ \{P^{\alpha}\ e^{(\beta\Delta T)}\ e^{(\gamma\Delta T)}\}\ (A/Ao),$

where V is the volcanic CO_2 emission rate (Vo the present rate), A the continental land area (Ao the present area).

Rewriting:

(3) $B = (A/Ao)(Vo/V)\ \{P^{\alpha}\ e^{(\beta\Delta T)}\ e^{(\gamma\Delta T)}\}$

Note that $e^{(\gamma\Delta T)} \leq 2$ (Pollack et al, 1987).

Thus for given values of D, B, A/Ao, V/Vo and t, P and Ts can be computed using equations (1), (2), and (3). For V as a function of time:

$V = Vo\ e^{(wt)};$

V parallels the decrease in radioactive heat generation in the Earth to present.

We assume a relation between the rate of continental growth and volcanic outgassing:

$(dA/dt)t = -c(dV/dt)t,$

giving $(A/Ao) = (1 + cVo) -cVo\ e^{(\omega t)}$

We constrain $(V/Vo)(Ao/A)$ at t = 3.8 b.y. for two extreme limiting models (models a and c) and a preferred model (b):

Model	$(V/Vo)(Ao/A)$ at t = 3.8 b.y.	cVo	ω
a	1 x 1 = 1	-	0
b	3 x 4 = 12	0.375	0.289
c	8 x 10 = 80	0.129	0.547

These models are consistent with limits derived from isotopic modeling of the evolution of the Earth's crust and mantle (Allegre and Jaupart, 1985; Armstrong, 1981; Des Marais, 1985). The parameters for the preferred model (b) are considered the most likely.

Model calculations are shown in Figures 1-7. Figure 1 shows Ts as a function of t for 3 values of B (D=1); note that for models (b) and (c) Ts decreases to present (except for B $\leq$ 5, t < 2 b.y.) in spite of increasing solar luminosity; increasing land area and decreasing volcanic emissions compensate forcing steady state Ts down. Ts for model (a) simply tracks solar luminosity since A and V are constant for this model. Dcritical is the minimum D for Ts = 50°C at t = 2 b.y. and Ts < 50°C from 2 billion years ago to now; for limiting model (a), Ts is a few degrees lower than 50°C at 2 b.y., and 50°C at present for Dcritical (Figures 2 and 3; Figure 3 shows a blowup of data from Figure 2). Figures 4, 5, 6 and 7 show the variation of Ts as a function of D for different B and t values (model (b) only). Figure 4 illustrates the effect of B variation at t = 0, Figure 5 at t = 2 b.y.. The patterns of variation shown in Figures 6 and 7 are the result of the competing effects of increasing solar luminosity, decreasing volcanic emissions and increasing land area to present.

We conclude that the inner boundary (Dcritical) of the habitable zone with potential for the emergence of human-like intelligence for an Earth surface governed by the purely geochemical climatic stabilizer (ie., abiotic) to be significantly greater than 1 AU, given the probable biotic enhancements of weathering and the most likely histories of volcanic outgassing and continental growth. For biotic enhancements greater than 200, Dcritical is > 1.4 AU, for enhancements of 500, Dcritical is > 3.8 AU, for the preferred model. Even for the extreme limiting model with constant outgassing and land area since 3.8 billion years ago, Dcritical is > 1.1 AU for biotic enhancements greater than 200.

Eucaryotes might have emerged even earlier than 2 b.y. ago; a Giardia-like protist may have appeared more than 3 b.y. ago, leaving no fossil record (Sogin, 1989). If so, Dcritical values would be pushed up even further for B $\geq$ 50, because of longer required durations with Ts $\leq$ 50°C (Figure 8 illustrates the effect of changing the age of origin of eucaryotes, "tcritical", on Dcritical for model (b)).

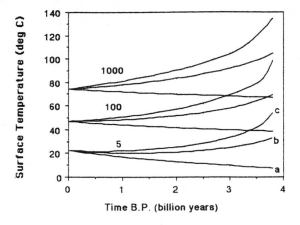

Figure 1. Computed surface temperature versus time for an abiotic Earth; models (a), (b) and (c) (described in text) for different assummed biotic enhancement of weathering factors (B), namely 5, 100, 1000. For B = 100 and 1000, the position of models (a), (b) and (c) is analogous to that for B = 5.

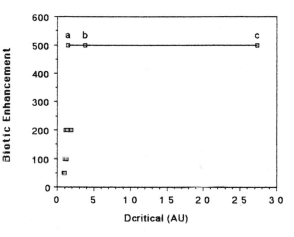

Figure 2. Biotic enhancement versus the minimum distance from Sun for surface temperature to have been less than 50°C for the last 2 billion years (Dcritical) for an abiotic Earth; models (a), (b) and (c) indicated.

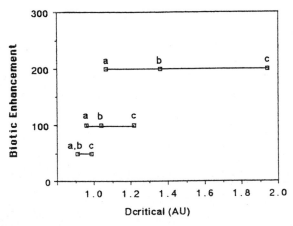

Figure 3. Biotic enhancement versus Dcritical; blowup of points from Figure 2 for B = 50, 100, 200.

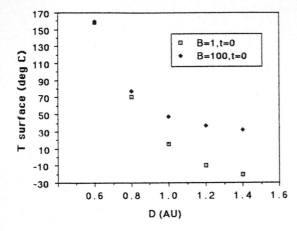

Figure 4. Surface temperature (Tsurface) versus distance from sun (D) for an abiotic Earth, B = 1, 100 at present (t=0); model (b) results only.

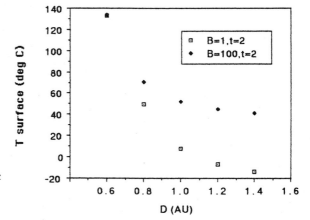

Figure 5. Tsurface versus D for an abiotic Earth, B = 1, 100, for 2 billion years ago (t = 2); model (b) results only.

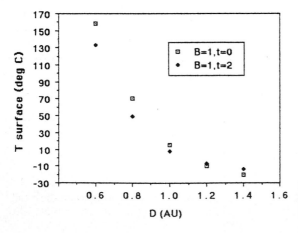

Figure 6. Tsurface versus D for an abiotic Earth, B = 1, for t = 0, 2; model (b) results only.

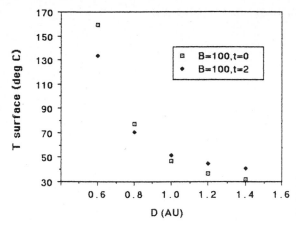

Figure 7. T surface versus D for an abiotic Earth, B = 100, for t = 0, 2; model (b) results only.

Figure 8. Dcritical versus time required for surface temperature to have been less than 50°C for the last (t) billion years (tcritical), with (t) assummed to be the time of origin of eucaryotes, for an abiotic Earth, B=1, 50, 100, 200 and 500; model (b) results only.

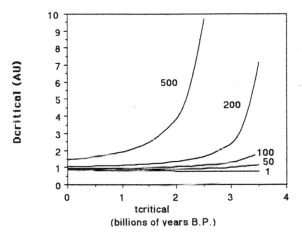

In summary, were it not for the biota's crucial participation in the geochemical climatic regulation of surface temperature, the Earth at 1 AU up to perhaps > 3.8 AU would not have evolved eucaryotes, including of course intelligent life. Since the outer boundary for the formation of terrestrial planets is on the order of 1.5 AU, intelligent life might well be absent altogether were it not for the emergence of the geophysiological mechanism outlined. Exogenic systems with similar geophysiologies to Earth (alien biospheres) should likewise emerge on terrestrial-like planets around sun-like stars, increasing the frequency of intelligent life in the galaxy.

References

Allegre, C.J., and Jaupart, C.: 1985, Earth Planet. Sci. Lett 74, 171
Armstrong, R.L.: 1981, Phil. Trans. Roy. Soc. Lond. A301, 433
Des Marais, D.J.: 1985, in The Carbon Cycle and Atmospheric CO_2: Natural Variations Archean to Present (Geophysical Monogr., 32), ed. E.T Sunquist and W.S. Broecker, Am. Geophys. Union, Washington, D.C., p. 602

Dole, S.H.: 1964, Habitable Planets for Man, Blaisdell Publ. Co., New York

Gough, D.O.: 1981, Solar Phys. 74, 21

Hart, M.H.: 1978, Icarus 33, 23

Hart, M.H.: 1979, Icarus 37, 351

Kasting, J.F.: 1987, Precambrian Res. 34, 205

Kasting, J.F.: 1988, Icarus 74, 472

Kasting, J.F., Toon, O.B., Pollack, J.B.: 1988, Scient. Am. 258, 90

Lovelock, J.E.: in The Geophysiology of Amazonia, ed. R.E. Dickinson, Wiley, New York, p.11

Lovelock, J.F., and Watson, A.J.: 1982, Planet. Space Sci. 30, 795

Pollack, J.B., Kasting, J.F., Richardson, S.M., and Poliakoff, K.: 1987, Icarus 71, 203

Sagan, C., and Mullen, G.: 1972, Science 177, 52

Schopf, J.W. (ed.): 1983, Earth's Earliest Biosphere, Princeton Univ. Press, Princeton, N.J.

Schwartzman, D.W.: 1981, SETI-81 Conference, preprint

Schwartzman, D.W., and Rickard, L.J.: 1988a, in Bioastronomy - The Next Steps, ed. G. Marx, Kluwer
    Academic Publ., Dordrecht, p. 305

Schwartzman, D.W., and Rickard, L.J.: 1988b, Am. Scientist , 76, 364

Schwartzman, D.W., and Volk, T.: 1989, Nature 340, 457

Schwartzman, D.W., and Volk, T.: 1990, V.M. Goldschmidt Conference, Program and Abstracts, May 2-4,
    Baltimore, MD , The Geochemical Society, p. 80

Shklovskii, I.S., and Sagan, C.: 1966, Intelligent Life in the Universe, Holden-Day, San Francisco

Sogin, M.L.: Amer. Zool. 29, 487

Tang, T.B.: 1982, J. Brit. Interplanet. Soc. 35, 236

Walker, J.C.G., Hays, P.B., and Kasting, J.F.: 1981, J. Geophys. Res. 86, 9776

Acknowledgments

    We thank Jim Kasting for his enlightening discussions and mini-tutorials on climatic
evolution, and James Lovelock for his review. The Scientific Organizing Committee of the Third
International Bioastronomy Symposium (IAU) provided travel support for D.W.S. to Val Cenis where a
preliminary version of this paper was presented.

Discussion

G. MARX: **Is your model detailed enough to give an answer to the question: how stable the present climate (temperature) is to sustain developed life?**

D. SCHWARTZMAN: We have calculated, as Lovelock had earlier, the time in the future when the habitability limit (for eucaryotes, i.e., T > 50°C) is reached because of increasing solar luminosity. Assuming the same biotic enhancement of weathering as now, the limits will be reached 3.2 billion years from now, not taking into account other effects. However, conditions are likely to become very unpleasant in only 1 billion years from now with the loss of the surface water of the Earth by photodissociation (Kasting, 1988).

EXOBIOLOGICAL HABITATS: AN OVERVIEW

Laurance R. Doyle (SETI Institute) and Christopher P. McKay (NASA/Ames)
Space Science Division, 245-7,
NASA Ames Research Center, Moffett Field, CA 94035

Paper Presented by L.R. Doyle

ABSTRACT

Exobiological habitats, defined here as a long-term ($\sim 10^9$ yrs) liquid water environment with available biogenic materials, can be examined from both a stellar and a planetary perspective. From the stellar perspective the solar system's galactic orbit may be fortuitous "straddling" galactic spiral arms, where stellar activity is high. However, interstellar molecular clouds, possibly enhanced in spiral arms, may also be a substantial source of organics to the protoplanetary disc. Expected ecospheres around various stellar spectral types can also be examined, in light of new atmospheric greenhouse models, to re-define the evolution of liquid water habitats with stellar evolution. From the planetary perspective of our solar system, perhaps four exobiological habitats have existed. In addition to Earth, we examine primordial Mars (early heating via greenhouse gases), the moon Europa (tidal heating via the orbital eccentricity), and possibly large cometary nuclei (radioactive heating from [26] Al).

1. THE STELLAR PERSPECTIVE

1.1 The Galactic Orbit of the Solar System

Although the galactic orbit of the Sun is fairly well known, constraints on the propagation of the spiral density waves that appear as galactic spiral arms vary widely (as do even the number of spiral arms) (Greenberg 1981, Dame et al. 1986, Marochnik and Mukhin 1988). The theory of spiral density waves (Lin et al. 1969) predicts that two types of density waves can develop in spiral galaxies, the so-called shortwave and longwave modes. While stellar systems should orbit the galaxy in a differential fashion (Keplerian orbits), the spiral density pattern should propagate with a constant angular phase velocity. If the galactic spiral arms are caused by the shortwave mode, then this velocity will be about 10 to 14 km/sec per kiloparsec. However, if the spiral structure is caused by the longwave mode, this velocity may be closer to 21 to 25 km/sec per kiloparsec (Marochnik and Mukhin 1988) for a solar distance from the galactic center of 10 kiloparsecs. This latter result changes the region where stars will be co-rotating between spiral arms from the outer galaxy, where the farthest visible HII regions can be seen (Feitzinger and Schmidt-Kaler 1980), to a region very   near to the Sun's orbital location (25 km/sec), a result perhaps not coincidental to the development of advanced biology in the solar system. However, there is a large uncertainty in the determination of the motion of this spiral pattern since the spiral pattern itself is difficult to define. An alternative theory is that the spiral pattern may be due instead to propagating small tidal perturbations (Toomre 1981), with most recent work suggesting either a combination of these two models or nonlinear density wave propagation (Shu et al. 1985).

Direct observations of this spiral pattern are also not well defined. The spiral pattern of our galaxy is generally traceable by observing the distribution of young systems like large mass stars, molecular clouds, open clusters, HII regions, CO cloud structure, and recently, pulsars (see Morgan 1990, for example). However, due to differential rotation, the location of these distributions remains highly uncertain. Consequently, estimates for the number of passages of the solar system through the galactic spiral arm regions vary from zero (Marochnik and Mukhin 1988, Balàzs 1988) to well over 40 (Greenberg 1981). The case for zero spiral arm crossings argues that the Sun presently lies in between the Sagittarius spiral arm of the Milky Way, in which it was formed about 4.6×10^9 years ago, and the Perseus arm, into which it will likely enter in about 3.3×10^9 years. As star formation in general, and large mass star formation in particular, seem to be enhanced along these spiral density waves, this approach argues that it would be necessary for a star to spend sufficient time in between spiral arms in a galaxy to be able to develop and evolve complex biological organisms at a safe distance from widespread supernovae radiation (Shklovskii 1976, Clark et al. 1977). Thus the largest-scale constraint on habitats could be that a parent star should have an orbit that allows significant time between galactic spiral arms.

Clark et al. (1977) have calculated that for the solar system's passage through a typical spiral arm region, one supernova within 10 parsecs would likely be encountered every 10^8 years. This would cause an increase in the total radiant energy of c and X-rays hitting the upper terrestrial atmosphere of from 20 ergs/cm^2/sec (solar flux at these wavelengths) to about 10^6 ergs/cm^2/sec. Atmospheric shielding could initially protect the terrestrial surface, but the atmosphere would become highly ionized, leading to effects such as the destruction of the ozone layer (Whitten et al. 1976, Clark et al. 1977). In addition, the expected cosmic ray flux would likely increase the background radioactivity by about two orders of magnitude (Shklovskii 1976, Marochnik and Mukhin 1988). A zone of co-rotation between spiral arms in a galaxy, therefore, may be an important factor in considering the advanced development of biological organisms.

There may also, however, be possible advantages of multiple solar system passages through spiral arm regions because these regions also show a marked increase in molecular cloud density (Dame et al. 1986), a possibly important source of organics to the early solar system. How might this passage of the solar system through molecular cloud regions quantitatively compare with other extraterrestrial prebiotic materials delivery mechanisms that have been suggested? One important source of post solar accretion delivery of organics that has been suggested is cometary impacts. This mechanism had initial difficulties in delivering organics intact to planetary surfaces because of expected impact velocities. Recent efforts modeling an early thicker terrestrial atmosphere (up to 100 bars of CO_2) and consequently aerobraking the impactors may, however, allow a significant amount of organic material (10^{17} to 10^{20} grams) to have been delivered intact to the early Earth (Chyba et al. 1989). Delivery of organic material by micrometeorites and by the passage of the Earth through cometary tails (Zahnle and Grinspoon 1990) could also have been a significant source. Scaling the micrometeorite data of Anders (1989), for example, would give about 10^{18} grams of organic material being delivered to the Earth over its history at the present flux rate, but could also have allowed values as high as 10^{20} grams to have been delivered depending on the micrometeorite flux rate at the early bombardment (before 4 Gyr ago).

For delivery of prebiotic material to the Earth by passage of the solar system through interstellar molecular clouds in its orbit through the galaxy, we can consider the following. There are probably about 2500 molecular clouds within the circle of the Sun's orbit (Dame et al. 1986) ranging in size from about 10^4 to 6 x 10^6 solar masses. Although these clouds tend to congregate along spiral arms, there is also an appreciable distribution of them in inter-spiral regions. In its orbit around the galaxy, the solar system may have passed through as many as from 15 to 50 molecular cloud complexes (lower non-zero estimates from Elmegreen 1985, and Solomon and Sanders 1985, and higher estimate calculated from data of Dame et al. 1986). The estimated density of star forming molecular clouds is about 10^4 H_2 atoms cm^{-3} (Elmegreen 1985), with organic molecular species (using IRC+10216 as an example) contributing about 0.002% to this density (Irvine et al. 1985). Although dependent on assumptions about the solar system's galactic orbit (Greenberg 1981) as well as the actual number, size, and type of interstellar molecular cloud complexes that the solar system has passed through, the total mass of prebiotic dust and gas accreted by the Earth appears to be in the range 10^{17} to 10^{19} grams, certainly a non-negligible contribution with respect to other delivery mechanisms. Again, the delivery rate during the early heavy bombardment (when the solar system was closer to the spiral arms) could have been higher. Interstellar grain sizes (Greenberg 1981) also appear to be just large enough to be compatible with intact delivery to planetary surfaces (Anders et al. 1973).

1.2 Habitable Zones Around Different Stellar Environments

It is important, in defining habitats, to examine in detail the various biospheric zones that stars of different individual types could allow. Present plans for a directed SETI are reasonably constrained to closely solar-type stars (Billingham 1973, Seeger 1977, Seeger and Wolfe 1985), but in light of recent advances in atmospheric modeling (Kasting 1988, Pollack et al. 1987, Zahnle et al. 1988, Kasting and Toon 1989), not only can the habitable zone around the Sun be broadened (for a terrestrial atmosphere from about 0.95 to about 1.5 AU) but habitable zones around a broader range of different stellar types than previous modeled (Hart 1978, 1979) may be possible. It is reasonable to argue that rapidly evolving high mass stars (spectral types earlier than about F4) would not be expected to allow enough time (10^9 years) for even primitive life-forms to have originated before evolving off the main sequence, (although recent convective models could double the age of early F-type stars; Andersen et al. 1990, for example). A preliminary lower limit on stellar mass around which a liquid water environment could exist has previously excluded the smallest M-type stars (less than 0.2 solar masses) due to the rotational tidal locking expected of a terrestrial planet within the ecosphere of such a small star and consequent freezing out of its atmosphere. However, this result depends highly on the density and thermal gradient of the planetary atmosphere being examined, and even a moderate atmosphere may be sufficient to allow an appreciable liquid water zone. Other considerations, of the M-type stars in particular, include their very high amplitude stellar flare activity and sunspot cycles as well. If certain planetary atmospheres can allow liquid water habitats around M-type stars, the number of such habitats would double, as M-type stars constitute the largest stellar population.

In addition to single main-sequence stars, earlier investigations have also preliminarily examined the possibility of biospheres existing around variable stars, giant stars, and binary systems (Gadomski 1961a, 1961b, 1965; Huang 1960, 1963; Dole 1964), but with very crude models of planetary atmospheres that, in general, do not reflect their adaptability to widely varying stellar luminosity and spectral characteristics that current, more advanced planetary atmospheric models show (Kasting 1988).

In addition, the expected location of extrasolar planetary orbital planes has been constrained (Doyle 1988) and the existence of certain stable planetary orbits in binary systems has recently been established (Pendleton and Black 1983). In double star systems, for example, a stable planetary orbit can exist if the orbital semi-major axis ratio is greater than about 7:1 (either the planet or the second star can be the most distant mass). It is of great interest to examine the possible biospheric zones for such double as well as multiple stellar systems.

Examples of modified environments due to detailed terrestrial atmospheric modeling for the solar system can be found (Kasting 1988, Whitmire et al. 1991) as well as for expected early martian atmospheric and geological conditions (Davis et al. 1991). The inner orbital distance for a terrestrial planet, for example, depends highly on whether a permanent cloud layer can be formed. Although a quantitatively reliable model is still being developed, it appears that such a cloud cover, by reducing the infrared albedo, can play a major role in shielding the planet. For example, a 100% cloud cover (if it can avoid being photodissociated into space) could allow liquid water to exist very near Venus' distance from the Sun. Considering only stellar heating sources at present, the outer boundary for liquid water habitats on planetary surfaces may be constrained by the condensation temperature of CO_2 and ice cover (Davis et al. 1991). Thus, although planetary environments such as present Venus seem to provide enough greenhouse heating to allow liquid water to exist on planetary surfaces as far out as perhaps the jovian orbit, this situation would not last very long as the CO_2 would be expected to quickly condense out. The present limit, then, might be expected to be somewhere just inside the orbit of Mars by these criteria (Davis et al. 1991, Whitmire et al. 1991). When we consider the various different stellar spectral types and luminosity classes, along with differing planetary atmospheric constituents and densities that could lead to liquid water environments, the work in modeling habitats from a stellar perspective has just begun.

2. THE PLANETARY PERSPECTIVE

Other than the Earth, there are three types of locations in the solar system which have been suggested to contain liquid water: Early Mars (and possibly Venus) (McKay and Stoker 1989, Kasting 1988), Europa (Cassen et al. 1979), and large cometary nuclei (Irvine et al. 1980). It is interesting to note that each of these liquid water habitats is the result of a distinct heating source. On early Mars the heating source is solar energy augmented by the CO_2 greenhouse; on Europa the source is tidal heating associated with Jupiter, and within large comets the heating source is radiogenic.

2.1 Liquid Water on Early Mars

The presence of fluvial features on Mars strongly suggests that liquid water had existed on its surface in the past (see for example, Carr 1987, Squyres 1989, McKay and Stoker 1989, McKay et al. 1990). The key question associated with the possible origin of life on Mars is the time interval that liquid water may have existed on early Mars. Several authors have pointed out that the apparent stability of liquid water on early Mars, as indicated by the runoff channels, implies warmer surface conditions which, in turn, implies a significantly thicker atmosphere at that time, (for example, Pollack et al. 1987). It is important to note here that it is the presence of liquid water per se that is of interest from a biological perspective and not the putative warmer, thicker atmosphere. If liquid water habitats can be maintained under a cold atmosphere that renders most of the Martian surface inimical to life, this would be of biological significance. This is analogous to what occurs in the Antarctic Dry Valleys. Mean annual temperatures in the dry valleys are -20°C, but liquid water habitats are found under perennial ice covers of 3 to 6 meters (McKay et al. 1985, McKay and Friedmann 1985) and are the major site for microbial life in the valleys (Parker et al. 1982a,b).

It has been suggested that the early atmosphere of Mars was composed of CO_2 in sufficient quantities to warm the surface above freezing despite the lower luminosity of the early sun, about 0.70 that of the present sun (Moroz and Mukhin 1977). This is similar to suggestions for keeping the early Earth warm (Owen et al. 1979, Kasting and Ackerman 1986). Subsequent studies by many authors have shown that one to several bars of CO_2 are needed to maintain the mean surface temperatures at or above freezing (Pollack 1979, Pollack and Yung 1980, Cess et al. 1980, Hoffert et al. 1981, Postawko and Kuhn 1986, Pollack et al. 1987). This initial CO_2 atmosphere would have been lost by reaction in water to form carbonate rock (Fanale et al. 1976, Kahn 1985, Pollack et al. 1987). The absence of tectonic activity precluded any long term recycling of carbonates and Mars eventually lost its atmosphere. Pollack et al. (1987) suggested that the thick atmosphere could be maintained by recycling of CO_2 due to increased volcanic activity and geothermal heat flow on early Mars. Combined with an extended outgassing model, they suggest that temperatures above freezing could have been maintained, in the optimistic limit, for 10^9 yr. Carr (1989) suggested that CO_2 was pumped into the atmosphere by impacts associated with the late bombardment, and that the thick atmosphere would have dissipated quickly after the cessation of impacts, 3.8 Gyr ago.

McKay et al. (1985) have developed a simple model which relates the thickness of ice on a perennially ice covered lake to climatological variables. In this annually averaged, steady state model, the thickness of ice is determined primarily by the balance between the conduction of energy from the ice, the inputs of energy via sunlight, and the transport of latent and sensible heat by the summer meltstream. The latent heat released upon freezing at the ice-water interface is the largest term in this equation. Because steady state conditions are assumed, the freezing rate of water at the ice bottom must be offset by ablation from the ice surface. Squyres (1989) has modeled the ablation rate on early Mars. Preliminary work in this area indicates that ice-covered lakes, similar to those found in the Antarctic Dry Valleys, could persist on early Mars for up to 700 million years after the mean temperatures fell below freezing (McKay et al. 1990, Davis and McKay 1990, Davis et al. 1991). However, these preliminary results also suggest that if the sun were brighter early in the history of Mars, the surface temperatures at a given pressure would be higher and the weathering rate, which is a strong function of temperature, would be increased. The result would be that conditions would become unfavorable for liquid water generation much sooner, reducing the probability that life could have evolved. Thus, indirectly the luminosity of a faint early sun helped maintain liquid water habitats on early Mars. Future work from these results will require the extension of this model to other stellar types at various stages in their evolution.

2.2 Liquid Water on Europa

Tidal heating can provide an alternative heating source for sustaining liquid water, and the existence of a possible ocean on Europa is based upon this heating source (Cassen et al. 1979, Reynolds et al. 1983, Reynolds et al. 1987). Tidal forces arise within a satellite because the gravitational force varies radially from the central body. Regions of the satellite closer to the central body feel a stronger gravitational pull than the corresponding regions on the other side of the satellite. In a steady state situation the system would reach a minimum energy configuration in which the orbit would be circular and the rotation of the satellite would be synchronous, that is, the orientation of the satellite is fixed with respect to the tidal forces. In this configuration the tidal forces are balanced by material and self gravitational forces in the satellite, and no net work is performed by the tidal forces. Thus there would be no tidal energy dissipation. This is inevitably the case for isolated satellites, such as the Earth's moon. However, in multiple satellite systems, such as characterize the giant planets of the outer solar system, mutual interactions between the satellites prevent the formation of synchronous states. The prime example of this is the so-called "Laplacian resonance" between Io, Europa, and Ganymede which maintains relatively high values for the eccentricities. Physically, the energy released by tidal dissipation comes from the orbital energy of the satellite and the rotational energy of the central body, and the entire system evolves to a lower moment of inertia, while conserving angular momentum.

Reynolds et al. (1987) has shown that the heating rate due to tidal forces (H_T) on a uniformly dense and rigid body with an eccentricity (e) much less than one, can be written as follows:

$$H_T = \frac{\dot{E_t}}{4\pi R_s^2} = \frac{21}{38} G^{\frac{5}{2}} \times \frac{R_s^5 \rho^2}{\mu Q} \times e^2 \left(\frac{M_p}{a^3}\right)^{\frac{5}{2}},$$

where ρ is the density, μ is the rigidity, E_t is the total rate of tidal dissipation, R_s is the satellite radius, Q is the dissipation factor analogous to the quality of a resonating system, M_p is the mass of the central body, and a is the semi-major axis of the satellite's orbit. The right-hand side of this equation has been grouped into three categories. The first term is a constant, the second term contains variables that depend on the nature and size of the satellite itself, and the third set of terms depend on the satellite's orbital parameters including the central gravitating planetary mass. This equation can thus be used to consider the effects of hypothetical changes in the configuration of the central body and satellite system, and how these changes would affect the extent and duration of a liquid ocean. This approach is being developed following initial work done by Reynolds et al. (1987). Recent models of planetary system formation (Pollack 1984, for example) can be used as a guide to the range and properties of central bodies and satellites that can be considered for general models of this new type of habitable zone.

2.3 Liquid Water in Comets

Comets are composed of water ice and are rich in organics. It is very unlikely that the comet nucleus presents a favorable milieu for organic molecules to react to form polymeric compounds and ultimately the structures of biology, such as membranes (Oró 1961, Ponnamperuma and Ochiai 1982, Lazcano-Araujo and Oró 1981, Oró et al. 1990). Perhaps the most serious difficulty preventing a continuing evolution of pre-biotic materials is the absence of liquid water. However, it has been suggested by Irvine et al. (1980, 1981) and Wallis (1980) that radiogenic heating of a cometary nucleus by the decay of an initial burden of ^{26}Al could have resulted in the formation of a liquid water core inside of large (radius > 10 km) comets. The existence of such a liquid core requires three conditions which are difficult to determine (Irvine et al. 1981): 1) The timescale for formation of comets must be shorter than the decay time of ^{26}Al, which is 7×10^5 years; 2) The cometary material must be a good insulator; and 3) The comet must be large. Given these uncertainties, Irvine et al. (1981) suggest that the duration of liquid cores in comets could range from 0 to 10^9 years. Since the upper limit is comparable to the time from the formation of Earth to the oldest definitive evidence for life 3.5 Gyr ago (see, for example, Schopf 1983), the possibility of the origin of life in liquid-core comets is not entirely out of the question. In fact, the high concentration of organics and the reducing conditions in a comet, coupled with the presence of liquid water, provides an environment that is actually closer to the conditions of the Miller synthesis than the early Earth, even in the case of a reducing atmosphere. One of the problems in the synthesis of biologically significant molecules in an early ocean on Earth is the polymerization of amino acids and nucleic acid monomers, and the higher concentration of organics in a comet would aid this polymerization (Irvine et al. 1980). The continuation of any comet life past the period of a liquid water core is unlikely, given the essential requirement of liquid water for all known life forms (for example, Kushner 1981). But more realistic models of this possible habitat, based on improved knowledge of cometary structure and composition from the Halley fly-bys, could be of significant exobiological interest.

3. CONCLUSION

We have examined exobiological habitats from both a stellar and a planetary perspective. We can examine these results in the context of one form of Drake's equation:

$$N = R_s f_s f_p n_e f_b f_i f_c L_c$$

where N = number of communicating civilizations; R_s = rate of star formation in the galaxy; f_s = fraction of stars capable of maintaining a stable ecospheric zone; f_p = fraction of those stars with planetary systems; n_e = number of ecospheres in that planetary system; f_b, f_i, and f_c = fractions of ecospheres where biology develops, becomes intelligent, and communicates, respectively; and finally L_c = lifetime of a communicating civilization. If exobiological habitats are constrained to lie within co-rotation zones between galactic spiral arms, then the term f_s can be considered only from a volume about 6% that previously considered. On the other hand, if stable ecospheres can be accommodated around M-type stars, in spite of their small ecoshells, their vast number could increase f_s to perhaps 0.4 or more. Considering recent developments in extrasolar planetary detection, f_p will likely be the next best constrained term in this equation, and if double stars can accommodate stable planetary orbits, as they appear to be able to do, this term too could be doubled. Finally, from a planetary perspective, it appears that the term n_e for our solar system (number of sustained liquid water habitats) could have been as high as 4. In conclusion, the application of detailed astrophysical and planetary models to the definition of exobiological habitats has just begun. We can expect a significant broadening of our ideas about possible locations for the origin and evolution of biological systems leading, as well, to a significantly better specification of targets in the dedicated search for extraterrestrial intelligence.

References

1)   Anders, E., R. Hayatsu, and M.H. Studier, (1973) "Organic compounds in meteorites ", Science 182, 781-790.
2)   Anders, E., (1989), "Pre-biotic organic matter from comets and asteroids", Nature 342, 255-257.
3)   Andersen, J., B. Nordstrom, and J.V. Clausen, (1990),"New strong evidence for the importance of convective overshooting in intermediate-mass stars", Ap.J. Lett. 363, L33-L36.
4)   Balazs, B., (1988), "The galactic belt of intelligent life" in Bioastronomy: The Next Steps, Kluwer Academic Pub., Dordrecht, Holland, G. Marx (ed.), 61-66.

5)   Billingham, J., (1973), Project Cyclops, CR 114445, Prepared under Stanford/NASA-Ames Research Center, Moffett Field, CA., pp. 7-20, 155-168.

6)   Carr, M.H., (1987), "Water on Mars", Nature, 326, 30-35.

7)   Carr, M.H., (1989), "Recharge of the early atmosphere of Mars by impact-induced release of CO_2", Icarus 79, 311-327.

8)   Cassen, P.M., S.J. Peale, and R.T. Reynolds, (1979), "Is there liquid water on Europa?", Geophys. Res. Lett. 6, 731-734.

9)   Cess, R.D., V. Ramanathan and T. Owen, (1980) "The Martian paleoclimate and enhanced carbon dioxide", Icarus 41, 159-165.

10)   Chyba, C.F., P.J. Thomas, L. Brookshaw, and C. Sagan, (1990), "Cometary delivery of organic molecules to the early Earth", Science 249, 366-373.

11)   Clark, D.H., W.H. McCrea, and F.R. Stephenson, (1977), "Frequency of nearby supernova and climatic and biological catastrophes", Nature 265, 318-319.

12)   Dame, T.M., B.G. Elmgreen, R.S. Cohen, and P. Thaddeus, (1986), "The largest molecular cloud complexes in the first galactic quadrant", Ap.J. 305, 892-908.

13)   Davis, W.L., L.R. Doyle, C.P. McKay, and D.E. Backman, (1991), "The Habitability of Mars-like Planets Around Main-Sequence Stars", (This volume).

14)   Davis, W.L. and C.P. McKay, (1990), "Duration of liquid water habitats on early Mars", Icarus, submitted.

15)   Dole, S.H., (1964), Habitable Planets For Man, Blaisdell Pub. Co., New York.

16)   Doyle, L.R., (1988), "Progress in determining the space orientation of stars" in Bioastronomy: The Next Steps, Kluwer Academic Pub., Dordrecht, Holland, G. Marx (ed.), 101-105.

17)   Elmegreen, B.L., (1985), "Molecular clouds and star formation: an overview" in Protostars and Planets II, University of Arizona Press, Tucson, D.C. Black and M.S. Matthews (eds.), 33-58.

18)   Fanale, F.P., "Martian volatiles, their degassing history and geochemical fate", (1976), Icarus, 28, 179-202.

19)   Feitzinger, J.V. and T. Schmidt-Kaler, (1980), "Contributions to the theory of spiral structure", Astron. and Ap. 88, 41-51.

20)   Gadomski, J., (1961a), "The ecospheres of pulsating stars" in Proc. 11th Int. Astronautical Congress, Springer-Verlag, 108-113 (in German).

21)   Gadomski, J., (1961b), "Ecospheres of giant stars and ecospheric evolution of Sun's planetary system", Astronautica Acta 7, 1-7 (in German).

22)   Gadomski, J., (1965), "Ecospheres of double stars", Postepy Astronomii 13, 241-243 (in Polish).

23)   Greenberg, J.M., (1981), "Chemical evolution of interstellar dust - a source of prebiotic material?" in Comets and the Origin of Life, C. Ponnamperuma (ed.), D. Reidel Pub. Co., 111-128.

24)   Hart, M.H., (1978), "The evolution of the atmosphere of the Earth", Icarus 33, 23-39.

25)   Hart, M.H., (1979), "Habitable zones about main sequence stars ", Icarus 37, 351-357.

26)   Hoffert, M.I., A.J. Callegari, C.T. Hsieh, and W. Ziegler, (1981), "Liquid water on Mars: an energy balance climate model for CO_2/H_2O atmospheres", Icarus 47, 112-129.

27)   Huang, S.S., (1960), "The limiting sizes of the habitable planets", P.A.S.P. 72, 489-493.

28)   Huang, S.S., (1963), "Life supporting regions in the vicinity of binary systems" in Interstellar Communication, A.G.W. Cameron (ed.), W.A. Benjamin Pub., New York.

29)   Irvine, W.M., S.B. Leschine, and F.P. Schloerb (1980), "Thermal history, chemical composition and relationship of comets to the origin of life", Nature 283, 748-749.

30)   Irvine, W.M., S.B. Leschine, and F.P. Schloerb (1981), " Comets and the origin of life", in Origin of Life, Proc. 6th Int. Conference on the Origin of Life, Y. Wolman, (ed.), pp 748-749. D. Reidel Publishing Company, Dordrecht, Holland, 27.

31)   Irvine, W.M., F.P. Schloerb, AA. Hjalmarson, and E. Herbst, (1985), "The chemical state of dense interstellar clouds: an overview" in Protostars and Planets II, University of Arizona Press, Tucson, D.C. Black and M.S. Matthews (eds.), 579-620.

32)   Kahn, R., "The evolution of CO_2 on Mars", Icarus 6 2, 175-190 (1985).

33)   Kasting, J.F., and T.P. Ackerman, (1986), "Climatic consequences of very high CO_2 levels in Earth's early atmosphere", Science 234, 1383-1385.

34)   Kasting, J.F., (1988), "Runaway and moist greenhouse atmospheres and the evolution of Earth and Venus", Icarus 74, 472-494.

35) Kasting, J.F., and O.B. Toon, (1989), "Climate on the terrestrial planets" in Origin and Evolution of Planetary and Satellite Atmospheres, University of Arizona Press, Tucson, S.K. Atreya, J.B. Pollack, and M.S. Matthews (eds.), 423-449.

36) Kushner, D., "Extreme environments: are there any limits to life?", (1981), in Comets and the Origin of Life, C. Ponnamperuma, (ed.), D. Reidel Publishing Co.

37) Lazcano-Araujo, A. and J. Oro (1981), "Cometary material and the origin of life on Earth", in Comets and the Origin of Life, C. Ponnamperuma (ed.), pp 191-225. D. Reidel Publishing Company, Dordrecht, Holland, 27.

38) Lin, C.C., C. Yuan, and F. Shu, (1969), "On the spiral structure of disk galaxies III. comparison with observations ", Ap.J. 155 721-746.

39) Marochnik L.S., and L.M. Mukhin, (1988), "Belt of life in the galaxy" in Bioastronomy: The Next Steps, Kluwer Academic Pub., Dordrecht, Holland, G. Marx (ed.), 49-59.

40) McKay, C.P. and E.I. Friedmann, (1985), "The cryptoendolithic mic robial environment in the Antarctic cold desert: temperature variations in nature", Polar Biol. 4, 19-25.

41) McKay, C.P., G.A. Clow, R.A. Wharton, Jr. and S.W. Squyres, (1985), "Thickness of ice on perennially frozen lakes", Nature 313, 561-562.

42) McKay, C.P. and C.R. Stoker, (1989), "The early environment and its evolution on Mars: implications for life", Reviews of Geophysics 27, 189-214.

43) McKay, C.P., W.J. Borucki, F. Church, and D. Kojiro, (1990), "Re-synthesis of organic material in cometary entry shocks", in preparation.

44) Morgan, S. (1990), "Pulsars as spiral arm tracers", P.A.S.P. 102, 102-106.

45) Moroz, V.I., and L.M. Mukhin, (1977), "Early evolutionary stages in the atmosphere and climate of the terrestrial planets", Cosmic Research 15, 774-791.

46) Oro, J. (1961), "Comets and the formation of biochemical compounds on the primitive Earth", Nature 190, 389-390.

47) Oro, J., S.W. Squyres, R.T. Reynolds, and T.M. Mills, (1990), "Europa: prospects for an ocean and exobiological implications", preprint.

48) Owen, T., R.D. Cess, and V. Ramanathan, (1979), "Enhanced CO_2 greenhouse to compensate for reduced solar luminosity on early Earth", Nature 277, 640-642.

49) Parker, B.C., G.M. Simmons, Jr., K.G. Seaburg,, D.D. Cathey, and F.T.C. Allnutt, (1982a), "Comparative ecology of plankton communities in seven Antarctic oasis lakes", J. Plank. Res. 4, 271-286.

50) Parker, B.C., G.M. Simmons, Jr., R.A. Wharton, Jr., K.G. Seaburg, and F.G. Love (1982b), "Removal of organic and inorganic matter from Antarctic lakes by aerial escape of bluegreen algal mats", J. Phycol., 18, 72-78.

51) Pendleton Y.J. and D.C. Black, (1983), "Further studies on criteria for the onset of dynamical instability in general three-body systems", A.J. 88, 1415-1419.

52) Pollack, J.B., (1979), "Climate change on the terrestrial planets ", Icarus 37, 479-533.

53) Pollack, J.B. and Y.L. Yung, (1980), "Origin and evolution of planetary atmospheres", Ann. Rev. Earth Planet. Sci. 8, 425-487.

54) Pollack, J.B. (1985), "Formation of the giant planets and their satellite ring systems: an overview" in Protostars and Planets II, D.C. Black and M.S. Matthews (eds.), U. of Arizona Press, Tucson.

55) Pollack, J.B., J.F. Kasting, S.M. Richardson, and K. Poliakoff, (1987), "The case for a wet, warm climate on early Mars", Icarus, 71, 203-224.

56) Ponnamperuma, C. and E. Ochiai (1982), "Comets and origin of life", in Comets, Laurel L. Wilkening (ed.), pp 696-703. University of Arizona Press, Tucson, Arizona.

57) Postawko, S.E. and W.R. Kuhn, (1986), "Effect of the greenhouse gases (CO_2, H_2O, SO_2) on Martian paleoclimate", Proc. 16th Lunar and Planetary Sci. Conf., J. Geophys. Res. 91, D431-438.

58) Reynolds, R.T., S.W. Squyres, D.S. Colburn, and C.P. McKay (1983), "On the habitability of Europa", Icarus 56, 246-254.

59) Reynolds, R.T., C.P. McKay, and J.F. Kasting, (1987), "Europa, tidally heated oceans, and habitable zones around giant planets", Adv. Space Res. 7, 125-132.

60) Schopf, J.W. (1983) (ed.), Earth's Earliest Biosphere: It's Origin and Evolution, Princeton University Press, Princeton, New Jersey.

61) Seeger, C.L., (1977), "Stellar Census" in The Search for Extraterrestrial Intelligence, NASA SP-419, P. Morrison, J. Billingham, and J. Wolfe (eds.), 143-146.

62) Seeger, C.L. and J.H. Wolfe, (1985), "The microwave search problem and the targeted search approach" in Bioastronomy: The Next Steps, Kluwer Academic Pub., Dordrecht, Holland, G. Marx (ed.), 391-395.

63) Shklovskii, J.S., (1976), Supernova, Nauka Pub., Moscow.

64) Shu, F.H., L. Dones, J.J. Lissauer, C. Yuan, and J.N. Cuzzi, (1985), "Nonlinear spiral density waves: viscous damping", Ap.J. 299, 542-573.

65) Solomon, P.M., and D.B. Sanders, (1985), "Star formation in a galactic context: the location and properties of molecular clouds" in Protostars and Planets II, University of Arizona Press, Tucson, D.C. Black and M.S. Matthews (eds.), 59-80.

66) Squyres, S.W., (1989), "Urey prize lecture: water on Mars", Icarus 79, 219-228.

67) Toomre, A., (1981), "What amplifies the spirals?", in The Structure and Evolution of Normal Galaxies, S.M. Fall and D. Lynden-Bell (eds.), Cambridge U. Press, New York, 111-136.

68) Wallis, M.K. (1980), "Radiogenic melting of primordial comet interiors", Nature 284, 431-433.

69) Whitmire, D.P., R.T. Reynolds, and J.F. Kasting, (1991), "Habitable zones for Earth-like planets around main sequence stars", (This volume).

70) Whitten, R.C., J.N. Cuzzi, W.J. Borucki, and J.H. Wolfe, (1976), "Effect of nearby supernova explosions on atmospheric ozone", Nature 263, 398-400.

71) Zahnle, K.J., J.F. Kasting, and J.B. Pollack, (1988), "Evolution of a steam atmosphere during Earth's accretion", Icarus 74, 62-97.

72) Zahnle, K. and D. Grinspoon, (1990), "Comet dust as a source of amino acids at the Creteous/Tertiary boundry", Nature 348, 157-160.

Discussion

D. BRIN: You mention the possibility of liquid interiors in comets, should they have formed quickly enough to incorporate sufficient Aluminum 26 early on. Have you also considered the proposal that electrical conductivity effects during the T-Tauri phase might also create liquid interiors? It is far-fetched but intriguing.

L.R. DOYLE: We haven't, but it is an interesting suggestion. I'd like to see how long such a heating phase could be sustained.

WM. IRVINE: Rather than hindering life, might nearby supernovae not actually stimulate the development of "advanced" life terms by creating environmental stress and perhaps extinctions that would open ecological niches for new biological "experiments" (as impacts may have done)? In any case, throughout most of the history of life on Earth only prokaryotes have existed, some of which are very resistant to ionizing radiation (e.g. Deinococcus radiodurans).

L.R. DOYLE: I agree that it is far from understood what the effects of supernovae on biology may have been. It may indeed have long-term benefits. It is interesting that such astronomical effects may be essential to understanding terrestrial biological evolution.

D.P. WHITMIRE: How did you estimate that only 6% of all disk stars are out of the spiral arms?

L.R. DOYLE: Comparing the rotation rate of the solar system around the galaxy with the long wave mode density wave propagation rotation rate, we can derive a delta radius for stellar orbits inward and outward from this radius that allows at least 5 billion years between spiral arm crossings. This leaves a narrow -- less than 500 parsecs wide -- circle of possible orbits for an ETI search if avoidance of spiral arms is necessary for the survival of biology.

HABITABLE ZONES FOR EARTH-LIKE PLANETS AROUND MAIN SEQUENCE STARS

D.P. Whitmire
Department of Physics
The University of Southwestern Louisiana

R.T. Reynolds
Theoretical Studies Branch
NASA Ames Research Center, Moffett Field, California 94035 USA

J.F. Kasting
Department of Geosciences
Penn State University

Paper Presented by D.P. Whitmire

ABSTRACT

As stars evolve and brighten, the radial zone within which liquid water can exist at the surface of an Earth-like planet expands outward. Using a new planetary climate model we have calculated the evolution of these habitable zones around several main sequence stars of masses between 0.50 and 1.25 solar masses. This evolution is presented in the form of a habitability continuum diagram for each star. We also give results for post main sequence evolution of the habitable zone around a 1.0 $M_\odot$ star. The inner radius of the habitable zone is determined by the moist greenhouse effect and the outer radius is taken to be the distance at which a dense CO_2 atmosphere begins to condense. Preliminary results indicate that the range of planetary radii for which liquid water can exist for > 4.5 Byr is considerably broader than previously calculated.

The habitable zone (HZ) around a star is usually defined as the radial band within which liquid water can exist at the surface of an Earth-like planet by virtue of the stellar luminosity. Huang (1959, 1960) pointed out the relation between the HZ and stellar luminosity or spectral type. He assumed that the HZ is determined primarily by the stellar flux and that the inner and outer radii corresponded to today's flux at Venus and Mars, respectively. Massive O and B stars ($L > 10^4 L_\odot$) have the largest HZ's but suffer from very short main sequence lifetimes ($< 10^7$ yr), whereas the most abundant low mass M type stars have extremely long lifetimes (> 30 Byr) but much narrower HZ's. This should decrease the a priori probability of occupancy by a suitable planet depending on the manner in which the distances of planetary formation scale with the mass of the system.

The mean planetary surface temperature (T_s) depends not only on the local stellar flux but also on albedo, clouds and atmospheric greenhouse gases. On Venus, for example, T_s is nearly twice the local black body temperature in spite of a highly reflecting cloud cover. Hart (1979) attempted to take these factors into account in determining the HZ around MS stars. This analysis was based in part on an earlier paper on the evolution of the Earth's atmosphere (Hart, 1978). He concluded that the HZ for which liquid water could exist on an Earth-like planet for the first 4.5 Byr (the continuously habitable zone, CHZ) was extremely narrow. For a 1 $M_\odot$ star it extended from only 0.958 AU to 1.0034 AU (sic), became even narrower for less massive G and early K stars and disappeared altogether for late K and all M type stars. This result has been cited as evidence that intelligent life is rare in the Galaxy and perhaps, given other constraints, unique to Earth. In view of the significance of this conclusion, we have re-examined the extent of the CHZ using a more recent planetary climate model.

The one dimensional radiative-convective model used in our analysis has been described elsewhere (Kasting et al., 1984a; 1984b; 1988a; 1988b). In addition to being radiative-convective, a significant feature of this model is the concept of the geophysical recycling of carbon dioxide. This process can lead to a negative feedback effect and the regulation of a planet's water supply and surface temperature over geological timescales, as has probably occurred on Earth. The carbon dioxide is cycled between the atmosphere and carbonate rock. In this process CO_2 is combined with mineral constituents ultimately producing carbonate rock. The carbonate rock is then subducted into the mantle and eventually the CO_2 is returned to the atmosphere by outgassing from volcanoes or other geological activity. Most of the CO_2 on Earth is now stored in carbonate rocks (as it has undoubtly been for most of Earth's history). If the solar flux increases, the surface temperature increases and more water is evaporated. This increases the rate of reaction and thus removal of atmospheric CO_2 which, in turn, reduces the surface temperature until equilibrium is established. A decrease in solar flux would reduce the surface temperature and rates of atmospheric CO_2 removal. The geological sources of CO_2 are connected to internal heating cycles and are thus unaffected by solar luminosity changes. Consequently, the CO_2 content of the atmosphere and T_s would eventually increase again. On Earth, life also enters into the regulation of CO_2 and surface temperature (Lovelock and Watson, 1982; Schwartzman and Volk, 1989, 1991). These negative feedback processes are probably responsible for maintaining Earth's surface temperature within a range suitable for life over geologic time (thus explaining the faint Sun paradox). There are of course limits to this sort of regulation as evidenced by Venus, which underwent either a runaway greenhouse or a moist greenhouse effect early in its history.

Venus is presumed to have formed with a water content comparable to that of Earth, but subsequently lost its water when it underwent either a runaway or moist greenhouse effect. In a runaway greenhouse, all water is evaporated and the surface temperature would soar to 1500 K. The thermal structure of the atmosphere is such that a cold trap exists at a very high altitude (> 100 km), allowing the water vapor to exist at high altitudes where it can be dissociated by the solar UV. Subsequently, the hydrogen escapes to space. Recent calculations (Kasting, 1988a) show that a runaway greenhouse would have occurred on a cloudless early Venus if the solar flux was f > 1.4 times the flux at Earth today (in the following we use flux units measured relative to the flux at Earth today). Since the actual flux at early Venus was about 1.4 and reflecting clouds may not have been negligible, it is possible that oceans existed for a short period. In this case, water would be lost by the moist greenhouse effect, which occurs at a critical flux of 1.1 in a cloudless atmosphere (Kasting 1988a; Kasting et. al, 1988b). Under these conditions, liquid oceans would initially exist but gradually evaporate. The atmospheric cold trap would occur at 100 km and H_2O dissociation by solar UV (and the subsequent loss of the hydrogen to space) would ensue as in the runaway greenhouse case.

We define an Earth-like planet as having a mass of roughly one Earth mass, terrestrial composition and significant near-surface quantities of CO_2 and H_2O (Earth has about 60 bars of CO_2 in carbonate rock). To estimate the extent of the HZ for such planets, we assume that the inner radius (r_i) is determined by the moist greenhouse effect and the outer radius (r_o) is determined by the critical stellar flux at which a dense CO_2 atmosphere starts to condense. Such an atmosphere (especially if augmented by some H_2O vapor and other greenhouse gases) could readily provide the necessary greenhouse to maintain a surface temperature $T_s > 273$ K out to distances well beyond Mars in the Solar System, assuming it does not condense. In using the cloudless moist greenhouse value for the critical flux to fix ri, we are being conservative, since clouds will probably reduce the actual critical flux for the onset of the moist greenhouse, and thus move the true r_i further inward. The early Sun had a luminosity of 0.74 $L_\odot$. Setting f = 1.1 gives r_i = 0.82 AU, which is slightly beyond Venus. The inner radius of the HZ has increased in time as the Sun has evolved on the main sequence and today is 0.95 AU.

In estimating the outer radius of the HZ, we have neglected clouds, greenhouse gases other than CO_2 and H_2O, and the non-ideal contributions to the equation of state. Preliminary calculations using a non-ideal equation of state indicate that the ideal gas approximation is probably adequate. To determine r_o, the mean surface temperature was conservatively fixed at 273 K. The partial pressure of CO_2 was then varied and the minimum flux required to maintain this temperature at the given pressure was calculated. It was found that for fluxes less than 0.37 it was not possible to maintain T_s = 273 K for any value of the CO_2 partial pressure. The minimum in the function f(PCO) occurred at PCO = 10 and f = 0.37. This minimum is due to an increase in Rayleigh scattering (and thus albedo) from condensing particles at high pressures. If the initial pressure was > 10 bar, the atmosphere would condense down to 10 bar at f = 0.37. For a flux less than 0.37, it would be impossible to maintain T_s = 273 K at any pressure. Terrestrial planets are expected to have dense (> 10 bar) CO_2 atmospheres after accretion is completed. Therefore ro can be computed on the basis of a critical minimum stellar flux of 0.37 at time t = 0. This critical flux is probably a conservative estimate since it would not allow for water on early Mars. There is in fact strong evidence for water on early Mars when the flux was only 0.32. Mars itself may have ultimately lost its liquid water because of the cessation of most geological activity, which is necessary to maintain CO_2 recycling.

If the zero-age main sequence flux is less than 0.37, the planet's H_2O will remain frozen until the luminosity increases. However, the flux must now increase above 0.37 to produce liquid water, since the partial pressure will be less than 10 bar and thus unable to maintain T_s = 273 K. Beginning with a frozen planet for which f < 0.37, the flux must increase to 0.6 before a surface temperature of 273 K can be reached. In the following habitability continuum diagrams we have calculated r_o at t = 0 by setting f = 0.37. For t > 0, we have assumed that at distances greater than ro(t = 0) a planet must wait until the flux is = 0.6 before T_s = 273 K can be reached. For an initially frozen planet, it might be possible that after the 0.37 flux level had been reached, a series of large volcanic eruptions (or a massive asteroid/comet impact) could generate a dense enough atmosphere so that liquid water temperatures could be reached and maintained before the 0.6 stellar flux level was reached. For this reason 0.6 is probably a conservative estimate.

In Figure 1, we illustrate how the luminosity of a 1 $M_\odot$ star varies throughout the star's evolution. The time dependence was based on the stellar models of Iben (1967a; 1967b; 1974; 1983). It can be seen in this figure that the luminosity increases by about a factor of 4 during the main sequence (MS) phase, a factor of 300 during the first red giant (RG) phase, and ultimately a factor of

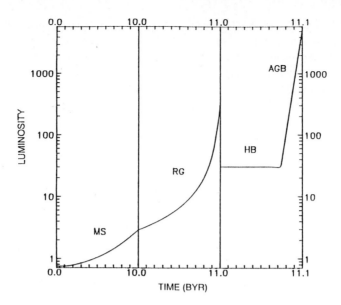

Figure 1: Luminosity in solar units as a function of time during the entire evolution of a 1 $M_\odot$ star with solar composition. Note the scale is broken twice at 10 and 11 Byr. The near discontinuity at 11.0 Byr is real and due to the helium flash. Details of the HB and late AGB evolution have been neglected. MS = main sequence, RG = red giant, HB = horizontal branch and AGB = asymptotic giant branch.

10^4 over the zero-age luminosity at the end of the asymptotic giant branch (AGB) or second red giant phase. We have approximated the luminosity as being constant during the horizontal branch (HB) phase. The lifetimes of the advanced phases are much shorter than the MS lifetime and therefore, although the HZ width will grow to 100 times its zero-age width, there is probably not much chance for the independent evolution of intelligent life during these phases. However, any pre-exiting MS intelligent life might find numerous outer worlds suitable for habitation during the post MS evolution, as illustrated in Figure 2.

In Figure 2 we show how the HZ evolves during the evolution of a 1 $M_\odot$ star. At a given time, the vertical height of the habitability band gives the radial width of the HZ and at a given distance from the star, the horizontal bandwidth gives the time interval for which an Earth-like planet at that distance could maintain liquid water. A habitability continuum diagram similar to Figure 2 was previously given for the more optimistic case where ro is determined at all times by the flux at early Mars, f = 0.32 (Whitmire and Reynolds 1990).

Figure 3 gives the MS habitability continuum diagrams for a range of stellar masses. Since the age of the Galaxy is approximately 10 Byr, the curves are terminated at 10 Byr, even though the 0.50 $M_\odot$ and 0.85 $M_\odot$ stars have much longer lifetimes. The zone within which an Earth-like planet would have liquid water for a period > 4.5 Byr sometime during the star's evolution, is broadest for stars in the mass range 1.0 - 1.2 $M_\odot$. (This zone is similar to Hart's continuously habitable zone). The width of this CHZ for a 1 $M_\odot$ star is about 0.5 AU for f(t >0) = 0.6, or about 1 AU for the more optimistic case of f(t >0) = 0.37, which is not shown in the figures. It is also evident from Figure 3 that all MS stars

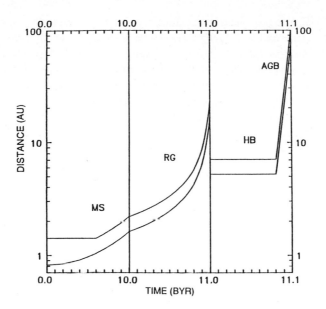

Figure 2: Habitability Continuum Diagram for the entire evolution of a 1 $M_\odot$ star. Note the scale is broken at 10 and 11 Byr.

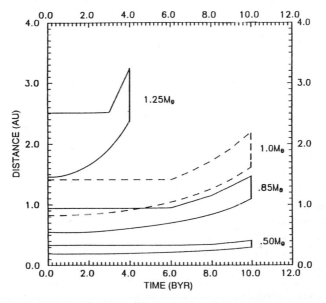

Figure 3: Habitability Continuum Diagrams for the main sequence evolution (< 10 Byr) for a range of stellar masses. Although not shown here, the main sequence lifetimes of the 0.85 $M_\odot$ star and especially the 0.50 $M_\odot$ star extend well beyond 10 Byr.

of mass < 1.2 $M_\odot$ have finite CHZ's, in contrast to Hart's results. Interestingly, although the CHZ is optimum for stars of approximately 1 $M_\odot$, our own planetary system would actually have had a better a priori probability of evolving intelligent life if it had formed around a less luminous star of mass between 0.85 and 1.0 $M_\odot$, in which case both Venus and Earth would have fallen within the CHZ.

In conclusion, we have used a recent climate model to calculate the evolution of the habitable zone for a range of stellar masses. Preliminary results for a 1 $M_\odot$ star indicate that the zone within which liquid water can exist for > 4.5 Byr is approximately 0.5 AU, which is an order of magnitude larger than previously calculated (Hart, 1979). We also find that this zone is finite for all MS stars less massive than 1.2 $M_\odot$, in contrast to the earlier results. Further modeling, taking into account a nonideal equation of state, parameterized-clouds, and additional greenhouse gases, is in progress. In view of several conservative assumptions and the apparent existence of liquid water on early Mars, the actual habitable zones could turn out to be somewhat broader than we have calculated here. In either case, it appears that habitable zones are not nearly as narrow as previously calculated, and the subsequent inference that intelligent life in the Galaxy must be exceedingly rare is premature.

Acknowledgment: We thank Dr. Pat Whitman for assistance.

References

Hart, M. H., The evolution of the atmosphere of the Earth, Icarus 33, 23-39 (1978).

Hart, M. H., Habitable zones about main sequence stars, Icarus 37, 351-357 (1979).

Huang, S., Occurrence of life in the universe, American Scientist 47, 397-402 (1959).

Huang, S., Life outside the Solar System, Sci. Amer. 202, 55-63 (1960).

Iben, I., Stellar Evolution. VI. Evolution from the main sequence to the red-giant branch for stars of mass 1 $M_\odot$, 1.25 $M_\odot$, and 1.5 $M_\odot$, Ap. J. 147, 624-649 (1967a).

Iben, I., Stellar evolution within and off the main sequence, Ann. Rev. Astron. Astrophys. 5, 571-626 (1967b).

Iben, I., Post main sequence evolution of single stars, Ann. Rev. Astron. Astrophys. 12, 215-256 (1974).

Iben, I. and Renzini, A., Asymptotic giant branch evolution and beyond, Ann. Rev. Astron. Astrophys. 21, 271-342 (1983).

Kasting, J. F., Pollack, J. B., and Ackerman, T. P., Response of Earth's atmosphere to increases in solar flux and implications for loss of water from Venus, Icarus 57, 335-355 (1984a).

Kasting, J. F., Pollack, J. B., and Crisp, D., Effects of high CO_2 levels on surface temperature and atmospheric oxidation state of the early Earth, J. Atmos. Chem. 1, 403-428 (1984b).

Kasting, J. F., Runaway and moist greenhouse atmospheres and the evolution of Earth and Venus, Icarus 74, 472-494 (1988a).

Kasting, J. F., Toon, O., and Pollack, J. B., Sci. Amer. Feb., 90 (1988b).

Lovelock, J. E., and Watson, A. J., The regulation of carbon dioxide and climate, Planet Space Sci., 30, 795-802, 1982.

Schwartzman, D. W., and Volk, T., Geophysiology and habitable zones around Sun-like stars, this volume (1991).

Schwartzman, D. W., and Volk, T., Biotic enhancement of weathering and the habitability of Earth, Nature 340, 457-460 (1989).

Whitmire, D. P. and Reynolds, R. T., The fiery fate of the Solar System, Astronomy 18, 20-29, April (1990)

TITAN'S ATMOSPHERE FROM VOYAGER INFRARED OBSERVATIONS: PARALLELS AND DIFFERENCES WITH THE PRIMITIVE EARTH

Athena Coustenis
Département de Recherche Spatiale,Observatoire de Paris (Section de Meudon),
92195 MEUDON Principal Cedex, FRANCE

INTRODUCTION

Titan was discovered by Huygens in 1655 and given its name due to the belief that it was the biggest satellite in our solar system. The existence of a substantial atmosphere around Titan was suspected since the beginning of this century, but the first "tangible" proof came only when Kuiper observed methane bands in Titan's spectrum in 1944. Numerous astronomers followed in observing Titan and trying to determine the methane abundance and the pressure/temperature conditions in the atmosphere and on the surface of the satellite. Two principal models, existing before the Voyager mission, favored either a thick atmosphere made essentially of nitrogen or a thin atmosphere with methane as the major component. Pre-Voyager knowledge of Titan is extensively discussed in Hunten (1977).

The Voyager 1 spacecraft encountered Titan in November 1980 and found it to be the second biggest satellite of our solar system. However, following the detection of nitrogen in its atmosphere by the UV experiment, Titan was still the only other object to possess an atmosphere in any degree similar to that of the Earth's. As a consequence, Titan attracted a special interest of scientists.

What follows is a brief summary of our current understanding of the physical processes that operate in the atmosphere and on the surface of Titan, to provide a foundation for a brief discussion on the resemblance of the satellite to the early Earth. By necessity, since this is not a review paper, details of the observations, of the analyses and of the existing models are omitted. The summary of these aspects is also somewhat selective and not exhaustive. Excellent review chapters exist, which are strongly recommended: Hunten et al. (1984), Atreya (1986), Morrison et al. (1986), Lunine et al. (1989), and others.

ATMOSPHERIC PROPERTIES

Disappointment answered the call for surface images from Voyager's cameras: Titan appeared totally covered by a thick stratospheric haze layer without "holes" through which a glimpse of the surface could be obtained. Thus, neither from ground-based observations nor from the Voyager cameras were we able to learn anything on the nature of the surface, which up to this date still remains a mystery. However, the radio-occultation experiment on board the mission gave information on the atmospheric temperature (Lindal et al., 1983), the radius of the satellite ($\sim$ 2575 km), and, in particular, it indicated the surface conditions: a temperature of ~ 94 K for a pressure of 1.5 bar. Other properties of Titan were more precisely determined (see Table in Hunten et al., 1984): the total mass of the satellite was found to be 1.4×10^{26} g, in other words almost a fiftieth that of the Earth's; the surface gravity is about seven times less than on Earth (1.35 m s^{-2}); Titan is at a distance of 1.3×10^6 km from Saturn and is assumed to orbit around the planet in ~ 16 days; at a distance of 9.5 AU from the sun, Titan receives only 1.1% of the solar flux that reaches our Earth; Titan's obliquity is assumed to be 26°.7; finally, the Bond albedo of the satellite is estimated to be around 0.20.

Our knowledge of the interior of Titan, based upon bulk density, is rudimentary and that of its surface mostly speculative. From the existing data models on Titan's surface have been developed (Lunine et al., 1983; Lellouch et al., 1989). These models predict the presence of an ethane-methane ocean covering the surface, but its extent, exact composition and depth of this ocean cannot at this point be determined with accuracy. On the other hand, radio ground-based observations tend to suggest the existence of at least some solid parts on Titan's surface (Muhleman et al., 1990). One can imagine ethane-methane "lakes" and icy landscapes. However, the real nature of the surface remains an open question and a most vital one, given that the atmospheres and surfaces of planets and satellites are highly coupled systems whose origins and evolutions are fundamentally interconnected.

Before the Voyager missions, Earth-based observations had shown evidence for the existence in Titan's atmosphere of components other than methane: hydrogen, ethylene, ethane, mono-deuterated methane and acetylene. We therefore knew already that a relatively complex chemistry was active on Titan. Besides helping establish nitrogen as the major atmospheric component, the infrared experiment

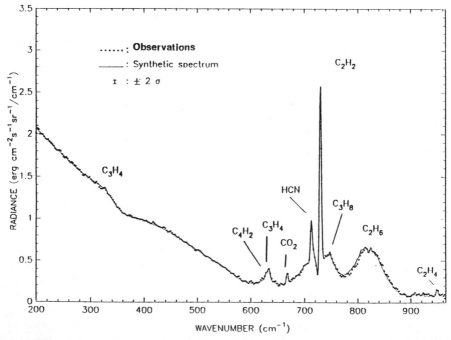

Figure 1: IRIS spectrum of Titan's equator, in which the emission bands of some gases present in Titan's atmosphere are identified in the 200-1000 cm^{-1} spectral region. Also shown for comparison is a synthetic spectrum corresponding to the best agreement with the observations obtained with the atmospheric model by Coustenis et al. (1989a).

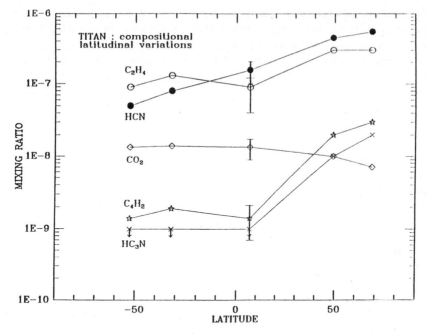

Figure 2: Examples of compositional variability with latitude in Titan's atmosphere. The nitriles and some hydrocarbons show an abundance increase towards the north pole. CO_2 on the contrary decreases with respect to the equator by a factor of > 2. The most abundant hydrocarbons after methane (ethane, acetylene and propane) seem to be homogeneously mixed in the atmosphere over all altitudes and are not shown in this figure. Some of the latitudinal variations are explained by current photochemical or seasonal effects models, others not. Error bars are shown for the equator. At other latitudes, the evaluation of uncertainties is in progress. Figure from Coustenis (1990).

on board Voyager 1 confirmed the presence of even more complex organic molecules on Titan (as, for instance, the nitriles and the oxygen compounds) and allowed the detection of a number of them. The presence of abundant organic gases in Titan's atmosphere is another exciting aspect of the satellite. The infrared data acquired by the Infrared Interferometer Spectrometer (IRIS), cover the 200-1500 cm^{-1} spectral region with a spectral resolution of 4.3 cm^{-1} and, of course, a very good spatial resolution. In this spectral range several emission bands of the atmospheric components were detected, with the exception of nitrogen, which does not possess a band in the infrared. Figure 1 shows an IRIS spectrum of Titan in the equatorial region, identifying the species present.

The components which are at this date known to be present in Titan's atmosphere list as follows: nitrogen ($\sim$ 90%), methane (0.5-3.4% in the stratosphere), hydrogen ($\sim$ 0.2%), trace amounts of hydrocarbons (C_2H_6, C_2H_2, C_2H_4, C_3H_4, C_3H_8, C_4H_2, CH_3D), nitriles (C_2N_2, HCN, HC_3N, C_4N_2) and oxygen compounds (CO --detected only from ground-based measurements-- and CO_2). The abundances of some of these gases are known today with good precision (Coustenis et al., 1989a, 1989b, 1989c) at various locations over the disk of Titan. This is obtained through an analysis of selected infrared spectra. Figure 1 is composed of the observed equatorial spectrum of Titan and of the associated synthetic spectrum. The latter is generated by a radiative transfer program which incorporates the temperature profile, spectroscopic data, cloud structure and gas abundances as parameters. By trial and error, the best possible fit of the observations is obtained, allowing us to derive information on the thermal structure and on the composition of Titan's atmosphere (see Coustenis et al., 1989a for more details). However, due to the medium spectral resolution of Voyager, the inferred mixing ratios of some molecules bear large uncertainties. Information is lacking for the following species: CO_2 (absent in the spectra for latitudes > 60°), HC_3N and C_2N_2 (present only in the north polar spectra), and C_4N_2 (present only in condensate form). Figure 2 gives some examples of the compositional variations in Titan's atmosphere as a function of latitude.

From a particular observational sequence over Titan's north pole, Coustenis et al. (1991) were able to derive vertical distributions for some of the components with spectral signatures in the north polar spectra between 200-1500 cm^{-1}. These vertical profiles are of great interest to the development of photochemical models to which they bring essential constraints. The only other vertical distribution available for a Titan component is for HCN which is derived from earth-based data obtained in the millimetric range (Tanguy et al., 1990) and is representative of the whole disk.

THERMAL BALANCE

The Voyager 1 mission provided information on the thermal balance of the atmosphere of Titan which actually confirmed pre-Voyager predictions based on simple physical concepts. The effective temperature at which Titan radiates to space is 83 K and corresponds to a ~ 73% absorption of the incident solar energy in the upper atmosphere (Neff et al., 1985). The 10 K increase at the surface level with respect to the effective temperature is due to a modest greenhouse effect (Samuelson, 1983). The thermal profile reaches a minimum value of about 71 K near an altitude of ~ 40 km (the tropopause). This temperature minimum, which represents a "cold trap" for methane, constrains the CH_4 mixing ratio in the stratosphere to about 2%. Above the "cold trap," temperatures increase with altitude attaining ~ 170 K at 200 km and even higher temperatures at lower pressures. The temperature profile of Titan's atmosphere is inverted because of sunlight-absorbing haze particles in the lower stratosphere which re-emit the thermal energy (resulting in the highly elevated asymptotic temperature as suggested by Danielson et al., 1973) and because of emitting stratospheric and absorbing tropospheric gases with bands in the infrared.

The thermal profile of Titan described above, was first obtained by Lindal et al. (1983) from the radio-occultation measurements. Referring to the equatorial region and the 0-200 km altitude range, it was calculated for a pure nitrogen atmosphere. However, thermal profiles at different locations on Titan's disk are now available, which correspond to more realistic atmospheric compositions. At the equator this information results from a re-analysis of the radio-occultation measurements, combined with examining of the observed emission in the ν_4 methane band at 7.7 μm (Lellouch et al., 1989). At other latitudes, temperature profiles were proposed by Coustenis et al. (1991) and Coustenis et al. (1989c) from investigations of IRIS data taken at various locations on Titan's disk. One such nominal profile, with the extreme temperature values allowed in the equatorial region, is shown in Figure 3, as suggested by Coustenis et al. (1989a).

The orange-yellow cloud deck that globally covered the satellite was not a total disappointment after all, in spite of its hindering the detection of the surface. It suggested the presence of aerosols in Titan's atmosphere, as on Earth. The molecular fragments and other components produced by impact of UV photons and energetic electrons in the upper part of the atmosphere form polymers, i.e.

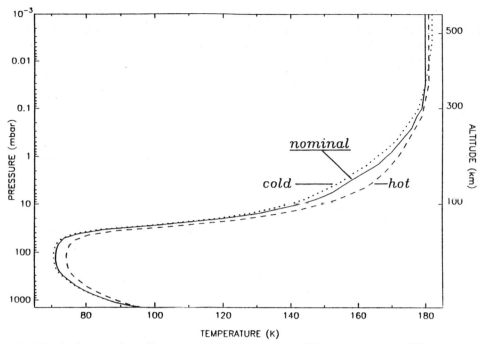

Figure 3: Nominal thermal profile suggested for the equator of Titan; extreme possible temperature profile in Titan's stratosphere is also shown. Each profile is associated to a certain atmospheric composition (Lellouch et al., 1989). Figure from Coustenis et al., 1989a.

molecular chains. All the hydrocarbons present in Titan's atmosphere, as well as the complex molecules produced by nitrogen and methane, condense in the coldest region of the atmosphere (between 50 and 100 km) and form solid suspended particles whose average radius is between 0.2 - 0.5 μm (Rages and Pollack, 1983). It is these particles that give Titan the characteristic orange color. They probably sink slowly into the deeper atmosphere, collide and form aggregates dropping even faster finally to land on the surface of the satellite.

The nature and size of the aerosols is not yet well defined, and more observations are needed to answer such questions as what is the effect of hazes on the thermal structure and what fraction of methane is converted to haze. A recent estimate of the aerosol mass production rate was given by Pope and Tomasko (1986): $\sim 8 \times 10^{-14}$ g cm^{-2} s^{-1}, that is to say one-fifth of the mass-loss rate of methane. The currently available observational studies of the aerosols are extensively reviewed in Hunten et al. (1984).

Some questions remain to be answered as concerns aerosols and the chemical evolution in Titan's atmosphere: for instance, how does the process in which particles form from gases occur in organic atmospheric systems, and what is the nature of the first condensed species? What is the complexity of chemical maturity in Titan's atmosphere? What is the role of aerosol growth in the development of complex organic compounds? A final question, synthesizing the previous ones would be: How may the answers to these questions be applied to the study of prebiotic chemistry in the primordial atmosphere of Earth?

Methane clouds could also be present at lower levels in Titan's atmosphere, where water vapor ones cover our own blue skies. A recent thermal model developed by McKay et al. (1989) is based on improved determinations of collision-induced gas absorption coefficients and constraints of the Voyager IRIS data. The calculations are based on the surface temperature found by Voyager, which corresponds to a balance between the amount of solar visible radiation deposit and atmospheric thermal radiation emitted (Samuelson, 1983). McKay et al. examined the greenhouse warming effects of radiation by gases and particles and methane clouds in Titan's atmosphere. The first factor was found to make important contributions to the greenhouse warming, whereas the latter can have a neutral or, even a cooling effect. McKay et al. (1989) find, however, a better agreement with the IRIS data when some cloud infrared opacity is added. These clouds could be limited in a certain atmospheric zone, formed of very large particles or broken in aereal coverage.

ORIGIN AND EVOLUTION

Another aspect of Titan which is still treated mainly in the form of questions and models is that of the origin and evolution of its atmosphere. Using the physical processes operating in the present-day environment (such as radiation-driven chemistry, atmospheric escape, photochemistry, surface-atmosphere coupling etc.) to "run the clock backward" would help modeling the origin and evolution of Titan.

McKay et al. (1989) developed a radiative-convective model of the thermal structure of Titan's atmosphere, as an extension to the work of Samuelson (1983). The photochemistry active in the satellite's atmosphere was modeled, among others, by Yung et al. (1984) and Yung (1987). Lunine and Stevenson investigated recently the evolution of Titan's coupled ocean-atmosphere system and interaction of ocean with bedrock (1985) and the origin of methane on Titan (1987). The present state and chemical evolution of the atmospheres of Titan, Triton and Pluto is explained by Lunine, Atreya and Pollack (1989) and references therein. Some of the questions related to the issue of Titan's origin and evolution, which still remain unanswered, are asked by these authors. A selected list would run as follows:

1. What is the origin of nitrogen in Titan's atmosphere? Was it derived from a gas containing molecular nitrogen or was it produced by chemical processing of ammonia?

2. How much carbon monoxide was introduced into Titan and its early atmosphere?

3. Was methane brought into Titan as frozen condensate or entrapped in clathrate?

4. Why does Titan have an atmosphere, while the Galilean satellites do not?

5. Could most of Titan's atmosphere have been added by impact of volatile-rich comets after accretion?

6. What would be the effect of an ocean in contact with the atmosphere, as opposed to a solid surface, in the evolution of Titan?

Several attempts have been made to answer these questions, and modeling is still under development. Lunine et al. (1989) address these questions and others and, in particular, they investigate surface models involving an ethane-methane ocean and its implications on the coupled evolution of Titan's ocean-atmosphere system. A schematic model of the ocean-atmosphere evolution on Titan is shown in Figure 4. The fundamental driving force in the long-term evolution of Titan's atmosphere is the photolysis of methane in the stratosphere to form higher hydrocarbons and aerosols. The current rate of photolysis, together with the inferred undersaturation of methane in the lower troposphere, suggests a reservoir of methane mixed with higher hydrocarbons. This "ethane-methane ocean" serves as both the source and the sink of photolysis, and contains a mass of nitrogen comparable to the atmospheric abundance. In the absence of outgassing or external resupply, the ocean composition evolves toward more ethane rich, as methane is photolyzed over geologic time. The atmosphere responds to the change in ocean composition with a corresponding change in gaseous composition and spatially averaged cloud composition (Lunine et al., 1989).

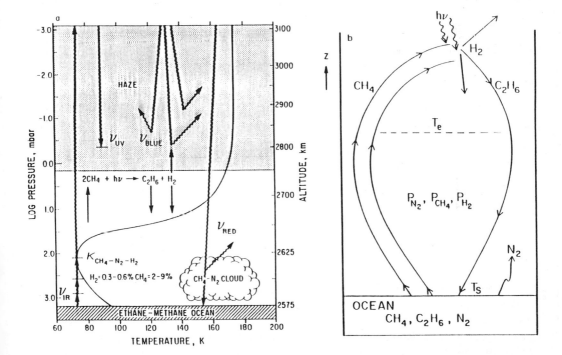

Figure 4: (a) The coupled system of atmosphere-ocean on Titan based upon Voyager data and inferred physical processes, from Thompson and Sagan (1984) (see text for more details); (b) Coupling between the photochemical cycle, atmospheric thermal structure and hydrocarbon ocean described in Lunine and Stevenson (1985). Methane evaporated from the ocean is photolyzed in the stratosphere to form predominantly ethane, acetylene and propane, with escape of hydrogen (only ethane is shown for clarity). These products condense and fall to the surface, changing the ocean composition and reducing its volume. As a result molecular nitrogen is evolved from the ocean into the atmosphere, altering the radiative balance.

Hereafter are summarized some of the tentative explanations proposed for some of the other questions in the above list.

Precise knowledge of the abundances of the noble gases, such as argon and deuterium, in Titan's atmosphere may be diagnostic tests of the competing hypotheses on the origin of nitrogen and methane on Titan. The outstanding problem of the origin of atmospheric nitrogen on Titan (was it introduced as molecular nitrogen or NH_3?), could find an answer in measurements of the Ar/N_2 ratio in the present atmosphere: a large argon abundance suggests N_2, while a small one favors NH_3. Unfortunately, argon has not yet been detected on Titan and the D/H ratio measurements bear large uncertainties.

The atmospheric CO and CO_2 observed in Titan require the existence of an exterior oxygen-bearing component in the atmosphere, such as water, which has not yet been detected. H_2O would then quickly dissociate into OH which combines with CH_4 photolysis products (such as CH_2, CH_3, etc..) and produces CO and CO_2 (Samuelson et al., 1983). The source for water on Titan can be found in the rings of Saturn, the meteorites or comets. CO could otherwise be directly captured in a clathrate-hydrate directly from the solar nebula. However, all of these mechanisms involve numerous uncertainties which have not yet been quantified and do not allow for a definite preference between the models.

There are several reasons which could explain why Titan is the only satellite with a large N_2 atmosphere in contrast with the Galilean moons of Jupiter and with Triton: (i) accumulation of an atmosphere on Titan could have proceeded unperturbed by losses from energetic charged particles, because Titan remains mostly outside Saturn's magnetospheric environment; (ii) Titan formed in the cooler part of the nebula from which it was accreted which permits a more favorable endowment of methane and ammonia ices (in addition to water ice) for the development of an atmosphere (Hunten et al., 1984); (iii) Titan became massive enough to have undergone some internal differentiation which concentrated ices towards the surface leading to substantial outgassing, but it still had enough gravity to retain all but the lightest gases; (iv) finally, the surface temperature on Titan (as opposed to that on Triton) is high enough to maintain significant amounts of CH_4 and NH_3 gas in equilibrium with their corresponding surface ices.

RELATIONSHIP BETWEEN TITAN AND EARLY EARTH ATMOSPHERES

Because molecular nitrogen is the major component in Titan's atmosphere, the satellite is evidently of general interest to the study of chemical evolution of such an atmosphere. On a planetary scale, Titan is probably the best laboratory that will ever be available to scientists and, as such, should be fully exploited. However, present day Titan is not a good model for the early Earth, because of important differences. Hereafter, we consider the possible relationship of the atmospheric composition and temperature of Titan to that of the primitive Earth.

Nitrogen- and carbon-containing gas species represent the major constituents of both types of atmospheres. Abundant organic gases are present in Titan's atmosphere, as well as aerosol layers, and, perhaps, methane clouds. The sources of organic reactants in Titan's atmosphere can be divided into three categories: (a) mother molecules, already present in the atmosphere at the time of its formation: N_2, CH_4, and possibly Ar; (b) pristine material deposited in the atmosphere by comets and interplanetary dust, or coming from the rings of Saturn: CH_4, CO_2, H_2O (undetected), NH_3 (undetected), etc...;(c) products of chemical processing in the atmosphere (hydrocarbons, nitriles, aerosols, polymers): (i) through photolysis of CH_4 and N_2, (ii) through irradiation by energetic particles, and (iii) through plasma-initiated chemistry (associated with lightning, cometary impacts,etc.). Thus, a variety of simple and necessary organic species is available, along with aerosols, for continued chemistry in Titan's atmosphere. A liquid ocean on the satellite's surface is consistent with the data. An orange-yellow cloud deck globally covers Titan, witnessing a complex chemistry active in the atmosphere, the products of which may have been preserved over all of the satellite's history.

On the other hand, conditions on present day Titan are not similar to those on primordial Earth. Here are some of the differences:

1. The composition of the satellite's atmosphere (CH_4-N_2) is different from that of the Earth's (CO-CO_2-N_2).

2. The surface temperature on Titan is $\sim$ 94 K (-180°C), and can only reach a maximum of $\sim$ 180 K (-93°C) in the upper stratosphere. This temperature range is lower than conditions on the early Earth.

3. Our planet was much closer to the sun and the UV radiation was about 100 times more than that reaching Titan.

4. Water, although suspected, has not yet been detected in Titan's atmosphere.

5. The state of oxidation between the inner and outer solar system atmospheres is very different. The Earth is an extreme case in this regard due to the high abundance of O_2. This apparently reflects the Earth's unique position in the solar system as an abode of life. C is present in its fully oxidized state CO_2 in all the terrestrial planets and in its fully reduced state in the outer solar system objects. This difference could reflect the ability of water to buffer the oxidation state of the carbon species in the atmospheres of the terrestrial planets, while the oxidation state of the C in the outer solar system atmospheres may be primordial, preserved because water buffering is due to dissociated water in magma chambers (Lunine et al., 1989).

6. Size may also have played an important role in the differing evolutionary history of these two atmospheres. Earth being larger, its atmosphere may have evolved significantly over its lifetime due to volcanic and tectonic changes with time. Titan, by way of contrast, may have undergone early dramatic changes followed by tectonic quiescence; subsequent evolution would be externally driven by photochemical processes. Since an ammonia-water magma beneath the surface of Titan could be maintained to the present day, on the other hand, the satellite may be tectonically active (Lunine and Stevenson, 1987).

7. The infall of carbonaceous material is smaller today than in the past. This source, considered as most important by models of Earth's evolution, is then less important for Titan.

In general, accretional and evolutionary models argue that the atmosphere of Earth was formed contemporaneously with accretion or shortly after (by a few 100 Myr) by impact degassing and/or highly vigorous overturning in a mantle heated by accretion and core formation. Subsequent evolution of the atmosphere and interior may have involved further degassing of volatiles retained within the planet at the end of accretion and regassing of the interior, if there existed mechanisms such as plate tectonics for cycling volatiles back into the mantle. Buoyant NH_3 and CH_4 magmas may have substantially degassed Titan's interior and determined the composition and state of its present atmosphere (Schubert et al., 1986).

Subsequent to a few 100 Myr after the end of accretion, the terrestrial planets underwent a slow and gradual cooling controlled by deep convective heat transport and near-surface conductive heat transport through a fixed or, in the case of the Earth, mobile lithosphere. Partitioning of carbon and nitrogen species between the atmosphere and lithosphere is another parallel between the inner and outer solar system atmospheres. In the case of the Earth, much more C resides in the lithosphere than in the atmosphere at present, due to weathering processes that readily convert CO_2 gas into carbonate rocks. For Titan, the buffering is largely due to the low surface temperature which puts most of the methane into surface frosts or oceans. Nitrogen is more volatile so that a much larger fraction of it is present in the atmospheres of Earth, Mars and Titan.

CONCLUSION

Although present day Titan may not be a good model for the early Earth, because of major differences discussed above, the satellite still provides a reasonable setting in which chemical and physical processes analogous to those on early Earth can be studied. The parallel study of our planet and Titan is undoubtly one of the best tools available to astronomers in search of scientific information on the origin of life.

Much about Titan has been revealed by ground-based and spacecraft observations, in particular those of Voyager. Nevertheless, more information is needed to secure our knowledge on various aspects of Titan: surface, lower atmosphere, origin and evolution, photochemistry, etc. Future missions should provide us with the data required to complement ground-based observations and teach us more about Titan. A specific mission planned toward the Saturnian system is Cassini. This combined ESA/NASA project is scheduled to be launched in 1996 and send back the first results on Titan early in the beginning of 2003. The spacecraft contains an orbiter which will make ~ 30 flybys around the satellite and a descent probe which will penetrate into the atmosphere of Titan and make in situ measurements. Another tool in the coming years will be the Infrared Space Observatory (ISO), an Earth-orbiting satellite which will observe the infrared spectrum of the Titan disk with high resolution and should allow the detection of new molecules.

The next decade will, no doubt, be essential to the investigation of Titan, since great opportunities will be made available to scientists. Full use of these opportunities will make available important clues to the understanding of our solar system.

References

Atreya, S. K. 1986. In Atmospheres and Ionospheres of the Outer Planets and Their Satellites, (New York: Springer-Verlag), pp. 145-197.

Cousténis, A., B. Bézard, and D. Gautier. 1989a. Icarus 80, 54-76.

Cousténis, A., B. Bézard, and D. Gautier. 1989b. Icarus 82, 67-80.

Cousténis, A., B. Bézard, and D. Gautier. 1989c. Bull. Amer. Astr. Soc. 21, 959.

Cousténis, A., B. Bézard, D. Gautier, A. Marten and R. E. Samuelson. 1991. Icarus, 89, 152-167.

Cousténis, A. 1990. Ann. Geophysicae 8, 645-652.

Danielson, R. E., Caldwell, J. J., and Larach, D. R. 1973. Icarus 20, 437-443.

Hunten, D. M. 1977. In Planetary Satellites, ed. J. A. Burns (Tucson: University of Arizona Press), p. 420.

Hunten, D. M., Tomasko, M. G., Flasar, F. M., Samuelson, R. E., Strobel, D. F., and Stevenson, D.J. 1984. In Saturn, eds. T. Gehrels and M. S. Matthews (Tucson: University of Arizona Press), pp. 671-759.

Lellouch, E., A. Coustenis, D. Gautier, F. Raulin, N. Dubouloz, and C. Frére. 1989. Icarus 79, 328-349.

Lindal, G. F., G. E. Wood, H. B. Hotz, D. N. Sweetnam, V. R. Eshelman, and G. L. Tyler. 1983. Icarus 53, 348-363.

Lunine, J. I., D. J. Stevenson, and Y. L. Yung. 1983. Science 222, 1229-1230.

Lunine, J. I., and Stevenson, D. J. 1985. In Ices in the Solar System, eds. J. Klinger, D. Benest, A. Dollfus, and R. Smoluchowski (Dordrecht: D. Reidel), pp. 741-757.

Lunine, J. I., and Stevenson, D. J. 1987. Icarus 70, 61-77.

Lunine, J. I., Atreya, S. K., and Pollack, J. B. 1989. In Origin and Evolution of Planetary and Satellite Atmospheres, eds. S. K. Atreya, J. B. Pollack, and M. S. Matthews (Tucson: The Univ. of Arizona Press), p. 605.

McKay, C. P., Pollack, J. B., and Courtin, R. 1989. Icarus 80, 23-53.

Morrison, D. M., Owen, T., and Soderblom, L. A. 1986. In Satellites, eds. J. A. Burns and M.S. Matthews (Tucson: University of Arizona Press), pp. 764-801.

Muhleman, D. O., Grossman, A. O., Butler, B. J., and Slade, M. A. 1990. Science 248, 975-980.

Neff, J. S., Ellis, T. A., Apt, J. and Bergstralh, J. T. 1985. Icarus 62, 425-432.

Pope, S. K., and Tomasko, M. G. 1986. Bull. Amer. Astron. Soc. 18, 816.

Rages, K., and Pollack, J. B. 1983. Icarus 55, 50-62.

Samuelson, R. E. 1983. Icarus 53, 364-387.

Samuelson, R. E., W. C. Maguire, R. A. Hanel, V. G. Kunde, D. E. Jennings, Y. L. Yung, and A.C. Aikin. 1983. J.Geophys.Res. 88, 8709-8715.

Schubert, G., Spohn, T., and Reynolds, R. T. 1986. In Satellites, eds. J. A. Burns and M.S. Matthews (Tucson: Univ. of Arizona Press), pp. 224-292.

Tanguy, L., Bézard, B., Marten, A., Gautier, D., Gérard, E., Paubert, G., and Lecacheux, A. 1990. Icarus 85, 43-57.

Yung, Y. L., M. Allen, and J. P. Pinto. 1984. Astrophys.J.Supp. 55, 465-506.

Yung, Y. L. An update of nitrile photochemistry on Titan. 1987. Icarus 72, 468-472.

Discussion

F. RAULIN: Athena, I am not sure that ASI and DISR experiments are the most important experiments of the Huygens probe, from a bioastronomical point of view. In this field, GC-MS and ACP experiments seem to me more important, since they will provide a detailed analysis of the organic compounds—both volatiles and polymeric materials—present in Titan's atmosphere.

A. COUSTENIS: I agree with your point of view. I did mention though that I was just giving some examples of the instruments on board Cassini and not an exhaustive list. I was also speaking from the point of view of improving Voyager measurements in terms of temperature and composition, and also surface conditions. That is my reason for mentioning CIRS and ASI and DISR.

LIQUID WATER AND LIFE ON EARLY MARS

C. P. McKay
NASA Ames Research Center, Moffett Field, CA 94035

L. R. Doyle and W. L. Davis
SETI Institute, Mountain View, CA 94043

R. A. Wharton
NASA Hq., Washington, DC 20546

Paper Presented by C. P. McKay

Summary

There is direct geomorphological evidence that in the past Mars had large amounts of liquid water on its surface. Atmospheric models would suggest that this early period of hydrological activity was due to the presence of a thick atmosphere and the resulting warmer temperatures. From a biological perspective the existence of liquid water, *by itself*, motivates the question of the origin of life on Mars. From studies of the Earth's earliest biosphere we know that by 3.5 billion years ago, life had originated on Earth and reached a fair degree of biological sophistication. Surface activity and erosion on Earth make it difficult to trace the history of life before the 3.5 billion years ago timeframe. If Mars did maintain a clement environment for longer than it took for life to originate on Earth, then the question of the origin of life on Mars follows naturally.

One interesting possibility that has grown out of recent work in the dry valleys of Antarctica is that the period on Mars during which biologically favorable habitats could persist may have been greatly extended by the phenomenon of perpetually ice-covered lakes. Although the mean temperature of the Antarctic dry valleys is -20°, deep lakes (>30 m) are formed by groundwater flow and by transitory melting of glacial ice. The ice-covered lake habitat is of major biological significance in the inhospitable cold desert valleys of Antarctica and provides a working model for the ultimate habitats for life on early Mars.

Investigating the record of prebiotic and biotic evolution on Mars will involve searching for many things, including: direct traces of life such as microfossils, organically preserved cellular material, altered organic material, morphological microstructures, and chemical discontinuities associated with life, isotopic signatures due to biochemical reactions, inorganic mineral deposits attributable to biomineralization, and hydrated minerals such as clays, as well as water-formed minerals such as carbonates. The fossil evidence of early life on Earth provides clues as to what form fossils on Mars might take. Of particular interest are stromatolites, macroscopic layered structures that result from the anchoring of sediments by microorganisms living in the photic zone. Since over two-thirds of the martian surface is more than 3.5 billion years old, the possibility exists that Mars may hold the best record of the events that led to the origin of life--even though there may be no life there today. Such insight may be the first step to a scientific assessment of the distribution of life in the cosmos.

NOTE: A complete discussion of this subject has been published by C.P.McKay, E.I. Friedmann, R.A.Wharton, and W.L.Davis in a paper entitled, "History of Water on Mars: A Biological Perspective," which appears in *Advanced Space Research*, 1990.

Discussion

C. CHYBA: Chris, I get a little nervous about stromatolite-like analogues as evidence for life on Mars. On Earth, you have the enormous advantage that there are contemporary, extant stromatolites, so that you can look at stromatolite fossils and say, "By God, these look just like stromatolites today." If you found something like a stromatolite fossil on Mars, how could you be sure that it had a biological, rather than abiological, origin? Even on Earth, there are "organized elements" in 3.8 billion year old rocks whose interpretation as microfossils or nonbiological structures has been controversial. It seems to me that this type of question is one of general concern to SETI as well.

C. McKAY: Good point. In order to determine that stromatolites on Mars are truly of biological origin, it may require additional information including: microfossils, isotopic analysis, the presence of organic material, etc. Even so, it may still be ambiguous.

D. BRIN: For lack of plate tectonics, the Mars surface looks unaltered since the end of the period of bombardment. Does this give you confidence [that] your lake sites can still be found after several billion years? And if so, what kind of orbital survey would you need in order to choose the very best sampling areas?

C. McKAY: Yes, over 2/3 of the martian surface dates back to the late bombardment. Already with the resolution allowed by the Viking Orbiter images, sites of possible lake-bed sediments have been identified.

F. RAULIN: Chris, in your model, have you considered the possible influence of dissolved CO_2 in the "Martian oceans" which could markedly decrease the freezing point of water?

C. McKAY: We have considered the reduction in atmospheric CO_2 due to CO_2 dissolved in water (assumed to cover 5% of Mars), but we do not include the effect of solutes (including CO_2) in depressing the freezing point. However, since we are primarily interested in the melting of glacial ice, the solutes will have already been forced out of solution during the freezing process and melting should still occur at $0°C$.

G. MARX: There are plans to send a mission to Mars, but with no return. Do you see a possibility to look for extinct life without bringing back samples?

C. McKAY: Yes, it would be possible to search for remnants of life, particularly organic material, with robotic missions.

M. PAPAGIANIS: Wasn't volcanism an important factor in maintaining a significant atmosphere on Mars, and hence, liquid water on Mars; which in the absence of a significant atmosphere, it would tend to evaporate very rapidly?

C. McKAY: Yes, volcanism could have played a key role in maintaining the initial atmosphere.

D. WHITMIRE: As you know, Jim Kasting is currently modeling the early dense Mars atmosphere and finds that CO_2 condenses out of the upper atmosphere. Do you have a comment on this?

C. McKAY: The question of condensation of CO_2 in the upper atmosphere of a thick martian atmosphere in problematic and certainly requires further analysis.

C. MATTHEWS: What happened to the nitrogen-containing compounds on Mars, nitrogen being an essential element of life?

C. McKAY: Based on the volatile composition of the Earth, one would expect Mars to have much more N_2 than observed today. Nitrogen could have been lost to impact erosion of the atmosphere, to nitrate formation, and possibly to biogenic organic material.

W. IRVINE: Do you think stromatolites could be identified without a manned mission?

C. McKAY: I think they could be since they are often quite large (>10 cm), however it may require humans on the surface.

R. BROWN: Can you define more explicitly just what constitutes "organic material" on Mars? Presumably you are interested in material of biological origin but this will have been degraded until it will merely be carbon-containing material (e.g., CO_2 is an extreme example of degraded "organic material"). When do you draw the line in deciding that the material is of biological origin?

C. McKAY: Any reduced carbon-containing compound would be of interest.

Editors Note: Referee's comment not addressed in time by the authors: "Weathering will be strongly influenced by precipitation and temperature by water vapor and clouds. These have not been adequately addressed for Martian paleoclimate."

NEW INTERPRETATION OF CRUSTAL EXTENSION EVIDENCES ON MARS

E. A. Grin

Laboratoire "Physique du Systeme Solaire"

Observatoire de Meudon, 92195 Meudon, France

Summary

The record of early evolution of life on Earth has been obscured by extensive surface activity. On the opposite, large fractions of the martian surface date back to an early clement epoch favorable to the needs of biological systems [1]. The upper martian surface reflects a wide variety of modifying processes which destroy the geological context. However, due to endogenic causes acting after the end of the primordial bombardment, abundant extensional structures display vertical sequences of stratigraphic units from late Noachian to early Hesperian periods [2]. Deep structural incisions in the upper crust provide unaltered strata, open flanks, and slope deposits that favor the use of an autonomous lander-rover-penetrator [3] that can perform geological investigations simultaneous with a biological search for fossils from various epochs.

The strategy for an exobiology search of such an optimum site should be guided by the recent attention devoted to extensional structures and their global significance [4]. Geological evidence supporting the martian crustal extension is suggested by abundant fractures associated with the dichotomy boundary northland-south upland, i.e., Aeolis Region, and peak igneous activity (Elysium bulge). As pointed out by [5], the system of fractures correlates with the endogenic origin of the dichotomy, as related to a major difference in the thickness of the crust. Perpendicular to this boundary, fractures of deep graben testify to a general tectonic crust relaxation. The opening of the graben, joined with compressive wrinkles, is the signature of a dynamical pervasive stress regime that implies a large scale roll-over of the upper crust over the ductile interface of a more dense mantle. This general motion is not a transport of material, as there is no thickening on the boundary of the dichotomy. The horizontal movement is due to the gravitational mechanism and differential thermal convection cells in the upper crust over the slope of the anti-flexure rigid interface consequential to Elysium bulge. The fracturation occurs as the neutral zone of the crust rises to the brittle surface of the crust. Deep extensional structures are logical sites for locating and sampling fossilized organisms from various epochs. Grabens suggest ancient lakes and the development of biological systems supported by bottom hot springs.

References:

(1) Cabrol N. (1989), Lunar Planet S. XX, p. 135
(2) McGille (1988), NASA TM, p. 549.
(3) Friedmann E. (1987), Space Life Proc.
(4) Morgan P. (1983), Tectonophysics 94, p. 1.
(5) McGille (1989), J. Geophys. Res.

MARTIAN PALEOHYDROLOGY AND ITS IMPLICATIONS
FOR EXOBIOLOGY SCIENCE

N. A. Cabrol*† and E. A. Grin†
*Laboratoire de Geographie Physique (UAO141 CNRS)
†Laboratoire de Physique du Systeme Solaire, Observatoire de Meudon, Meudon 92195 France

Paper Presented by N.A. Cabrol

The controversy about the genesis of martian channels is based upon the main hypotheses of formation, one endogenetical, and the other exogenetical. The endogenetical hypothesis includes three processes of formation leading to:

(a)   Terrestrial fluvial-like features, corresponding to martian channels originating from running water flowing from high basins towards lowlands. Observed runoff traces should be the result of regular flows with seasonal variations.

(b)   Catastrophic features, which are related to a sudden release of high discharges by melting of ground ice or release of aquifers.

(c)   Seepage features, which correspond to a sudden melting of underground ice (permafrost) during volcanic activity or during exogenetical events, such as meteorite impacts or modification of the solar energy intensity by variation of the eccentricity and inclination of the planet.

The exogenetical processes could be divided into two categories:

(a)   Processes linked to a paleoatmosphere. These imply the presence of a pre-Noachian or Noachian atmosphere, allowing a complete water cycle from a mega-ocean (hypothesis of the "Oceanus Borealis"), with a relevant water evaporation. The climatic variation effects are critically marked by meteorical impacting. All these external events lead to a broad resurfacing of the planet and a new apparent distribution of the networks.

According to this hypothesis, the martian networks (of supposed meteorological origin, i.e., rainfalls) were generated during the Ocean phase and disappeared progressively with the water content of the atmosphere. Owing to the cooling of the climate and the pressure decrease, the water was trapped in the substratum or evaporated.

The last results of the Phobos-Mars mission (1989) and Termoskan surveys experiments led us to set up a new model for the martian paleoatmosphere defined as the "greenhouse effect."

The information on the atmospheric Deuterium content shows that during the post-accretional period, Mars contained nearly six times more water than the present residual amount (Owen et al, 1988). To generate the observed erosion of the supposed river-bed features on Mars, this amount should correspond to a water depth of 200 m to 500 m (nearly 7.10^7 km^3) covering the whole surface. Impact craters plotted on the martian surface show that this erosion was maximum 3.5 to 4 billion years ago.

Today, there is no more liquid water at the surface of Mars because of the low atmospherical pressure (mean value of 6 millibars) with a mean temperature of 220 K.

If the residual polar caps are considered, their volumes correspond to a water depth of 10 m to 20 m covering the whole surface. The difference between the previous estimated volume of 200 m to 500 m and the present should be trapped underground. But, the presence of hydrogeological traces implies another climate than the actual one. This past climate could correspond to a CO_2 atmosphere creating a "greenhouse effect" corresponding to an increase of the pressure/temperature system allowing the presence of liquid water. The coexistence in the past of liquid water and CO_2 led to the dissolution of the latter and consequently to the modification of conditions over the entire planet up to the present.

(b)   The second exogenetical process is linked to the accretion of meteorites containing large amounts of high-volatile-rich components and the position of the planet Mars in the primary Nebula.

After considering the possible morphogenetical processes for martian channels, a new problem is raised. The importance of water in the genesis of channels is obviously established, but the existence of rainfalls and a complete water cycle must be demonstrated.

The physiographic characteristics of the martian networks point out hydrogeologic specificities implying different evolution regarding terrestrial drainage basins. Therefore, are these networks of meteorological or hydrogeological origin? The answer could influence the possible existence of a water-dependent life. A meteorological regime of rainfalls could lead to a stabilized basin feeding during a long period of time. A hydrogeological origin of this water (seepage or aquifers) implies accidental releases and discharges of short period of time, inconsistent with complex life development.

The first traces of flow are located in 4 billion year old terrains, with a maximum of intensity around 3.5 billion years. These observations argue in favor of a short period of favorable conditions until 3 billion years before the present (Fig. 1).

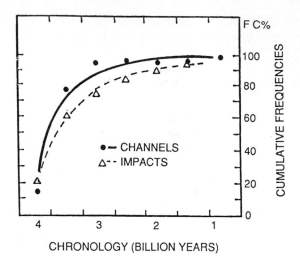

Fig. 1: Correlation between impact crater flux and channel apparitions during the different geological periods.

But, a global survey of the martian drainage basis shows a distribution of headwater densities varying from 100 to 1 per 10^6 km² for two adjoining regions. In addition, the maximum percentage of drained area is 16% for Margaritifer Sinus Region, which can be considered as an arid area in terrestrial conditions. Consequently, the results are not in favor of an organized meteorological feeding.

Other results of the physiographic analysis allow us to determine the structure of the martian networks and could give a new light on their origin. This analysis was established on 44 martian networks (Cabrol, 1990) and give the following results.

The compacity (Kc) of a network allows us to estimate the drainage efficiency. A low value indicates a short time of flowing between the sources and the outlet. Kc is defined by the equation Kc = $0.28.(\sqrt{S})^{-1}$. where S is the basin area. The terrestrial value for Kc is around 1.3.

The drainage density (Kd) characterizes a hydrographical basin. Kd is defined by the relation between its area (S), and the mean value by km² of the whole branch network systems (L_i). So, Kd is expressed by: Kd = $\Sigma L_i.(S)^{-1}$ On Earth, Kd varies from 0.02 for karstic terrains to 300 for badlands.

The martian networks are generally of mean or small sizes and divided into two categories:
- basins of $(5\pm2)\cdot10^3$km² (nearly 86% of the whole population)
- basins of $(30\pm10)\cdot10^3$km².
Some larger basins are identified but remained exceptional.

The martian values for Kc vary from 1.2 to 4, which indicates a drainage capacity three times less efficient than on Earth. Finally, the drainage density on Mars has only small variations (0.03 to 0.1) (Fig. 2).

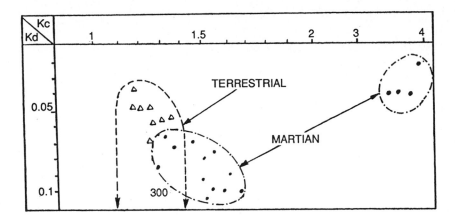

Fig. 2: Comparison between terrestrial and martian drainage basis by their respective physiographic coefficient of drainage density (Kd) and their area compacity (Kc).

Therefore, the conclusion is that the martian drainage is very less efficient than the terrestrial one, and the values plotted between -43° Lat. and +19° Lat. (craterized uplands) indicate a very porous substratum corresponding to basins weakly drained by tributary systems.

These results added to the dimensions and the dispersion of martian networks give good arguments in favor of a non meteorological origin for these channels and lead to support the underground water hypothesis and the seepage origin. In addition, the laboratory experiments provide obvious indications of the similarity for martian networks and seepage basins.

These results have important consequences for the hypothesis of a water-dependent life on Mars. The model of seepage implies only accidental or periodical phases of melting and release. But, the hypothesis of underground water can be associated to the possibility of specific life evolution.

In terrestrial views, it is supposed that martian chemistry--in good climatic conditions and with solar energy interaction--could evolve towards a primitive life system. But, this suggestion causes a major dilemma: in the best hypothesis, the favorable period for life development extended only during some thousands of years, then changes quickly into critical atmospherical and climatic conditions.

Consequently, if martian life was present in the first billion years after the planet formation, it must have been preserved against every kind of cycle to insure a sufficient stability for its survival and development. So--and still in the supposition of a water-dependent life system--the hypothesis of an underground life apparition can be considered, near hot springs. The question of hydrothermalism must be supported by some elementary parameters such as magmatic activity close to the surface, the presence of fractures, and underground water. The only restrictions are in the necessity for the life to be preserved of cyclic variations, which means that this life developed its own energy underground, out of solar processes. So, in this hypothesis, the best chances to find it--may be--today (active or fossilized) could be to explore regions of high densities of volcanic pit or low depression areas.

References

Cabrol, N.A.: 1989, *Lunar and Plan. Inst. Houston* XX, 136.

Cabrol, N.A.: 1990, *Lunar and Plan. Inst. Houston* XXI, 134.

Cabrol, N.A., Grin, E.A., and Dollfus, A.: 1990, *Adv. in Space Res. and Cospar Col. series. Sopron-Hungaria*

Carr, M.H.: 1979, *Jour. Geo. Res.*, 84, 2995-3007.

Mouginis-Mark, P.: 1985, *Icarus*, 2.

Discussion

N. EVANS: **Do you have evidence that the channels were cut by liquid H$_2$O, as opposed to other possible liquids?**

N. CABROL: We have no absolute evidence that water carved the martian channels. The only evidence is that water can exist on Mars under the gaseous, solid and liquid forms (even if this last form is limited by the low atmospherical pressure). But our analysis of the martian channels shows that water was not absolutely necessary to cut the fluvial-like features. Our hypothesis is that martian channel networks are probably fracture patterns, generated by tectonics and impactism (cf answer to Mr. Leger), that water released from aquifers used and shaped (with other erosion types) them only partly. On the other hand, it is obvious on Viking imagery that lava flowed also inside some large valleys.

A. LEGER: **I have some problem understanding how a very small amount of water can have made impressive river patterns which, at first glance, seems similar to earth patterns that have been dug by the flow of abundant rivers during long periods.**

N. CABROL: First, martian channels are not similar to terrestrial rivers. The disposition of drainage basins, the angles of branch junctions, the drainage density and capacity are not the same and do not correspond to a terrestrial-like hydrological system. Second, Mars probably had an important underground water content under permafrost or aquifer forms. The results of our hydrogeological analysis of the channel networks show that the drainage basin could not be of meteorological origin (low drainage capacity). The headwater systems correspond to a seepage process of underground water. The drainage of aquifers is triggered by the fracture networks generated by tectonics and impactism and the thermal and mechanical destabilization of the permafrost. The fluidized permafrost is released in the preexisting fractures and does not shape them.

BOUNDARIES OF THE EARTH'S BIOSPHERE

M. D. Nussinov
125195 Moscow, Belomorskaja Str. 22, Bldg. 3, Apt. 351, USSR

S. V. Lysenko
Institute of Microbiology, Academy of Sciences, USSR.

Paper Presented by M. D. Nussinov

Summary

We estimated the boundaries of the Earth's biosphere using the presentations developed earlier by authors of the new physical thermo-vacuum mechanism which causes irreversible ("explosive") damage to vegetative cells and spores of microorganisms under space vacuum conditions around the Earth (M. D. Nussinov and S. V. Lysenko, JBIS, $\underline{42}$, 431 (1989)). As the consequences of this mechanism, which are based on experimental testing, the conditions of microorganisms may be written as set of inequalities: $Pe \wedge Ps_{H_2O}(1)$, $Pe > Ps_{H_2O}$ (2) and $T_e \wedge S_c(3)$, where: Pe-environmental (ambient) pressure; Ps_{H_2O}-water vapour pressure; Se-environmental (ambient) temperature and Sc-critical temperature for H_2O (Sc~$160^{\circ}C$). These inequalities allowed us to estimate roughly the upper (in atmosphere) (H atm) and the lower (in Earth's interior) (H int), boundaries of biosphere. The estimation gives the H atm value <100 km. This estimation was confirmed from two series of measurements of microorganisms sampled in the upper atmosphere with high-altitude sounding rockets [(S. V. Lysenko, Advanced Microbiology, $\underline{16}$, 231, (1981) (in Russian)]. Approximate agreement between predicted and experimental results is observed. From H > 80 to 90 km no microorganisms were observed in the rocket traps. Therefore, to a first approximation H atm \ 100 km. The lower boundaries for Pe and Se conditions clearly correspond to H int ~ 30 km below the Earth's surface, at the Moho layer and some deeper.

These conditions may help to explain the existence of thermophilic microorganisms discovered recently in the "black smokers" of the deep-sea bottom, i.e., in regions at high pressure and temperatures. We concluded that the cells of microorganisms able to reach Earth orbit due to light pressure action or other means, e.g., from space or on outer surfaces of space probes, would be heated quickly by IR sunlight up to S ~ 400 K (considering a microbial cell as a black body) and would quickly explode. Therefore, both space vacuum and ionizing space radiations are serious obstacles to radiopanspermia and throw doubt on the comet hypothesis.

Conditions (1) to (3) may be considered suitable for a rough estimation of the possibility of primitive microbial life on other worlds. For example, this fact may be indicative of the absence of microbial life on the surface of Venus where Te ~ 500°C (i.e. higher than Tc) and pressure is high (Pe ~ 100 atm). As to Mars then, the conditions (1) to (3) also give reason to doubt the possibility of primitive microbial life on its surface because the diurnal temperature difference is ΔT ~ 100°C with T_{max} ~ 27°C. Then with the pressure on Mars surface P ~ 5 torr, conditions (1) to (3) are not satisfied.

IV. ADVANCED EVOLUTION - SEARCHING FOR EVIDENCE

THE IMPACT OF TECHNOLOGY ON SETI

Michael J. Klein and Samuel Gulkis
Jet Propulsion Laboratory, California Institute of Technology
Pasadena, CA 91109 USA

Paper Presented by M.J. Klein

INTRODUCTION

The probability of contact by interstellar communication depends upon the technology of both the transmitting as well as the receiving civilization. Until very recently, technology here on Earth had not advanced sufficiently to permit interstellar communication. Only during the last 50 years has it become possible to contemplate realistic searches for evidence of extraterrestrial intelligent life. We would now be capable of communicating with a hypothetical twin technology over distances of thousands of light-years if we knew the location, distance, and signal characteristics such as frequency and bandwidth.

The technical challenge for Searches for ExtraTerrestrial Intelligence (SETI) is often characterized by analogy with the problem of searching for a very small needle in an enormous "cosmic haystack". To conduct a comprehensive search program, SETI scientists must consider a variety of dimensions that include frequency, direction, intensity (which is related to transmitter distance and power), polarization, signal bandwidth, and time duration. The cosmic haystack is multidimensional and very large because nothing is known about the properties of putative signals. To keep search programs affordable, SETI researchers usually constrain the problem by reducing the "haystack" to just three dimensions. This is accomplished in practice by designing the SETI system to receive two orthogonal polarizations simultaneously and by making reasonable assumptions about the bandwidth and the duration of putative signals. Figure 1 is an example of a particular three-dimensional search volume (ref. 1) that describes the parameter space covered by the two components of the NASA Microwave Observing Project (ref. 2).

Throughout the following discussion we assume that (a) all dimensions of the cosmic haystack are equally important to search and (b) the bandwidths and time durations of putative signals will arbitrarily match the corresponding parameters in the receiving systems. We believe these assumptions are valid for the purpose of this paper, which is to illustrate how technological advances have affected our capability to search a particular set of parameters.

PROBABILITY OF DETECTION

Following Drake (ref. 3) and Gulkis (ref. 4), we calculate the relative probability of success of various SETI search programs to demonstrate the growth in search capability during the past few

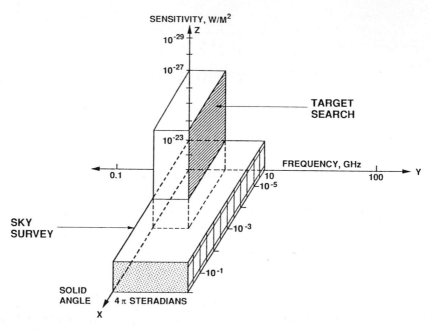

Figure 1. A three dimensional representation of the two 'volumes' of search space that will be examined by the Sky Survey and the Targeted Search components of the NASA Microwave Observing Project (from Klein and Gulkis, 1985).

decades. The derivation is based on the assumption that a given search is able to detect a minimum radio flux S_m radiated by an ETI transmitter at distance R, where

$$R = [P_t / 4\pi S_m]^{1/2} \tag{1}$$

and P_t is the equivalent isotropic radiated power (EIRP) of the transmitter. The total number of detectable signals within the spherical volume at distance R is given by

$$N_{det} = \left(\frac{4\pi}{3}\right) \rho R^3 \tag{2}$$

where ρ is the (unknown) density of transmitters assumed to be uniformly distributed within a few kiloparsecs from the sun. At greater distances the star density is concentrated in the plane of the galaxy, which is a highly flattened disk, but this fact is not considered in this discussion.

Drake (ref. 3) then shows that the relative probability of success of each SETI program is proportional to the volume of the haystack that is searched; i.e., the detection probability P_d is proportional to (a) the total frequency searched, (b) the solid angle searched, and (c) the three-dimensional volume defined by a radius R. Gulkis (ref. 4) discusses how the detection probability varies as a function of the fraction of the sky that is searched.

These proportionalities lead to the following expression of the probability, P_d, for a search over M different directions on the sky using an antenna with beam-solid-angle equal to Ω_a and a receiving system having N discrete frequency channels, each with detection bandwidth B:

$$P_d \propto \text{ß} \; [R^3] \; [NB] \; [M\Omega_a/4\pi] \tag{3}$$

where ß is a proportionality constant, R is the distance of an ETI transmitter, [NB] is the span of frequencies searched and $[M\Omega_a/4\pi]$ is the fraction of solid angle of sky that is searched.

The relationship between the parameter R and the SETI system parameters is controlled by the minimum flux S_m, which is given by the expression

$$S_m = \frac{\alpha \, (kTB)}{\pi \dfrac{D^2}{4}} \left[\frac{1}{\sqrt{B\tau}} \right] \tag{4}$$

where α is a numerical multiplier that is chosen to control false alarm rates in the SETI detection system, $k = 1.38 \, e^{-23}$ watts per Kelvin is the Boltzmann constant, T is the equivalent-noise temperature of the receiving subsystem operating on an antenna with diameter D, B is the detection bandwidth of one receiver channel, and τ is the integration time of a single receiver output sample. Substituting (4) into (1), we have a new expression for R:

$$R \propto \frac{D}{4} \sqrt{\frac{P_t}{kTB}} \; [B\tau]^{1/4} \tag{5}$$

For dish antennas with circular apertures, the solid angle of the antenna beam, Ω_a, is proportional to $[1/fD]^2$, where f is the received frequency.

Upon substitution, equation (4) can be rewritten:

$$P_d \propto P_t^{3/2} \left[\frac{D^3 \, (B\tau)^{3/4}}{(TB)^{3/2}} \right] \; [NB] \; \left[\frac{M}{f^2 D^2} \right] \tag{6}$$

It is clear from this expression that P_d increases as the three-halves power of the transmitter power, P_t. Therefore, it follows that detection will be easier if "they" decide to use their technology to transmit very powerful signals. We surely hope that is happening!

POSITIVE IMPACT OF THE RECEIVING TECHNOLOGY

With the exception of P_t, the detection probability can be enhanced if the parameters in equation (6) are optimized by the receiving civilization, i.e., by the application of our technology. The three factors in brackets can be visualized as "dimensions" of the search volume, i.e., range, frequency and beam solid angle, in that order. Note that some of the parameters are mutually independent while others are interrelated, which often means that tradeoffs must be considered. For example, reducing the system temperature, T, increases the sensitivity without affecting the other two search parameters. Likewise, increasing the number of channels (N) directly expands the search in the frequency domain. No compromises are required. On the other hand, decreasing the channel bandwidth (B) will increase the range at the expense of the frequency parameter. This occurs because P_d varies as $B^{-3/4}$ in the range "dimension" while P_d varies as B^{+1} in the frequency "dimension". Similarly, the range is profoundly expanded as the antenna diameter increases; but the amount of sky coverage is reduced. The net effect on P_d is the product of $[D^{+3}] \star [D^{-2}]$, that is, $P_d \propto D$.

This part of the analysis is summarized in Table 1, where we focus on five parameters, indicate the design goals for each parameter, and list a few of the independent and/or interrelated impact statements. Positive impacts (+) as well as negative impacts (−) are noted and briefly explained. In the final column we give a subjective opinion of the potential improvement to be gained if future design goals are realized.

Table 1.

[P]	DESIGN GOAL	IMPACT STATEMENTS	FUTURE PROSPECTS
T	Minimize	(+) Always helps (−) Approaching practical limit	WEAK
N	Maximize	(+) Always helps (+) Capitalize on high-speed signal processing	VERY STRONG
D	Maximize	(+) Increased sensitivity is attractive (+) Might be critically important to reach minimum sensitivity required for strongest ETI signal. (−) Sky coverage is reduced unless observing time is increased or multiple antennas or phased arrays are required. (−) Significant gains are costly if new antennas are required.	STRONG
M	Maximize	(+) Increased sky coverage (limit $\leq 4\pi$ steradians) (−) Reduces integration time in specific direction unless time to complete search is increased.	MODEST
B	Multiple Values	(+) Capitalize on high speed signal processing to design systems with multiple outputs to increase likelihood matching transmitter bandwidth.	MODEST

Several aspects of the NASA Microwave Observing Project can be identified in Table 1. The NASA SETI project will use state-of-the-art receiving systems to minimize T. The parameter D will be maximized by using the largest and most sensitive antennas, but the project will only use existing antennas to avoid the cost of building new (and perhaps larger) antennas. The project will capitalize on the burgeoning technology of high-speed digital processing by building megachannel spectrum analyzers with large input bandwidths to maximize NB. The two complementary searches that will be conducted are examples of the tradeoffs that were discussed above. The Targeted Search emphasizes sensitivity and provides time to search for complex signal patterns while the Sky Survey emphasizes complete sky coverage (more than 1000 times the area of the Targeted Search) over a wide range of frequencies (more than 4 times that planned for the Targeted Search) but with reduced sensitivity.

It is informative to assess the impact of technological advances that have been made over the past four decades. Using equation (6), we are able to compare the relative search capability of a few representative SETI observing programs with the capability of project OZMA, the first search. conducted by Drake in 1960 (ref. 5). The values of P_d were calculated for each observing program over a common observing period of 200 hours, which was the total search time of Project OZMA. Some of the searches are wide area surveys and others are targeted searches. Note that our intent is to establish a very specific, but quantitative, basis to contrast the capability of the OZMA search with recent searches, which have benefited from advances in receiver technology and high-speed digital signal processing. We do not wish to compare the overall merits of one search with another because so many subjective factors must then be considered. For example, it is entirely possible that two searches with similar technological capabilities could have unequal detection probabilities if one strategy guessed the correct "preferred" frequency (or direction, or pulse period, etc.) and the other did not!

Figure 2 displays the results of the ratio Pd/Pd(OZMA) that was calculated for the SETI program conducted at the Ohio State University; the META and SENTINEL projects conducted at Harvard University under the sponsorship of The Planetary Society; and the Sky Survey and Targeted Search components NASA's Microwave Observing Project. The relative capability ratio is plotted on the vertical

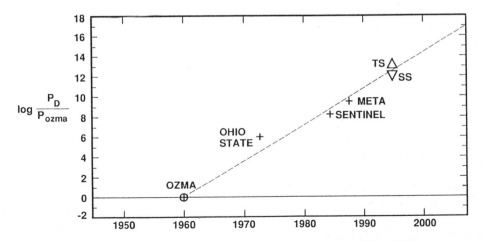

Figure 2. A representation of the increase in relative detection probability of SETI searches with date. The positive slope of these data is correlated with the technological enhancements that have benefited SETI search systems from one decade to the next.

(logarithmic) scale against the initial observation date measured on the horizontal (linear) scale. The result is dramatic; the relative capability has increased approximately 3.5 orders of magnitude per decade. Furthermore, if signal processing speed and bandwidth improve as expected, the trend may continue for another decade.

NEGATIVE IMPACT OF OUR TECHNOLOGY

On the negative side, advancing technology also tends to make the search more difficult from the Earth's surface. The ever increasing use of the microwave spectrum for terrestrial and satellite communications greatly complicates the detection process. Presentations by E. T. Olsen, J. W. Dreher, and D. J. Werthimer at this symposium (refs. 6, 7 and 8) describe the results of new observations and special surveys of radio frequency interference at microwave frequencies of interest to SETI researchers. Significant resources are now required to develop detection algorithms that will mitigate terrestrial radio frequency interference which threatens to block out very weak signals that may have originated at interstellar distances.

The research described in this paper was carried out by the Jet Propulsion Laboratory, California Institute of Technology, under contract with the National Aeronautics and Space Administration.

References

1. Klein, M.J. and S. Gulkis (1985) "SETI: The Microwave Search Problem and the NASA Sky Survey Approach" in The Search for Extraterrestrial Life: Recent Developments, IAU Symposium No. 112, D. Reidel Pub, Ed. M.D. Papagiannis, pp 397-404.
2. J.H. Wolfe, et al. (1981) "SETI - The Search for Extraterrestrial Intelligence: Plans and Rationale" in Life in the Universe, NASA Conference Publication 2156, Ed. J. Billingham, pp. 391-417.
3. Drake, F.D. (1983) "Estimates of the Relative Probability of Success of the SETI Search Program," SETI Science Working Group Report, NASA Technical Report, Eds. F. Drake, J. H. Wolfe, and C. L. Seeger, October 1983, p. 67-69.
4. Gulkis, S. (1985) "Optimum Search Strategy for Randomly Distributed CW Transmitters", IAU Symposium No. 112, D. Reidel Pub, Ed. M.D. Papagiannis, pp. 411-417.
5. Drake, F.D. (1985) "Project Ozma" in The Search for Extraterrestrial Intelligence, NRAO Workshop No. 11, Eds. Kellermann & Seielstad, pp. 17-26.
6. Olsen, E.T., E.B. Jackson and S. Gulkis,(1991) "RFI Site Surveys at Selected Radio Observatories," paper published in these proceedings.
7. Dreher, J.W., (1991) "The Targeted Search High Resolution RFI Survey," paper published in these proceedings.
8. Werthimer, D.J. , (1991) "SERENDIP II SETI Program: Observations and RFI Analysis," paper published in these proceedings.

Discussion

S. ISOBE: Detection probability in your equation is proportional to the third power of telescope diameter. Since cost for building a telescope is also nearly proportional to the third power of its diameter, the difference in detection probability comparing small telescopes and large telescopes Is proportional to the cost. However, I feel there should be some optimum value of telescope diameter to have high-detection probability.

M.KLEIN: For the NASA SETI Project we decided we could increase our probability of detection most cost-effectively by using existing radio telescopes and putting our research money into the development of very fast wideband signal processing systems. In the future, it might be important to design arrays of antennas to increase sensitivity as your suggest.

D.WERTHIMER: To point out another complication, the probability depends on the bandwidth of the signal. Currently, META is best for signals with bandwidth <0.1 Hz if it can guess the magic reference frame, Serendip II has the highest probability for signals between 0.1 Hz and 100 Hz (and needn't make reference frame assumptions), and the Ohio State program has the highest probability for signals with bandwidths greater than 100 Hz.

M.KLEIN: Perfectly true—I should emphasized that we assumed each search system is matched to the bandwidth of the transmitter; that is, we make no assumptions about transmission bandwidth since we have no way of knowing what it might be. The NASA spectrometers are designed to span a range of resolution bandwidths from about 1 Hz to 100 Hz.

D.BLAIR: I think your probability calculations also assume a random choice of transmission frequency. This does not give them credit for an intelligent choice of frequency, nor does it give us credit for the intelligence to identify the correct interstellar contact frequencies. Is it not true that if we first choose to search several such magic frequencies, then follow on to fill in the gaps, we can only increase the probability of detection?

M.KLEIN: The design goal of the NASA system is to systematically cover a very wide range of frequencies; 1-3 GHz and 1-10 GHz for the two parts of the program. Our design assumes no one is smart enough to be sure to guess the correct magic frequency. On the other hand, the order of frequencies to be searched might be sequenced to search certain frequencies before others as you suggest.

RFI SURVEYS AT SELECTED RADIO OBSERVATORIES

Edward T. Olsen, Earl B. Jackson, and Samuel Gulkis
Jet Propulsion Laboratory, California Institute of Technology, Pasadena, CA 91109

Paper Presented by E.T. Olsen

Radio frequency interference (RFI) is one of the most vexing problems confronting any endeavor to search the microwave spectrum for signals of extraterrestrial intelligent origin. Algorithms which detect faint signals from civilizations at interstellar distances will also detect the much stronger and more numerous signals of intelligent origin which are transmitted from the surface of the earth and from nearby and distant spacecraft of human manufacture. Unless great care is taken to automatically recognize and excise these signals from the data stream, the search for extraterrestrial intelligence (SETI) is destined to become entangled in an endless thicket of false alarms.

The scope of the problem is not well understood at present, for very little observational data exists which quantifies the proliferation of RFI throughout the terrestrial microwave window. There are databases which catalog the licensed transmitters, but this information does not give a true picture of the electromagnetic smog through which a SETI program must peer at the distant stars. The databases do not contain information vis-a-vis the duty cycle of these transmitters, nor do they contain information on the out-of-band transmissions (sidebands and harmonics) which are inadvertently emitted. Thus an entry in such catalogs does not give a true picture of the amount of RF bandpass which is obscured by a transmitter, nor does it inform the investigator as to how often or when the transmitter is used.

Radio frequency (RF) ranges which are infiltrated by ubiquitous RFI are permanently closed to SETI programs which are earth-based. These are lost frequencies which cannot be assayed for extraterrestrial signals unless an observatory is built on the back side of the moon.

The majority of the terrestrial microwave window is still free of such omnipresent emissions. However, commercial interests are rapidly moving to utilize the unprotected RF bands for a wide variety of services, ranging from direct broadcast television from geosynchronous satellites to air navigation to cellular telephone communications to remote control of garage doors. Many of these transmitters will be sporadic or periodic in their emissions. These transmitters are sources of RFI which must be recognized and excised by a SETI program whose goal is to observe as much of the terrestrial microwave window as possible before it is lost forever to ground-based observation.

Automatic recognition and removal of these emissions requires a combination of carefully crafted observational strategies and sophisticated and expensive real-time hardware. RFI mitigation impacts all facets of the SETI program. For example, emissions which exhibit a diurnal periodicity (such as sidebands of commercial television broadcasts) may be avoided by scheduling the SETI observations of those frequencies during the time of day that the emissions are minimal. On the other hand, emissions which are strictly periodic may be excised in hardware by blanking those frequencies at the times they are obscured. Sporadic transmissions require that dynamical avoidance techniques be employed. These may include automatic movement of the antenna and shifting of the receiver frequency and finally blanking of the frequency until the emissions terminate. The cost-benefit tradeoff decisions which a SETI program must make, in observational time and frequency coverage and hardware expansion, depend upon the nature of the RFI emissions and the importance of observing the partially obscured frequencies.

In support of the future NASA SETI Microwave Observing Program, the Jet Propulsion Laboratory (JPL) is conducting omni-directional, low-sensitivity, coarse-resolution surveys of RFI at selected radio observatory sites (1) to assess the site RFI environment and (2) to test the utility of this type of survey to predict the RFI environment which will be experienced by SETI prototypes on the main beam. To date, measurements have been made at (1) DSS 13 (Venus Station) in Goldstone, California, (2) Arecibo Observatory in Puerto Rico, (3) Algonquin Radio Observatory in Ontario, Canada, (4) Ohio State University Radio Observatory in Columbus, Ohio, (5) NRAO in Green Bank, West Virginia and (6) Meudon Observatory in France. Planned sites for future observations over the next two years include Owens Valley Radio Observatory in Big Pine, California, and Tidbinbilla in Australia.

The JPL Radio Spectrum Surveillance System (RSSS) is a computer-controlled, single-channel scanning spectrometer with programmable resolution, bandwidth, sweep frequency range, integration time, and threshold. It is a stand-alone system capable of unattended operation with its own antenna, front-end amplifiers, spectrum analyzer, data recorder and controlling computer. A noise diode at the front end provides a calibrated reference signal from which the detected signal levels can be deduced. The antenna can be either a 60-degree included angle inverted discone or a 1-meter paraboloid. In the latter case, the azimuth of the antenna is controlled by computer as well.

The standard operational strategy utilizes the inverted discone and sweeps the RF band between 1.0 GHz and 10.4 GHz repeatedly at 10 kHz resolution. The accumulation constant is chosen so that $\beta \times \tau = 5$, and a threshold 10 dB above the local noise floor is applied to the data. The sensitivity achieved by this strategy is ≈ 120 dBm and the 9.4 GHz band is swept at ≈ 45 minute intervals. An event is recorded on disk whenever the detected power exceeds the threshold. The disks are returned to JPL for analysis in a customized relational database. The database entries are reports containing time-tagged power, frequency and (if appropriate) azimuth.

The sensitivity of this survey does not approach the sensitivity which will be achieved by the future NASA SETI Microwave Observing Program, even for transmitters which appear in the sidelobes of the antennas. Only observations with prototype SETI hardware can duplicate the ultimate SETI sensitivities. However, this survey does detect the strong emissions which will definitely impact the future SETI observations and thus the design of the strategies, algorithms, and hardware which must be employed to mitigate the RFI.

The standard operational strategy is designed to rapidly and repeatedly sample the RFI environment over the 1.0 GHz → 10.4 GHz RF band. In addition, the surveys were carried out in a manner which ensured that the observations were equally spaced over weekdays, weekends, and all hours of the day. Thus there should be no selection effects which might arise from the interplay of the duty cycles of RFI transmitters and the observation strategy of the surveys. Each day of the week was observed continuously for at least two succeeding weeks, and each frequency resolution element was observed at least 10 times in each one hour interval of each day of the week.

The JPL database is analyzed to determine (1) the time variability of RFI activity, (2) the distribution of incident power of detected events, (3) the probability that a frequency will be obscured by RFI, (4) the fraction of the RF band obscured as a function of detected power and (5) a catalog of the RF occupancy as a function of maximum detected power and duty cycle.

In this paper, we present the abridged results obtained from surveys conducted at DSS 13, Arecibo, Algonquin, Ohio State University, NRAO and Meudon (Nancay). The detailed analysis will appear elsewhere.

Table I: Particulars of the surveys.

Observatory	Frequency Span Swept	Dates of Observations	Number of Sweeps	Sweep Repeat Period (Minutes)	Number of Observations of 4 Hour Periods
Goldstone	1.0 → 4.0 GHz	12/21/87 → 01/04/88	1,479	14	35
(Survey #5)	4.0 → 7.0 GHz	01/07/88 → 01/21/88	1,622	12.5	38
	7.0 → 10.4 GHz	01/21/88 → 02/01/88	1,196	15.5	30
	7.0 → 10.4 GHz	02/29/88 → 03/03/88			
Arecibo	1.0 → 10.4 GHz	06/25/89 → 07/18/89	685	50	18
Algonquin	1.0 → 10.4 GHz	10/15/89 → 10/30/89	527	40	12
OSU	1.0 → 10.4 GHz	01/20/90 → 02/06/90	567	43	13
Goldstone	1.0 → 10.4 GHz	03/19/90 → 04/03/90	449	47	11
NRAO	1.0 → 10.4 GHz	06/13/90 → 06/27/90	482	41	12
Meudon	1.0 → 10.4 GHz	09/18/90 → 10/03/90	541	40	12

The first Goldstone survey reported here differs from all the other surveys in that the RF band was partitioned into roughly three equal frequency intervals which were swept separately. This strategy was an early one, and was carried out in an attempt to detect transmitters which had very low duty cycles or whose activity schedules were very restricted. Comparison with the second Goldstone survey reported here, which was carried out according to the standard strategy, indicates that the standard strategy is sufficient at the sensitivity levels achieved in the surveys.

The Arecibo survey was extended to three weeks duration over the standard two weeks. This was due to the fact that the fourth of July holiday appeared during the second week of the survey, and we felt it was desirable to extend the observations in case there were significant differences in activity due to this holiday. Indeed we found that there were significant differences over the holiday.

The "Number of Sweeps" in Table I is the number of times the RF span was swept over the course of the observations. Nominally, these sweeps followed one upon another continuously throughout the survey period, with no interruptions.

The "Sweep Repeat Period" in Table I is the amount of time required for a single sweep to be executed. This time includes the overhead imposed by data transfer to the spectrum analyzer and switching among the low noise amplifiers. For a fixed range of swept frequency, this time becomes longer if there is more RFI activity.

The "Number of Observations of 4 Hour Periods" in Table I is the average number of sweeps which were completed during each 4 hour period in each day of the week. It is an indication of how many times a particular time of day was sampled over the course of a survey.

Since these surveys randomly sampled each frequency at least 449 times, they are capable of detecting all transmitters whose incident power exceeds the sensitivity of the surveys and whose random transmission duty cycle is greater than 1%. Additionally, since each 4 hour period in a day was sampled at least 11 times, the surveys are capable of detecting all transmitters whose emissions are scheduled to take place during a restricted time interval only once a week if their duty cycle within that time period is greater than 10%.

Table II: Overall results of the surveys.

Observatory	Total Number of Events Reported	Total Number of Unique Frequencies Reported	Portion of RF Band Thus Obscured (%)	Greatest Detected Power Reported (dBm)	Number of Unique Frequencies with Probability $\geq 1\%$	Portion of RF Band Thus Obscured (%)
Goldstone (Survey #5)	397,579	14,120	1.50	-76.9	2,644	0.28
Arecibo	370,373	31,657	3.37	-87.8	8,966	0.95
Algonquin	1,469	1,029	0.11	-93.2	4	0.00
OSU	39,367	4,492	0.48	-92.5	1,538	0.16
Goldstone	97,578	8,787	0.93	-80.8	1,716	0.18
NRAO	26,126	3,778	0.40	-93.5	1,523	0.16
Meudon	37,252	3,885	0.41	-93.3	1,389	0.15

The "Total Number of Events Reported" in Table II gives the number of times a detection report was generated in the course of a survey. This gives an indication of RFI activity and is the product of the spectral occupancy of the RF band, the duty cycle of the transmitters and the number of times the band was swept.

The "Total Number of Unique Frequencies Reported" in Table II gives the number of 10 kHz resolution frequency elements in the 1.0 GHz → 10.4 GHz RF band which were reported at least once in the course of a survey and is a measure of the spectral occupancy of the RF band.

Column 4 in Table II, the "Portion of RF Band Thus Obscured", is the worst-case calculation of the obscuration of the 1.0 GHz → 10.4 GHz RF band. It identifies a 10 kHz frequency resolution element as being obscured if a signal (no matter how narrow) was detected in the frequency resolution element at least once. This results in an overestimation of the fraction of the RF bandpass which is obscured, but cannot be avoided due to the inherent frequency resolution of the surveys.

The "Greatest Detected Power Reported" in Table II gives the strongest signal which was detected in the course of the survey. The signals which the NASA SETI Microwave Observing Program will be capable of detecting will be in the range, -170 dBm → -200 dBm, over frequency resolutions of 20 Hz to 1 Hz.

The "Total Number of Unique Frequencies Reported with Probability ≥ 1%" in Table II gives the number of 10 kHz resolution frequency elements in the 1.0 GHz → 10.4 GHz RF band which were reported at least once in every 100 scans during the course of a survey.

Column 7 in Table II, the "Portion of RF Band Thus Obscured", is the next worst-case calculation of the obscuration of the 1.0 GHz → 10.4 GHz RF band. It identifies a 10 kHz frequency resolution element as being obscured if a signal (no matter now narrow) was detected in the frequency resolution element at least once in every 100 observations of that element. This results in an overestimation of the fraction of the RF bandpass which is obscured at this duty cycle, but cannot be avoided due to the inherent frequency resolution of the surveys.

At all sites, RFI is mostly concentrated in the 1 GHz → 3 GHz frequency range. This RFI is due to a variety of sources, whose mix changes from one site to another. Most, however, is due to air navigational radars and communications transmissions and the higher harmonics of commercial broadcasting. In fact, the commercial broadcasting harmonics appear to be concentrated in the band 1.0 GHz → 1.2 GHz, and they exhibit a strong diurnal variation which is tied to the civil time of the observatory location. Between 11:00 pm and 6:00 am local time, most of the RFI in this narrow band disappears at all sites. Above 3 GHz, each site displays its own characteristic RFI signature, although all appear to have some sporadic RFI above 9 GHz.

Goldstone, for example, exhibits additional RFI near 4 GHz (due to commercial cross-country microwave links) and near 8 GHz (due to a microwave communication link utilized at the Deep Space Network complexes which is scheduled to be deactivated in early 1991).

Arecibo exhibits very little RFI outside of the very heavily infiltrated 1 GHz → 3 GHz RF range. There is some RFI around 4 GHz and in the 6 GHz → 7 GHz RF band.

Ohio State University exhibits additional RFI near 4 GHz and just below 7 GHz. During the observations at Ohio State, we discovered that nearby cellular telephone repeaters operating around 800 MHz broadcast signals which were strong enough to render our lowest frequency amplifiers useless. We were forced to obtain high-pass filters which provided an additional isolation of 50 dB below 1 GHz before we could continue with the survey. These filters were made part of the standard configuration of the RSSS from that time onward.

NRAO is nearly free of RFI over the entire RF range surveyed. The most active RF ranges are 1.05 GHz → 1.15 GHz and two small bands around 1.35 GHz and 1.57 GHz. This site, which has an active RFI transmitter mitigation program and has been declared a radio quiet zone, is the only one in which we did not detect any interference in the radio astronomy protected bands.

Algonquin is nearly free of RFI over the entire RF range surveyed. The most active RF range is 1.05 GHz → 1.15 GHz, and the duty cycle is even lower than we experienced at NRAO.

Meudon (Nancay) exhibits the usual concentration of RFI in the band 1.0 GHz → 1.2 GHz, and a widely spaced set of transmissions which are localized in frequency up to 2.3 GHz. The RF band is quiet thereafter save for 4.2 GHz → 4.5 GHz and scattered activity above 9.2 GHz.

Changes in the RFI environment at DSS 13 are detectable in the two surveys at Goldstone which were taken almost exactly two years apart. The gross difference is that the amount of the RF band obscured by RFI decreased overall from the first to the second survey (mostly due to reduced utilization of the DSN broadcast microwave communication link near 8 GHz). However, the low probability RFI spread to involve a greater range (although a lesser number) of unique frequencies.

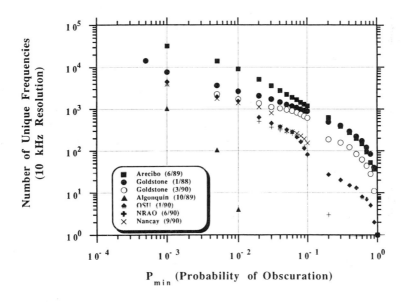

Figure 1: Obscuration of the 1.0 GHz → 10.4 GHz Band by RFI as a function of observed probability.

Figure 1 displays the level of obscuration of the RF band at the various observatories as a function of the probability of obscuration. It shows how many unique frequencies are obscured by RFI at an observed probability greater than or equal to P_{min}. Since there are 9.4×10^5 10 kHz frequency resolution elements in the range 1.0 GHz → 10.4 GHz, full scale on the Y-axis corresponds roughly to 10% of the observed RF band. Thus we see from the figure that at Arecibo, $\approx$ 1% of the RF band between 1.0 GHz and 10.4 GHz is obscured $\geq$ 1% of the time. All curves have the same general form, suggesting that there may be a general population model for transmitter distribution and power which can fit the data. The two Goldstone surveys indicate that some high duty cycle transmitter(s) decreased in activity between the times that the earlier and the later surveys were carried out. Comparison with other analyses show that this was the microwave link between DSN sites near 8 GHz.

The research described in this paper was carried out by the Jet Propulsion Laboratory, California Institute of Technology, under a contract with the National Aeronautics and Space Administration.

Discussion

M. KLEIN: The RFI system at Arecibo was installed on the platform 500 ft. above the antenna surface. From that vantage point the results show the very high density of radio frequency interference events reported. Will you comment on the shielding against these signals that the Arecibo receivers experience from their location far below the platform?

E. OLSEN: The Arecibo telescope, fortunately, is of a unique design which is very non-receptive of RFI whose origin is on the horizon. Thus when we placed the RSSS receiver on the main beam, we did not detect the high level of RFI which we experienced using the horizon-scanning discone antenna. We are currently comparing simultaneous observations (RSSS on omni-directional antenna, and narrowband systems on main beam) and the simultaneous detections. Preliminary analysis suggests that the isolation of the main beam from the horizon is more than 44 dB.

P. BOYCE: Some frequencies look very black on your plots. Are you going to give up on these frequencies or use some extraordinary measures to combat the interference - such as the use of an interferometer for the targeted search.

E. OLSEN: The plots look very black because of the limiting resolution of the laser writer used to generate the hard copy. There is a lot of "white space" which is not visible at this resolution. Less than 1% of the RF band pass at Arecibo is obscured more than 1% of the time. However, the RF regions which are heavily infiltrated by RFI will likely be masked by digital filters (in blocks of 1 kHz resolution) from further consideration by the detection algorithms. An interferometer is not appropriate to the sky survey, since so little time is spent on each beam area.

F. BIRAUD: Can you tell whether the interference observed at 1420 is emitted at this frequency, or some harmonic of a transmitter outside the protected band? Is it important that the receiver has a very steep frequency response to have a good rejection?

E. OLSEN: The interference detected at Arecibo in the band, 1420 → 1430 MHz, appears to be a very high harmonics of (usually) commercial transmitters. We have found that a steep frequency cutoff below 1 GHz is desirable. For example, at Ohio State the RSSS ran afoul of cellular telephone repeaters. Even though these transmitters operated in the range 800 → 900 MHz, they were so strong that they caused intermodulation in our FETs until we added a high-pass filter in the RF. This filter suppressed the signal below 1 GHz by an extra 50 dB (beyond the FET roll-off).

N. EVANS: The RFI problems at 1-10 GHz will spread to higher frequencies pretty soon. Shouldn't we make some effort to begin a search, with small telescopes perhaps, at higher frequencies (up to ~ 300 GHz) before they too are hopelessly polluted?

E. OLSEN: We desire to observe first in those bands where we can achieve the greatest sensitivity. We can now use very large telescopes and achieve $T_{sys} < 25$ K routinely over the RF range 1 → 10 GHz. Yes, RFI is greater in this range than at higher frequencies. However, these microwave frequencies are still unobscured for the most part. They will be lost to us over the next 50 years as utilization throughout the world increases. Thus we believe it is logical to search the microwave band first. Most of our development is in signal processing, and so we can attach our instruments to any front end which provides us an IF in the range 1 GHz ± 500 MHz.

A 1 HZ RESOLUTION RFI SURVEY: PRELIMINARY RESULTS

J. W. Dreher
NRC Senior Research Associate
NASA Ames Research Center, Mountain View, CA USA

J. Tarter
SETI Institute and University of California, Berkeley
Mountain View, CA USA

D. Werthimer
University of California, Berkeley CA USA

Paper Presented by J. W. Dreher

In support of the nascent SETI Microwave Observing Project at NASA, radio-frequency interference (RFI) data with a resolution of 1 Hz and an integration time of 1 second were collected at several radio observatories during the last year. At each site, we surveyed a frequency range of > 10 MHz over the course of a few days, with the antenna stowed. The threshold for the detection of RFI was −164 dBm into 1 Hz for the first two telescopes and −173 dBm for the NRAO telescope. Data with less sensitivity but much better frequency and time coverage, obtained with a scanning spectrum analyzer at kHz resolutions and a horizon-sensitive surveillance antenna, are presented in an accompanying paper in this Symposium by Olsen et al.

1. METHODS

The high-resolution spectra were obtained using the Serendip II+ system, which is described by Werthimer, Tarter, and Bowyer (IAU Symposium 112, 1984). Serendip accepts IF frequencies near 30 MHz, with the RF amplification and remaining IF conversions being provided by the receiver system of the host telescope. The system samples its baseband IF for ≈1 second at the Nyquist rate, then Fourier transforms the data to 65536 complex coefficients spanning 62.5 kHz. For this survey the system was upgraded with an optical disk drive so that the entire spectrum could be recorded, rather than just thresholded powers. The rate at which spectra were obtained was limited to about 16 per minute by the speed at which the data could be written to disk.

For the data sets used here, the IF frequency was stepped so that Serendip recorded 16 spectra at one frequency range, then shifted upward by 50 kHz. A typical set of observations took a few hours to span 6.95 MHz, stopping when the optical disk cartridge was full. Observations were usually obtained at frequencies in the 1.4 GHz protected band, which we expected to be fairly quiet, and in the vicinity of 1.7 GHz, which we expected to be fairly noisy. These frequencies were chosen because they are generally available at all radio-astronomy telescopes. Data from Green Bank were obtained last month at 1.7 GHz, with the 1.4 GHz observations still to be made this summer.

The complex spectra were later converted to powers, corrected for the average bandpass response of the system, and normalized using a median-based estimator of the noise power of each spectrum. Regions roughly 2 kHz wide were blanked at both ends of each spectrum because of roll-off caused by the pre-sampling anti-aliasing filter. A region ±1000 bins (±954 Hz) around DC was also blanked to avoid multiples of the AC power line frequency that were visible on long averages.

2. RESULTS

Observations were made at the Hertzberg Institute's 150-foot steerable paraboloid antenna located in Algonquin Park, about three hours drive north of Ottawa, Canada, from October 30 to November 1, 1989. During the RFI scans, the antenna was pointed at the zenith and had a measured system temperature of roughly 130 K. Observations with the Ohio State University's large Kraus-type transit telescope, located on the outskirts of the city of Columbus, Ohio, were made from January 18 to 20, 1990. The antenna main beam was directed at 0^h HA and $+20°$ δ; the system temperature was estimated to be $\sim$ 150 K. The National Radio Astronomy Observatory's 140-foot steerable paraboloid antenna is located in Green Bank, West Virginia, the center of the U.S. National Radio Quiet Zone. During the RFI scans this antenna was stowed at the zenith and the measured system temperature ranged from 19 to 20 K. In all, during the survey scans, about 3×10^4 spectra were obtained, with 1.5×10^9 powers.

The accompanying table summarizes the strongest RFI found during the survey. The signals reported from Algonquin and Ohio State were selected because they had at least one 1-Hz bin in one spectrum ≥ 21 p_{noise}, where $p_{noise} = kT \Delta\nu$. For pure random noise, the powers would be distributed as $\exp(-p/p_{noise})$ and there would be ~1 event over 21 times the mean in our total set of data. To facilitate comparison, the threshold for the tabulated RFI from the Green Bank observations has been set to the same level in absolute power as that of the other two observatories, despite the much lower system temperature at this site. As was expected, the observations at Ohio State, which is surrounded by the suburbs of a large city, showed the most RFI. Indeed, the surprise is that the RFI was not much worse! Somewhat surprisingly, although there was not much RFI found at Algonquin, the two examples were both very strong. Finally, based on the partial observations at Green Bank, the National Radio Quiet Zone seems to have been effective in suppressing RFI. One should note, however, that most of the NRAO's efforts go to suppressing RFI within the commonly used radio-astronomy bands; observations outside these bands might be worse.

Figures 1 and 2 show an example of one of the RFI signals from Algonquin. It is not feasible to plot 60,000 bins, so a compressed representation is used in Figure 1, with the data grouped into sets of 200 adjacent bins and the maxima being plotted. This RFI signal is a comb in frequency (corresponding to a pulse train in time), the strongest component of which is very close to 1424.150000 MHz (unfortunately, this component is not plotted because it lies within 1000 bins of the origin). Figure 2 shows a plot of the individual bins around one of the comb elements. Note that these elements are as narrow as 1 Hz. Unlike most of the RFI found in the survey, this example is very transient. From the narrowness of the comb features, the signal must have persisted for a least one second, but on the spectra before and after, the signal is not present.

Figures 3 and 4 show an example of a simpler RFI signal, in the same format as above. This signal was clearly present (above about 13 times the mean) for 10 out of 16 spectra at this frequency step. Note the excellent stability of the transmitter.

```
================================================================================
                    RFI Observed with Peak Power Exceeding -164 dBm
--------------------------------------------------------------------------------
Observatory      Frequency     # of    --------RFI Found in Scan---------  Comment
                 Scan Range   Spectra  Spectra  Peak   Frequency   Width
                    MHz       in Scan  Affected dBm       MHz        Hz
--------------------------------------------------------------------------------
Algonquin 150'

              1400.00 1406.80   2235      16    -147  1399.999846    20
              1406.80 1413.70   2235       0
              1413.50 1420.45   2235       0
              1420.00 1426.95   2235       1    -149  1424.150006  >60000   Transient

Ohio State

              1402.00 1408.10   1825       6    -151  1408.047146  >60000
                                           1    -164  1404.513079     1     Just random noise?
              1414.80 1421.75   2235      12    -161  1421.593780    20
                                           8    -161  1415.001810    20
                                          11    -162  1416.003909    30
              1421.60 1428.63   2235      15    -161  1427.935297    15
                                          15    -161  1426.753817    17
                                          10    -162  1424.603877     3
                                          10    -163  1423.601776    16
              1691.00 1697.95   2235      16    -155  1690.999330  20000    Scan of "bad"
                                          16    -148  1691.254964   5000    area on survey
                                           3    -163  1692.196968    30     of Dixon et al.
                                           7    -164  1693.965572   200
                                           3    -164  1694.033388    40
                                          13    -163  1697.251955    30
              1661.00 1667.95   2235      16    -153  1663.754799    43

Green Bank 140'                                                             *

              1605.00 1611.95   2235       0                                Includes GLONASS bands
              1663.00 1669.95   2235       0
              1716.00 1720.55   1461       0

================================================================================
```

* The Green Bank spectra had an intrinsic noise threshold of -173 dBm, but for comparison only
  events above -164 dBm are reported in this table.

Now we consider the results for Green Bank at full sensitivity, with a threshold of −173 dBm. The 1663-1670 and 1716-1721 MHz scans remain free of any RFI. The 1605-1612 scan is another matter entirely. Figure 5 shows an example of the abundant RFI found in this scan. This comb is characteristic of RFI from the Glonass satellites. We were able to detect Glonass at the other two observatories as well, but only by looking many times especially for them. The much greater sensitivity afforded by the NRAO receivers made observing Glonass satellites all too easy: we appear to have detected Glonass transmissions in slots 5 and 8 strongly, as well as possibly slots 9 and 14 (the combs are so wide it is difficult to determine the center frequencies).The Glonass comb was present at 17% of the frequency steps used in the 1605-1612 scan. Fortunately, one can probably observe around the RFI from these satellites by waiting until the offending satellite at a particular frequency is below the horizon.

3. CONCLUSIONS

The small amount of RFI found in this survey is encouraging. At least at the relatively high power threshold of −164 dBm, only a tiny portion ($< 10^{-3}$) of the frequencies surveyed had RFI contamination at the time of observation. The Targeted Search will use multichannel spectrum analyzers (MCSAs) that will provide 1-Hz resolution bins over a 10-MHz bandwith. The results summarized in the table then suggest that at a signal level of −164 dBm, and at a quiet site such as Green Bank or Algonquin, there will be many frequencies for which there will be no RFI present in the

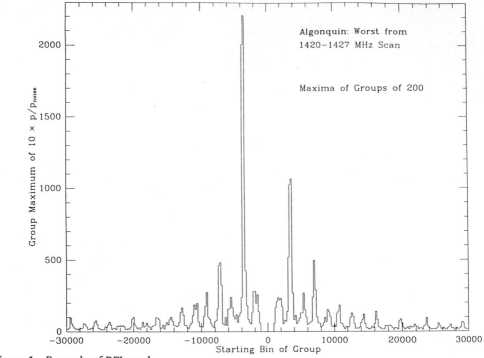

Figure 1: Example of RFI comb.

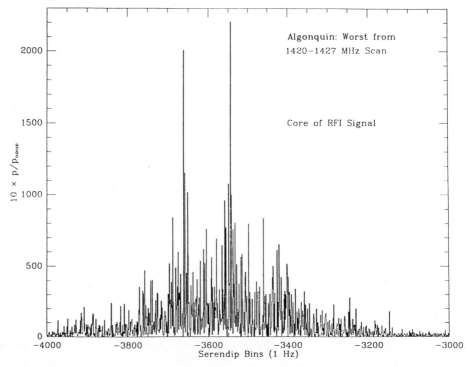

Figure 2: Expanded view of comb element.

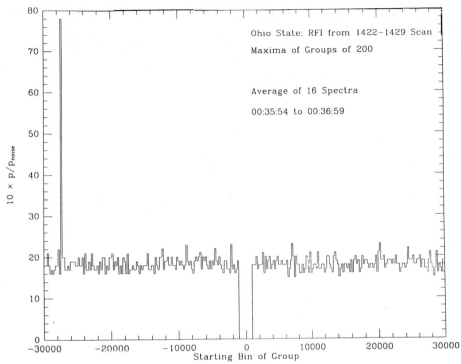

Figure 3: Example of CW RFI.

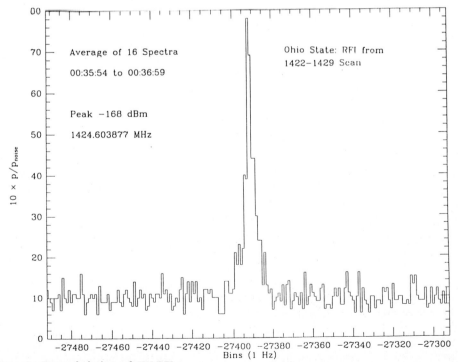

Figure 4: Expanded view of CW RFI.

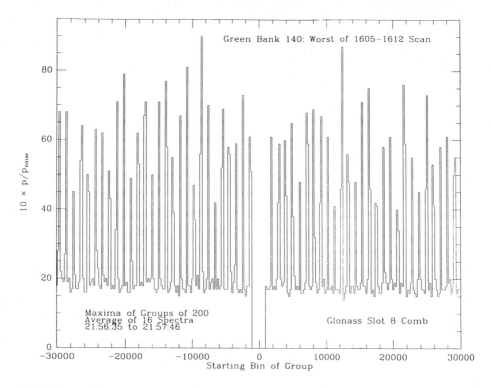

Figure 5. RFI from Glonass.

MCSA spectrum. Less is known, however, about how the incidence rate of RFI will increase as the sensitivity increases. The partial data set from Green Bank suggests that at a threshold of −173 dBm, RFI is still sparse: of three data sets, only one showed RFI contamination. Since some of the RFI is expected either to be predictable, such as Glonass, or recognizable, such as the frequency comb of Figure 1, it seems likely that the fraction of spectra containing RFI that needs to be diagnosed by reobservation or other techniques, will not be so large as to severly affect the observing schedule. The Targeted Search CW detectors are expected to work to a threshold of ~ −189 dBm; however, and it remains unclear how much more RFI will be found at this level. Another uncertainty lies in the temporal behavior of the RFI. We observed at each frequency for only a minute; over the course of the 5 to 20 minutes planned for each Targeted Search observation, a population of rapidly variable RFI could also increase the total amount of RFI found. Despite these uncertainties, however, our observations still render implausible those scenarios in which the presence of numerous RFI signals in most observations would overwhelm the processing capacity of the Targeted Search systems.

Acknowledgement

This work would not have been possible without the support of the Space Astrophysics Group, University of California, Berkeley, California

THE SERENDIP II SETI PROJECT: OBSERVATIONS AND RFI ANALYSIS

C. Donnelly, S. Bowyer, W. Herrick, D. Werthimer,
M. Lampton, and T. Hiatt
University of California, Berkeley, CA USA

Paper Presented by D. Werthimer

The Berkeley Search for Extraterrestrial Radio Emission from Nearby Developed Intelligent Populations (SERENDIP) is a piggyback search for narrow-band radio signals of extraterrestrial origin. The original SERENDIP system is discussed by Bowyer et al. [1983]. The SERENDIP instrument utilizes the regular schedule of observations of an existing radio astronomy observatory. In contrast to a dedicated search, we are not able to select the observing frequency or polarization, and we do not control the pointing of the radio telescope. In piggyback mode, however, SERENDIP operates autonomously, collecting large quantities of high-quality data in a cost-effective manner. Figure 1 shows the broad range of frequencies SERENDIP has monitored in the period from December 1986 through November 1988 as a piggyback experiment. Note that SERENDIP spends a significant amount of time within frequency ranges considered "likely" to host interstellar communication. Figure 2 shows SERENDIP's cumulative sky coverage during this same period.

Data are sent from the observatory on floppy disk to the Space Astrophysics Group, Space Sciences Laboratory at the University of California in Berkeley, where they are analyzed for coherent signals having high probability of extraterrestrial (ET) origin. Signals that are identified as ET candidates are logged for further study. Finally, a targeted search program utilizing dedicated telescope time allows us to study further those signals which are not categorized as radio frequency interference (RFI) or white noise.

The SERENDIP instrumentation is based on a high performance personal computer (PC), an array processor (for signal processing), and some front-end electronics that interface the PC to the observatory's receiver. Figure 3 shows a block diagram of the SERENDIP instrument. The SERENDIP hardware is discussed in detail by Werthimer, Tarter, and Bowyer [1984] and Werthimer et al. [1988]. The PC receives a digitized signal from the front-end system and passes it to a high-speed array processor, which performs a 65,536-point complex Fast Fourier Transform (FFT) on the input data. After FFT processing, the complex amplitudes of adjacent bins are added using a technique suggested by Cullers et al. [1984] in order to increase sensitivity to signals falling between the 1-Hz frequency bins. All peaks in the power spectrum that pass over a mean-power detection threshold are transferred back to the PC and recorded onto hard disk for later off-line analysis.

The intermediate frequency signal from the telescope is mixed down to baseband by mixing it with a signal produced by a frequency synthesizer. To increase frequency coverage during search operations, the synthesizer steps along the RF spectrum in 20 kHz increments covering a total frequency range of 1.2 MHz before returning to its starting frequency and repeating the scan process. The entire

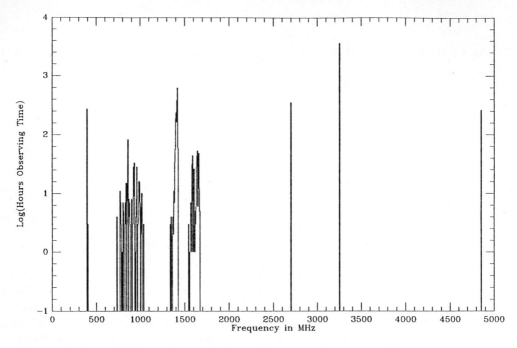

Figure 1: Amount of time SERENDIP has logged at each frequency. Note that a significant amount of observing time has been spent around the spectral lines of hydrogen and the hydroxal radical (1.420–1.662 GHz), commonly dubbed the water hole. Noise is minimized at around 1.5 GHz, making this spectral region optimal for interstellar communication.

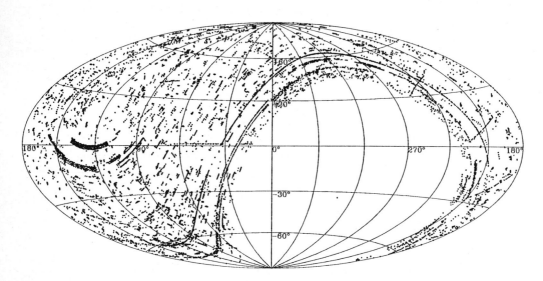

Figure 2: Galactic coordinates of SERENDIP's observations between 1986 and 1988.

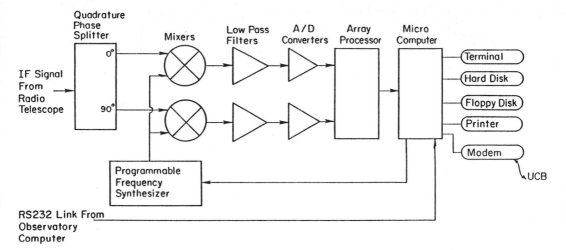

Figure 3: Block diagram of the SERENDIP real-time hardware system.

frequency scan takes about 60 seconds and is repeated continuously during normal operations. Figure 4 illustrates the synthesizer steps and how an ET signal might appear if detected repeatedly by SERENDIP over a typical observation of several minutes.

To keep up with the large volumes of data collected by SERENDIP, we developed an automated off-line data analysis system that first identifies and rejects RFI and then searches the remaining data set for candidate ET signals. Of the several million events recorded by SERENDIP over the last few years, only 126 were selected as candidate ETI signals. All other events were discarded as RFI or white noise. Among the RFI events rejected were those that (1) persisted in a narrow range of frequencies over multiple pointings of the telescope's beam, (2) persisted in a narrow range of Fourier Transform bins over multiple pointings of the telescope's beam, (3) persisted in a narrow range of Fourier Transform bins over multiple settings of the frequency synthesizer, and (4) were detected in large groups over a broad band of the RF spectrum in a single FFT (spread spectrum RFI).

The first three rejection criteria require a cluster analysis on the input data set. Significant deviations from the expected Poisson distribution of events are detected, and all events within the spectral area under question are rejected if they fall into one of the rejection criteria above. By rejecting events over spectral regions, we risk losing ETI signals that happen to fall within the suspect areas. However, normally less than 1% of the observed spectrum is eliminated in this manner.

Spread spectrum RFI is rejected when the following condition is detected

$$S < \frac{\sum d^2}{x}$$

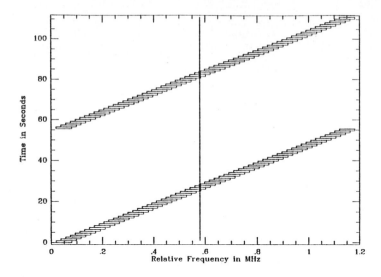

Figure 4: The rectangles represent SERENDIP's 65,536-channel spectrum analyzer as it steps across a portion of the RF spectrum. A signal (represented by the vertical line) could be detected up to four times during each scan.

where d is the number of bins (0.98 Hz/bin) between adjacent events and x is the number of events in the spectrum. S, as the limiting value, is usually set at $S = 10{,}000$. Nearly all spread-spectrum RFI is identified using this condition. Figure 5 shows a particularly severe period of spread-spectrum RFI that was identified by this condition.

Our automated off-line data analysis system includes a two-pass RFI filter with built-in statistics algorithms which are dynamically adaptive to the changing size of the data set as the filter process proceeds. In this way, RFI sensitivity is increased as large RFI clusters are identified and removed from the calculations for the expected Poisson distribution of events over Fourier or frequency bins.

Candidate ET signals are selected in the last phase of the data analysis process from those events surviving RFI rejection. Signal selection criteria include (1) events recorded above 60 times the mean spectral power, (2) three or more events recorded over a short time period in any reference frame with drift rates up to 10 Hz/sec, and (3) two or more events recorded in the solar system's barycentric reference frame, the geocentric reference frame, or the topocentric reference frame at different times. Time periods between event detections in selection criterion (3) range from several seconds to several months.

SERENDIP operated at a sensitivity of 4 x 10 $^{-24}$ W/m^2 on the 300-foot radio telescope at NRAO for about two years of nearly continuous observation until the telescope collapsed in November 1988. The sensitivity of SERENDIP at NRAO is discussed further by Bowyer et al. [1988]. During this period, SERENDIP analyzed about 3.2 trillion spectral bins, recording several million events with power above the detection threshold.

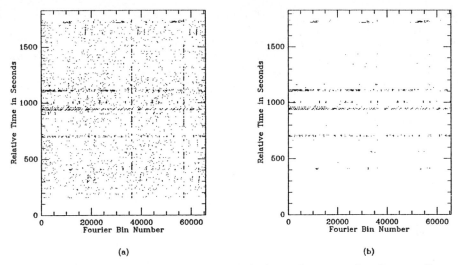

(a)   (b)

Figure 5: Real data from a particularly severe period of spread-spectrum interference. Dots represent thresholded events recorded by SERENDIP. Part (a) shows the data-set before any RFI filtering. Part (b) shows those events which were identified by our off-line analysis system as spread-spectrum RFI.

We intend to install a SERENDIP system at the Arecibo Observatory in Puerto Rico in the future. The large size of the Arecibo telescope (1000 feet) will increase our sensitivity to 1×10^{-24} W/m^2. Table 1 compares SERENDIP's sensitivity at Arecibo with other major SETI projects.

Table 1: Comparison of SERENDIP's sensitivity with other major SETI projects

	Diameter (ft)	Temp K	Gain K/Jy	b Hz	t sec	Sensitivity W/m$^2 \cdot 10^{-24}$
SERENDIP at Arecibo	1000	40	8	1	1	1
SERENDIP at NRAO	300	20	1.4	1	1	3
META	85	80	0.1	0.05	20	8
Proposed META in Argentina	100	90	0.13	0.05	20	7
NASA all sky search						100
NASA targeted search						0.002

Acknowledgments

The authors wish to acknowledge the many contributions by Jill Tarter and Vicki Lindsay, and the excellent support from the staff at NRAO. This work was supported by NASA Grant NAGW-526 and the Research Corporation.

References

Bowyer, S., Zeitlin, G. M., Tarter, J., Lampton M., and Welch, W. J.: 1983, Icarus 53, 147-155.

Bowyer, S., Werthimer D. and Lindsay, V.: 1988, "The Berkeley Piggyback Seti Program: SERENDIP II," in Bioastronomy-The Next Steps , ed. G. Marx. Dordecht, Holland: Kluwer.

Cullers, D. K., Oliver, B. M., Day, J. R., and Olsen, E. T.: 1984, "Signal Recognition," in SETI Science Working Group Report , NASA Technical Paper No. 2244, eds. F. Drake, J.H. Wolfe, and C.L. Seeger. Washington, D.C.: Scientific and Technical Information Office.

Werthimer, D., Brady, R., Berezin, A., and Bowyer, S.: 1988, Acta Astronautica 17(1) , 123-127.

Werthimer, D., Tarter, J., and Bowyer, S., "The SERENDIP II Design," in The Search for Extraterrestrial Life: Recent Developments, ed. M. D. Papagiannis, IAU Symposium 112, 421-424.

Discussion

W. SULLIVAN: Are you keeping long-term records of the anomalous burst events that you reject as ETI signals: The only way we will ever detect periods longer than a few days is by retrospective examination of a long-term archive. It may well be that we will recognize a long-term periodicity only in the manner that resulted in Edmund Halley having a comet named after him. He was not the first to see the comet, but he was the first to recognize the long-term 76-year periodicity and to predict the return.

D. WERTHIMER: We archive all the "hits" above our threshold - several million. We have also started an RFI database - groups of events we believe are interference; and of course we keep all our interesting candidates even though we are not able to confirm them when we re-observe.

Normally we do not store the entire spectrum - only the hits; but when NASA uses Serendip for their RFI surveys they store the complete complex spectra - so far about 200 Gbytes of data recorded on optical disks.

A. LEGER: Have you any explanation for the "powerful" events that you have rejected?

D. WERTHIMER: Most are probably short-term interference, e.g., nearby transmmitters that are only on while the telescope is pointing in one beam. The most misleading interfering signals are unstable transmitters that drift in frequency with a rate that matches the doppler drift due to the earth's motions.

M. HARRIS: Have you searched for straight-line trajectories formed by a number of the high-power events you described, which might be consistent with a rapidly-moving object such as a spacecraft?

D. WERTHIMER: No. We assume the transmitters are stationary on the sky. If we find the same signal in different parts of the sky, we reject it as interference.

SETI: ON THE TELESCOPE AND ON THE DRAWING BOARD

Jill Tarter
SETI Institute and U.C. Berkeley
Mountain View, California USA

Michael J. Klein
Jet Propulsion Laboratory
Pasadena, California USA

Paper Presented by J. Tarter

On Columbus Day 1992, NASA will initiate the observational phase of its SETI Microwave Observing Project that will continue to the end of the century. Paul Horowitz and the Planetary Society will soon be operating META SETI in both the Northern and Southern hemispheres. Smaller scale projects at the Nancay Observatory in France, the Ohio State University Radio Observatory in the US, the Algonquin Observatory in Canada, the Parkes Observatory in Australia and an observational strategy being employed at optical and radio wavelengths in the USSR round out the list of near-term SETI observing programs for this generation. This paper discusses these searches as well as the strategies and technologies that have been suggested for the next generation of searching.

INTRODUCTION

Over the past 30 years, since Frank Drake's original Project Ozma, there have been at least 60 searches, in 8 countries, looking for evidence of distant technologies. Most have been at radio wavelengths, but optical, infrared, UV and gamma-ray observations have also been tried (see archive of searches compiled by Tarter, June 1990). Almost all of these searches have made use of existing instrumentation, primarily designed for astrophysical research, rather than SETI. As a result, the searches have been confined to frequencies that are of interest to the astronomical community, where receivers have already been built. In the extreme, some of the searches have confined themselves to "magic frequencies" that might be mutually guessed by both sender and receiver. Although the past 30 years has seen a growing number of telescope hours devoted to this search, the fraction of the observable universe that has actually been explored for evidence of extraterrestrial technologies is infinitesimally small. This is beginning to change.

ON THE TELESCOPE

OSURO

The SETI observations at Ohio State University Radio Observatory represent the longest running SETI program anywhere in the world. Begun in 1973, this program continues today on the basis of volunteer effort organized through a consortium of local colleges and universities. The radio telescope provides some "hands-on" training for science and engineering students, as well as an intro-duction to managing a scientific program on a shoe-string budget. The antenna is a unique design by John Kraus, consisting of a fixed standing spherical segment and a tiltable flat mirror, and provides the equivalent collecting area of a 53 m parabolic dish. The current receivers are uncooled FETs, providing a system temperature of about 300K in a single vertical polarization. The observations consist of decli-nation scans over all of the sky visible from Ohio and take place at the 21 cm HI line frequency, shifted to be at rest with respect to the galactic center. The system covers a bandwidth of 500 kHz, with frequency resolutions of either 10 kHz or 1 kHz. In addition to trying to monitor the local interference environment with a discone antenna mounted on the main feed horns, OSURO is attempting to implement an automated follow-up signal verification scheme. In this mode of operation, the sidereal rate sky scan will be interrupted whenever a signal is detected with the proper response pattern from the dual beam horn. The system will select the narrow resolution and continue to track the source until it sets or is identified as known RFI.

META SETI: North and South

Since 1985, Paul Horowitz has operated his META (mega-channel extraterrestrial assay) SETI system at the Oak Ridge Observatory in Harvard, Massachusetts. The project is funded by the Planetary Society, and enjoyed a one-time donation from filmmaker Steven Spielberg. The antenna at Oak Ridge is a 26 m parabola that performs a sidereal rate sky survey, operating with uncooled FET amplifiers and a system temperature ranging from 50K to 85K. META is perhaps the ultimate "magic frequency" observing program, as it assumes that any signal is being deliberately beamed to the Earth with all relative accelerations compensated by both sender and receiver. Under this hypothesis, the best sensitivity is achieved with ultra-narrow frequency resolution. META consists of 8 million spectral channels, each 0.05 Hz wide. Every two minutes it investigates the patch of sky moving through its beam sequentially in two orthogonal polarizations, at a particular frequency relative to three possible reference frames: the Earth-Sun barycenter, the galactic center and the 3K background. Information on signals exceeding about 20 times the rms fluctuation in the system temperature are automatically recorded to tape for later reobservation. To date META has searched 400 KHz surrounding the 21 cm HI line, twice that frequency, and is once again searching the sky at the HI line. Because of the need to rapidly chirp a local oscillator in order to compensate for the diurnal acceleration of the Earth, META turns out to be relatively insensitive to terrestrial interference since it is spread over many channels. Over the past year, two young engineers from Argentina have collaborated with Prof. Horowitz to build a replica of the META system. META II is scheduled to be inaugurated on a 30 m dish in Argentina in October of 1990. It will be the first sky survey of the southern hemisphere.

TARGET

The Algonquin Observatory near Ottawa, Canada, is currently in a "moth-balled" state, awaiting a funding source with a valid scientific project. One of the handful of staff who maintain the facility is Bob Stephens. He has moved his collection of secondhand and home-built electronic gear from the Northwest Territory to an 11 m dish at Algonquin to establish Project TARGET. In between sleeping and maintenance of the 46 m site, Bob improves upon his equipment and conducts declination drift scans of the sky at 21 cm.

ZODIAC

Both the 6 m optical telescope at the Special Astrophysical Observatory in Zelenchukskia and the large RATAN 600 radio telescope in the Soviet Union have begun participating in a SETI project called ZODIAC, each observing a few of the selected target stars to date. The idea, as you will hear from Dr. Strelnitskij later in this symposium, is to observe the selected targets when they are close to opposition. In this way, one can perhaps improve the chances of intersecting an ultra-narrow beamed SETI signal generated by an extraordinarily large, high gain transmitter. This project will continue with a list of 21 solar-type targets near the ecliptic, operating at frequencies where receivers currently exist. It will soon be augmented by AILETA, an amateur effort of the Young Pioneers and radio astronomers. Its 3 m telescope will be operated by the youngsters at 21 cm, scanning the sky when not making targeted observations of the selected stars near opposition. Later in this symposium, Dr. Likachev will elaborate upon plans for future SETI projects using a 64 m antenna near Moscow.

Australian SETI

We were delighted to learn, upon arriving in Val Cenis, that the Australians have once again initiated SETI observations. The Parkes 64-m antenna is being used by a consortium, headed by David Blair, to conduct a preferred frequency search at π times the HI frequency. Three rest frames are considered: geo-, galactic- and barycentric with 500 channels of 100 Hz spectral resolution devoted to each. Dr. Blair will tell us more about the selection of target stars and the rationale for this search later in this symposium.

RFI Studies

The Jet Propulsion Laboratory has built a stand-alone system for surveying the radio frequency interference environment from 1 to 10 GHz at sites of interest to SETI. You will hear more about the results of surveys with the Radio Spectrum Surviellance System from Dr. Olsen later in this symposium. As a complement to the RSSS studies, the SETI Office at NASA Ames Research Center has funded Dr. Bowyer's group at UC Berkeley to replicate and upgrade their SERENDIP system to permit a high spectral resolution survey of RFI over those frequency bands where a given observatory site has receivers. SERENDIP II$^+$ continuously records 65,536 channels of 1 Hz data onto an optical disk. These data are later analyzed for drifting and non-drifting CW signals using algorithms based on those developed for the NASA SETI Microwave Observing Project. On occasion, this system is actually used to do SETI while conducting RFI studies. Last June we made 21 cm observations at Arecibo, targeting two of Bruce Campbell's candidate planetary systems and two stars which had previously produced unexplained "Birdies". The observations lasted for approximately 1000 seconds each in a single circular polarization. This unsuccessful search achieved the highest sensitivity to date for drifting CW and pulsed signals, although for a very limited set of targets.

SETI ON THE DRAWING BOARD

The groups at UC Berkeley and Harvard University have begun planning for 100-million channel, high-resolution spectrum analyzers in order to expand their frequency coverage. These systems will be based on commercially available digital signal processing chips and are projected to be very affordable. You will hear more about some of these plans later from Dan Werthimer.

Far beyond the planning stage, NASA is completing production prototypes of the observing systems it intends to deploy for its SETI Microwave Observing Project on October 12, 1992. The search strategy will be bi-modal. The Targeted Search Systems will provide two polarizations covering 10 MHz with 16 million channels of 1 Hz resolution, and simultaneously, 5 coarser resolutions ranging from 2 to 30 Hz. This spectrometer will be combined with low noise feeds and receivers and real-time signal recognition capability to permit fully automated searches of approximately 1000 nearby solar-type stars over the frequency range 1 to 3 GHz. These systems will be deployed at Arecibo, Parkes, Tidbinbilla, Nancay and a dedicated site in the Northern Hemisphere starting in 1992 and continuing until the end of the century, in a manner that does not place undue burden on any of the large radio astronomy facilities around the world. The real-time signal recognition software will examine the frequency-time domain for non-drifting and drifting CW signals as well as narrowband pulses. In a complementary search, one Sky Survey System will be deployed on the new 34 m DSN antennas at Goldstone and then Tidbinbilla in order to survey the entire sky from 1 to 10 GHz in right and left circular polarizations. To accomplish this search, the antennas will be driven rapidly across the sky in a race track pattern at 0.2 degree per second. The spectrum analyzer produces 10 million channels of 30 Hz resolution and the real-time signal recognition system records any narrowband signals above threshold and compares them with subsequent detections both along a given scan line and between successive scan lines.

A unique feature of the NASA Microwave Observing Project is that candidate ETI signals are automatically re-observed as soon as possible; i.e., in the Targeted Search, immediately after their detection and in the Sky Survey, at the end of surveying a pre-set pattern on the sky. In this way, it is hoped that most sources of interference can be recognized and eliminated without the need for human intervention. Any signal that continues to look promising following the automated discrimination tests triggers an alarm to summon an observer and initiates recording. This real-time analysis is a new capability in SETI and the data processing requirements it imposes are daunting. Table I compares the data rates that are generated by more familiar astronomical observatories with the data rates generated by a single Targeted Search System (Sky Survey data rate is similar.) In the astronomical examples, the data are often analyzed long after the observations, in processes that often require much more time than the data acquisition. In SETI this data rate must be handled "on the fly," in no more time than it took to acquire. Thanks to some custom VLSI development by Stanford University, some new commercially available parallel processing capability, and breakthroughs in rapidly accessible large memory devices, we believe that we can meet this processing challenge with workstation-class machines. Columbus Day 1992 is not very far away, but if our funding is approved, we expect to launch the next wave of exploration with systems that are both extraordinarily powerful and affordable. In times past, we have often discussed the unit of the "kilocray"; our processing requirements would either keep 1000 Crays very busy, or kill a single one! Fortunately, we believe that special-purpose processors can replace the need for the "kilocray" unit for SETI. At the next Bioastronomy Symposium, you will see whether we have lived up to our words.

Table I. Data Rate Comparison

VLA: Worst Case $= 4 \times 10^4$ bytes/sec

- 256 channel spectral line observation with all 27 antennas (351 baselines)
- 2 x 16 bit complex frings
- 10 second integration time

HST: Worst Case $= 3 \times 10^7$ bytes/sec

- WF/PC with four 800 x 800 CCDs
- 12 bit digitization
- 0.12 seconds (shortest exposure)

SETI Targeted Search System: Normal Case $= 7 \times 10^8$ bytes/sec*

- 15 x 10^6 channels
- 2 x 16 bit complex amplitudes
- 2 polarizations
- 1 second integration
- 6 resolutions simultaneously, all with same data rate.

SETI Sky Survey System: Normal Case $= 2.5 \times 10^9$ bytes/sec

- 16 x 10^6 channels
- 32-bit power values
- 2 polarizations
- 0.05-second accumulation

---

* Targeted Search will deploy up to 6 such Targeted Search Systems around the world.

Abbreviations: VLA - Very Large Array in Socorro, New Mexico   WF/PC - Wide Field/Planetary Camera
                 HST - Hubble Space Telescope

SETI AND HAYSTACKS

SETI is often likened to "looking for a needle in a haystack." Just how hard a job is that? Is it something that an individual can do in a working lifetime of 50 years? A good reference library can tell you that there are:

- 370 billion acres of solid land on Earth,
- 1.8 billion acres of which are potentially arable
- on average, 22% or 390 million acres are under cultivation,
- and 5% of that, or 19.5 million acres produce hay.

If haystacks are 5 feet x 6 feet, then 1 acre yields 20 haystacks per year. If you could acquire the annual crop of haystacks around the world and examine them for 50 years, you would be able to spend 4 seconds searching each one. That's not a lot of time, but it isn't infinitesimally small either. It is a time scale commensurate with human activity and indeed, if the task is too big for one person, it is the sort of enterprise that could be accomplished by many people in a cooperative effort. SETI is much bigger and harder than that! To accomplish a systematic SETI Project requires not individuals, but technology and some real technological breakthroughs.

We have described how technology is being applied to SETI; more is taking shape every day. With luck, and continued fiscal support, we will master the challenges of a systematic SETI, begin searching on October 12, 1992, continuing until we succeed.

Discussion

B. CAMPBELL: I am wondering about the selection of haystacks in which to look for needles. Concerning the targeted search list of stars, are you going to ask the optical astronomy community to look at the stars and gather various data, such as age estimates from CaII emission?

Okay, can I suggest that this is a good opportunity to involve some optical astronomers who are not already in the SETI community. What about issuing a NASA Announcement of Opportunity to have them look at these stars?

J. TARTER: The current target list consists of F,G, K, V stars from the Wooley "RGO Catalog of Stars within 25 pc." No prioritization has been made of that list except for distinguishing single from multiple stars (as listed in catalog). Much work needs to be done, starting with modern classifications and multiplicity from the SIMBAD database. Ages will be incorporated where available.

We would very much like to include the optical community. I suggest that the current NASA Research Announcement for the SETI Microwave Observing Project is precisely the invitation needed, and I hope it will be heeded.

S. VON HORNER: I am very impressed by this progress of technology, but I am also afraid about omitting the human mind too much. My Chinese fortune cookie said "The elves' thing is to be prepared for the unexpected," and no computer can do this. For example, you spoke of searching for a needle in a haystack; but is it really a needle we should look for? This question is illustrated by a nice definition of serendipity: "looking for a needle in a haystack, but finding there the farmer's daughter."

J. TARTER: Sebastian, I will not comment specifically on the "farmer's daughter." It is perfectly true that our automated systems will only detect what we have instructed the computers to look for. That is why we have spent so much effort on creating instruments that are sensitive to as wide a class of signal types as we can afford. This is far more comprehensive than what has been done in the past. As for signals that are not in this class, we can only encourage observers to keep their eyes open for unexpected results.

A. SUCHOV: Is there among the first targets to observe any globular cluster? And if not, wouldn't it be reasonable to include some of them since among hundreds of thousand stars in a cluster, the oldest in the Galaxy, one may expect to find the most advanced intelligence?

J. TARTER: The target list is drawn from the RG0 catalog with measured stellar parallaxes. It only extends to 25 pc ($\approx$ 80 light years) and so it contains no globular cluster stars. I expect that we will augment this list with other good candidate targets. The globular clusters give a lot of targets within a single beam. However, there is still a debate about whether the low metallicity, greater stellar tides and high-energy activity at the cores of some clusters make terrestrial type planets unlikely to be found in the clusters. This is an open question currently being debated.

A. LEGER: What are the main criteria of your selection of RF signals?

J. TARTER: We have tried to consider what the natural universe produces in the way of emission and to try and find a type of signal that as far as we know is never produced by nature. This leads us to consider signals that are band-limited in the frequency domain. The narrowest natural signal is ~300 Hz. We will look for signals narrower than that. In addition, if the signals are pulsed, we require that the signals be no broader than required by their duration i.e., the time-bandwidth product should be approximately unity (BT~1).

PROJECT OF ETI SIGNAL SEARCH AT THE WAVELENGTH 1.47 MM

S. F. Likhachev
Special Design Department, Power Institute
Moscow, USSR

G. M. Rudnitskij
Sternberg Astronomical Institute, Moscow State University
Moscow, USSR

Paper Presented by G. M. Rudnitskij

INTRODUCTION

Starting from the article of Cocconi and Morrison (1959), the decimeter-wavelength range was usually assumed to be the optimum one for CETI and SETI. The frequencies of the 21-cm line of neutral hydrogen, the 18-cm OH lines, or the band between these two lines (the "Water Hole") etc., was used for the search (see review by Tarter, 1986).

However, during the last years, another approach to the selection of the search frequency was developed. Kardashev (1979) showed that the millimeter-wavelength range presented a number of advantages (minimum background noise temperature, high antenna directivity, etc.). According to Kardashev, in the millimetric range, a natural frequency standard could be the hyperfine-structure transition $1^3S_1 - 1^1S_0$ of the positronium (PS) atom, which is a short-lived system consisting of an electron and a positron. This line at 203.385 GHz is analogous to the 21-cm line of neutral hydrogen.

In this contribution, we present, for the first time, a practical scheme to realize Kardashev's idea (see Figure 1). The novel idea is to use simultaneously two parallel channels, and to search both for narrow-band ETI beacon signals and for broad-band informative ones.

THE NARROW-BAND CHANNEL (Δf = 100 MHz)

The narrow-band channel would be used for the search of signals which could have a frequency band as small as a fraction of a Hz. The frequency is down-converted to the band 0–100 MHz. Further, digital spectral analysis by means of the Fast Fourier Transform (FFT) is applied. We consider it possible to obtain a 1-Hz frequency resolution in the entire 100-MHz band, i.e., to have a 10^8-channel spectrum analyzer. However, even with modern electronics, the necessary buffer memory capacity is millions of Gigabytes to have reasonable times of processing. For an illustration, we give below some numerical estimates concerning the signal processing in a simplified case of a relatively small number

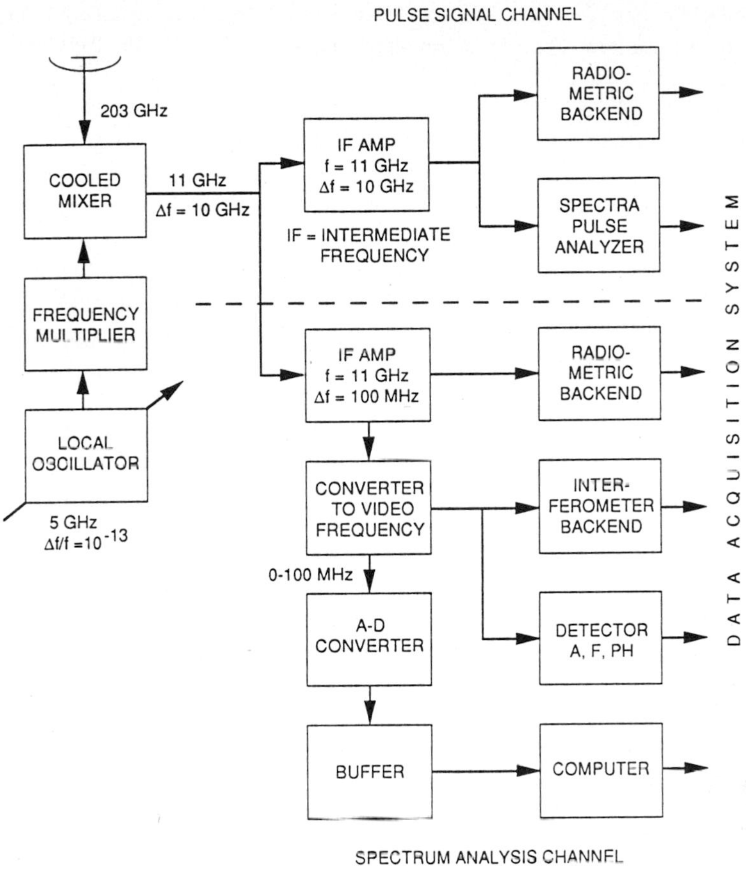

PULSE SIGNAL CHANNEL

Figure 1. The dual broadband narrow-band receiver concept

of spectral channels (N = 1024). With the sampling rate 200 MHz, the duration of one realization for the buffer memory of 1024 samples is 5.12 x 10^{-6} seconds; with 8 bits of signal digitization, the total amount of information in it is 8.2 kbit.

The signal processing software includes:

(a)   Preliminary filtration by a Kalman filter; this is a fresh idea which allows one to adapt the computer medium to the manifold of input signals and, when properly used, to increase the signal-to-noise ratio. Earlier, Kalman filtering was successfully applied to processing of speech signals (Likhachev, 1987) and radioastronomical images (Zheng Yi and Basart, 1990). Kalman filtering requires $2m^3 + 3m^2 + 3m + 1$ multiplications (and division) per one signal value, where m is the order of the filter. With m = 1 and N = 1024, we have 9216 multiplications per realization.

(b)   Fast Fourier Transform (see, e.g., Brault and White, 1971), requiring $Nlog_2N = 10240$ multiplications. Thus, in the simplest case (N = 1024), $19455 \sim 2 \times 10^4$ multiplications are needed per realization.

For the integration time $\tau = 1$ hour, the number of realizations will be;

$$n = 3.6 \times 10^3 \text{ s} / 5.12 \times 10^{-6} \text{ s} = 703,125,000.$$

The integrated circuit Intel 386 performs one multiplication in $\sim 10^{-6}$ s; one realization could be processed in 0.02 s. The necessary buffer capacity for realization of the conveyor (lattice) method of spectral analysis is 4×10^6 Gbytes.

The quoted estimates of computational complexity for the given facility show that the procedure can be realized *uniquely* on dedicated lattice (parallel) processors, fabricated in the USSR. In this case, the volume of information (for an integration time of 1s) is 50 Mbytes. For a 6 to 8-digit A/D converter, channel number $\approx 10^6$, $\Delta F_{input} = 25$ MHz ($t_{digital} = 20$ ns), two types of signal treatment are possible:

1)   On-line processing, $\sim (4\text{-}8) \times 10^3$ channels,

2)   Off-line processing, $\sim 10^6$ channels, computational time ~ 5 min. (The time of processing in this case is determined by the type of computer used.)

Viewing the difficulties of technical realization of such a buffer memory, we suggest to form the backend of the proposed receiver as a network of computers which would deal with the realizations averaged over the ensemble (with ergodicity assumed and with parallel computations).

THE PULSE BROAD-BAND CHANNEL ($\Delta f = 10$ Ghz)

Kardashev (1979) argues that an ETI informative transmission may have a frequency band as board as 8 GHz (corresponding to the reciprocal value of the Ps $^1S_0$ State annihilation time). This channel would include a pulse analyzer with a sub-nanosecond time resolution. More detailed considerations for automated search of pulses will be given in a subsequent publications.

To extend SETI and astrophysical possibilities of the receiving system, radiometric and interferometric backends are foreseen in both channels. In the narrow-band channel, a set of various analog-signal detectors (amplitude, frequency and phase) is used at video frequencies to detect possible slow modulation of a beacon signal (Petrovich, 1986).

Main search targets at the initial stage could be nearby solar-type stars. Further, the direction towards the galactic center may be investigated. According to Kardashev's estimates, an ETI transmitter with $P_m = 100$ MW and a receiving antenna with an effective area of 4000 m^2 can provide (with the noise temperature of the considered 1.47-mm receiver $T_n = 200$ K) communicating to distances of about 10 kpc.

The experiment can be realized at the new 70–m antenna which is being constructed now in Soviet Central Asia (Slysh, 1986). The adaptive reflecting surface of this antenna will provide a minimum working wavelength of 1.3 mm.

With the same receiver, numerous astrophysical observations can be proposed, among them high-spectral resolution observations of radio lines (i.e., searches for the positronium 1.47-mm spectral line in astrophysical objects, for known and new molecular spectral lines with the frequencies close to 203 GHz, and for radio recombination lines). Thus, the proposed SETI experiment can also yield a large amount of astrophysical information.

The authors are grateful to V. P. Davydov for illuminating discussions.

References

Brault, J.W., White, O.R.: 1971, Astron. Astrophys, 13, 169.
Cocconi, G., Morrison, P.: 1959, Nature 184, 844
Kardashev, N.S.: 1979, Nature 278, 28
Likhachev, S.F., Nazarov, M.V., Prokhorov, Yu.N.: 1987, in Proc. of the 11th Internat. Congress Phonetical Sci. Tallinn, August 1-7, 1987. Vol. 3, p. 267
Petrovich, N.T.: 1986, in The Problem of Search for Life in the Universe. Proc. Tallinn Symp. Moscow, Nauka, p. 152
Slysh, V.I.: 1986, in Proc. 18th Soviet Conf. Radio Astronomy. Abstracts of papers. Irkutsk. Pt ii, P. 222.
Tarter, J.: 1986, in The Problem of Search for Life in the Universe. Proc. of the Tallinn Symp. Moscow, Nauka, p. 170
Zheng Yi, Basart, J.P.: 1990, Astron. J. 99, 1327

Discussion

D. BLAIR: What is the maximum flux sensitivity of such a millimeter wave system? Is there not a serious problem of atmospheric absorption?

G. RUDNITSKIJ: The flux sensitivity depends on what receiver noise temperature could be obtained. In the favorable case of $T_n = 100$–200 K, the sensitivity in a 1-Mz frequency band would be a few Janskys. I agree that at millimetre wavelengths, atmospheric absorption is important and can significantly increase T_n, reducing the sensitivity.

A SETI SEARCH TECHNIQUE: MONITOR STARS TO WHICH WE HAVE SENT SIGNALS

Peter B. Boyce
American Astronomical Society, Washington, DC, 20009, USA

Lawrence H. Wasserman
Lowell Observatory, Flagstaff, AZ, 86002, USA

Paper Presented by P. B. Boyce

ABSTRACT

After the appropriate light travel time, we should monitor stars which have been illuminated by earth's high power radar systems. We know the position of the star, the frequency and the time after which a reply, if sent, could arrive. The problem of computing which stars have been illuminated by the Arecibo planetary radar has been solved by using occultation predictions. It is possible that four stars have been illuminated. Because of planetary motion, radar signals beamed toward any given star have a maximum duration of a few hours, making verification difficult. In order for our searches to be sensitive to an alien civilization's radar signals, immediate verification of any suspicious signal is essential.

1. THE CONCEPT

One of the problems with any SETI program is the great number of stars or the vast amount of sky to be searched. Anything which narrows the phase space to be searched is worth trying. In the course of the past two decades powerful microwave radar signals of known frequency have been sent out from the earth in many directions. It is possible that some of these signals will impinge upon nearby solar-type stars and could possibly trigger a reply. Since we know, in principle, when such signals have been sent, we gain a tremendous advantage. We know the position of the star and the frequency of transmission. We also know the distance, and light travel time, to these stars. If one of these stars has an advanced civilization around it, it is possible that our signals might be detected and a return signal might be sent back to earth.

We suggest that it is worth monitoring any nearby stars which have been unwittingly illuminated with one of our strong radar beams. We propose that, after a period of time equal to the round trip travel time, any such stars be monitored at the transmitting frequency for possible return signals. The odds of having our signal detected and returned are certainly very small, but by removing the uncertainty in position, frequency, and arrival time of a possible return signal it becomes a small undertaking to add a few nearby stars to any comprehensive search program.

2. THE ARECIBO RADAR

The most powerful transmitter on the earth is the S-band planetary radar at Arecibo (Morrison et. al., 1977) which went into operation in 1976 and produces an EPIR of 7.1 x 10 e 12 W with a half power beamwidth of about 2.2 minutes of arc. Within 25 parsecs the Arecibo signal would be easily detectable by modest sized radio telescopes. Prior to 1976, radar observations were made at 430 Mhz at a lower power and a half power beamwidth of about 9 minutes of arc. Although the probability of the Arecibo beam intersecting a nearby star is low, it is possible that one of the radar transmissions might, by chance, have been directed at a nearby star. Unfortunately, an easily searchable log of the planetary radar pointings does not exist. It would be a prohibitive expenditure of time to manually search the logs and compute the telescope positions for checking against the positions of nearby stars.

3. USING OCCULTATION PREDICTIONS

However, it is possible to work backward by computing occultation predictions for all the planetary radar targets for the Arecibo site. The occultation program at Lowell Observatory was used to find close approaches of the planets to the nearby stars. The catalog of RGO solar type stars within 25 parsecs of the sun was checked for possible appulses by Mercury, Venus, Mars, Jupiter and Saturn. Within the 1960-1990 time frame there was only one appulse in which a star would have approached within the 2.2 minute of arc (full width at half maximum) size of the S-band Arecibo radar beam and also satisfied the 20 degree zenith angle limitation of the Arecibo telescope. On July 4, 1968 Mars came within 91.4 seconds of arc of the RGO star 241, a K6 dwarf at a distance of 37.9 l.y.

Table 1: Stars with planetary appulses as seen from Arecibo, 1960 - 1990.

Planet	RGO Star	Date			Dist l y	Closest Apprch (arcsec)	HPBW (arcmin)
Mars	241	1968	Jul	47.630	37.9	91.4	2.2
Mercury	449	1967	Sep	7.738	32.6	374.9	9
"	174	1970	Jun	20.673	39.8	346.4	9
"	9027	1973	Apr	24.630	75.8	378.3	9

At this time, Arecibo was using a lower frequency transmitter with a half power beam width of about nine minutes of arc. Extending our search to include the larger beamwidth yielded appulses of Mercury with three additional candidate stars (see Table 1). We do not yet know if the radar transmitter was in use at the time of the appulses. Unfortunately, all the appulses we have found occurred over seventeen years ago and radar records may not be available -- three of the appulses took place while Arecibo was a defense department installation. The closest star has a round trip light travel time of 64 years and no response can occur before the year 2031. Monitoring these stars is a job for the next generation.

4. LIMITS ON DURATION OF PLANETARY RADAR SIGNALS

What are the characteristics of a planetary radar signal? Is there any chance at all that our planetary radar would be recognized as an ETI signal by another civilization? Billingham and Tarter (1990) concluded that the planetary radar is an effective signalling device. It has all the right characteristics. It is extremely stable, as narrow band as technology will allow and with a repetitive pseudo-random phase coding that is obviously an intentional signal (Davis, 1989). However, the radar is only used about five percent of the time so only a few hundred beamwidths on the sky are illuminated in one year.

There is another major drawback to the recognition of planetary radar transmissions as ETI signals. Because of the motion of the planet which is being illuminated, this is a one-time signal for the star in the background. Recognition and verification have to be done by the other civilization within a period of a few hours. Unless they are specially set up to deal with such events, it is unlikely that another civilization could verify a radar signal that quickly. Also, after a few minutes the radar is turned off for an equal amount of time in order to receive the return echo.

Mercury's apparent motion can be quite large, approximately 2 degrees per day at its most rapid. For Arecibo's older radar with a nine arcminute beam this gives a maximum of 1 3/4 hours illumination of a given star for a central occultation, less for off-center appulses. For the two arcminute beam of the present S-Band radar the duration is only twenty minutes. The other planets move more slowly but the duration of the signal is limited by the Arecibo telescope's constrained tracking capability. Our planetary radar signals on any given area of the sky will, in general, only last for a few hours and will not be directed at the same area of the sky on succeeding days.

5. IMPLICATIONS FOR OUR MICROWAVE SEARCHES

This last point is particularly important in deciding how to handle our own ETI searches. If we assume dimensions in other solar systems are similar to ours, then other civilizations' radar signals made for the purposes of examining and monitoring other bodies in their planetary systems will be subject to the same constraints in duration. This may cause a problem on our end. There is no chance to see a signal the next day. All verification will have to take place immediately. Given the characteristics of present microwave searches, many of which are not analyzed for days or months, it is not likely that we would currently recognize another civilization's planetary radar as a real signal.

Should we be setting up our ETI searches to detect another civilization's radar transmissions? Judging from our own experience we can expect that other civilizations might use powerful microwave radars with reasonable frequency. The high-gain, high-power characteristics of these transmissions will make them detectable at distances significantly greater than other types of leakage radiation. It is easy to imagine scenarios under which such "unwitting" radar transmissions, as infrequent as they may be, would be among the most numerous of the detectable ETI signals. We believe that radar leakage from other civilizations may turn out to be an important class of ETI signal.

Search strategies which count on having an hour to get back to the same region of sky to check a signal will stand a good chance of missing one-time radar signals. Modern, computer-controlled searches should be set up to immediately recognize and verify suspicious signals.

6. FUTURE DIRECTIONS

This present exercise is just the first step. In recent years Ostro (1988, 1990a, 1990b) has used the Arecibo radar for extensive studies of asteroids. Our next step will be to run these same calculations for the asteroids which have been studied. The Goldstone radar is also a powerful transmitter. For instance Muhleman et. al. (1990) have recently observed Titan for four nights. Searching the records at Goldstone and even extending the concept to other powerful transmitters could result in a list of additional candidate objects for monitoring.

References

Davis, M. (1989) Personal Communication.

Morrison, P., Billingham, J., and Wolfe, J., eds. (1977) The Search for Extraterrestrial Intelligence SETI, NASA SP-419.

Muhleman, D.O., Grossman, A.W., Butler, B.J., and Slade, M.A., (1990) Science, 248, 975.

Ostro, S.J., Connelly, R., and Belkora, L., (1988) Icarus, 73, 15.

Ostro, S.J., Rosema, K.D., and Jurgens, R.F., (1990a) Icarus, 84, 334.

Ostro, S.J., Campbell, D.B., Hine, A.A., Shapiro, I.I., Chandler, J.F., Werner, C.L., and Rosema, K.D., (1990b) AJ, 99, 2012.

Billingham, J.B., and Tarter, J.C., (1990) Proceedings of 1989 IAF Congress, Torremolinos, Spain.

Discussion

E. OLSEN: Would you really believe a signal which was detected for 10 minutes and never seen again?

P. BOYCE: The short answer is no. But, I think it would depend upon the characteristics of the signal. A strong (20 sigma) signal from one specific area of the sky, with 1 Hz bandwidth and switched in a repeating, non-random pattern would certainly be tempting to believe. It would, in any case, indicate a region of the sky which should be monitored and searched intensively for other signals.

D. WERTHIMER: As Jill Tarter briefly mentioned in her talk, the new Ohio State search does some real time verification of candidate signals by tracking the source for several minutes, going on/off source and recording the potential signal for later analysis, so at least we've got one system along the lines of what you're thinking.

P. BOYCE: That is very good. Any analysis will help in our understanding of the nature of any peculiar signal.

RADIO SEARCH FOR ALIEN SPACE PROBES

A.V. Arkhipov

Institute of Radio Astronomy, Acad. Sci. of the Ukr. SSR

Kharkov, USSR

ABSTRACT

It is shown that the search for alien space probes by occasional interceptions of their radio communication beams appears to be a promising task not beyond the ability of amateur radio astronomers and all-sky monitoring systems.

The existence of alien space probes in our solar system and beyond could become a natural consequence of the ETI space activity (Freitas, 1983b). The probe's radio communication and active radio location (for the flight safety) must be accompanied by the narrowband radio emission, radiated in narrow beams. The occasional interception of such a radiobeam by an antenna could be registered as a radio flash. Apparently, sky monitoring for such events appears to be a useful addition to the other SETI strategies (Arkhipov, 1989).

Before searching for alien probes by radio eavesdropping, it is necessary to consider the conditions for success. In general, such analysis is not identical to estimates for probability of communication by high-gain antennae with the ETI, located in the other planetary systems. So, if the beam orientation is quite accidental relative to the Earth antenna, the observation time, sufficient for a search with w probability of success, is:

$$t = \left[\tau G_e \ln(1 - w) \right] / \left[N \ln(1 - G_p^{-1}) \right] \tag{1}$$

where: τ is the time scale of orientation variability of the beam, relative to the Earth-probe direction; N is the all-sky number of radio beams accessible for detection; G_e, G_p are the gains (relative to an isotropic radiator) of the Earth and probe antennae, respectively. Hence t is minimized if $G_e \to 1$. A single dipole antenna cannot localize a signal source and has a small effective area. That is why for the discussed search the "Obzor"-like SETI systems are most promising. The "Obzor" system, proposed by Troitskii (1986), consists of about 100 autonomous radio telescopes (the gain of each is ≈ 200) monitoring the total hemisphere of sky in sum. For two such systems the total antenna complex gain can reach 1. There are some situations favorable for employment of "Obzor"-like devices.

One of them is the search for alien space probes located in the vicinity of our planet. The possible presence of alien reconnaissance satellites is admitted by a good many scientists (Bracewell, 1960; Deardorff, 1986; Freitas, 1983 a,b; Ksanfomaliti, 1981; Lawton and Newton, 1974; Lunan, 1973; Suchkjin et al., 1986). Alien probes could monitor human activity and send information to proprietors located in other planetary systems. Then:

$$\tau \sim 2\sqrt{\pi}\, r \,/\, (V_{\perp}\sqrt{G_p})$$

where: r is the Earth-probe distance; $V_{\perp}$ is the tangential component (with respect to the Earth direction) of probe velocity. In the case of a circular geocentric orbit and reasonable parameter estimates (w=0.9; N=1; $G_p=10^6$; $2 \leq r/R \leq 20$, where R is the earth radius) the observations time is: 0.6 yr x $G_e \leq t \leq 18.7$ yr x G_e. For w=0.5:0.2 yr x $G_e x \leq t \leq 5.6$ yr x G_e. Therefore, the search for a radio communication beam has some sense if $G_e \sim 1$. On the other hand, for a typical gain of the antenna planned for the NASA SETI program ($G_c \sim 10^6$; Papagianis, 1985) success is almost unattainable. Apparently, search success is more probable if N>1 and/or $\tau < 2\sqrt{\pi} r/(V_{\perp} \sqrt{G_p})$. Such a situation can be realized in the case of radar beams of a probe. Moreover, the probe radio emission intended for interstellar communication could be discovered by means of leakage in the side lobes of the probes's antenna radiation pattern. Indeed, if the radio flux at a distance of 2 pc is equal to 10^{-24} W/m^2 (50-kW transmitter; $G_p=10^6$), then at distance $r/R \leq 20$, it is $\geq 2 \times 10^{-7}$ W/m^2. In this case:

[alien radio flux] / [flux limit of the "Obzor"] $\geq 2 \times 10^{12}$

Hence, even the side lobes of -120 dB level would be observable.

The probe, orbiting the Moon, is too slow for direct interception of its radio communication, because t>30 yr (w=0.9; $G_e=1$; N=1; $r=4 \times 10^5$ km; $V_{\perp} < 3.4$ km/s; G=10). Nevertheless, it is more probable to receive the reflection of a probe's emission from the Moon. Then, for stable selenocentric circular orbits (3×10^3 km $\leq a \leq 5.8 \times 10^4$ km; Freitas, 1983a): $I^d \leq t \leq 5$ yr. The diffuse backscattering from the Moon decreases the radio flux at the Earth to $\sim 1\%/G_p$ level (at frequency range from 26.5 MHz to 35 GHz; Krupenio, 1971) or to $\sim 10^{-16}$ W/m^2 for a 50-kW transmitter and $G_p=10^6$.

For preliminary identification of direct interception of an alien communication channel, the following criteria can be proposed:

-   The high amplitude of a transient radio signal, apparently of non-interference origin, received from a discrete source in the sky;
-   The orientation of a radio emission to a certain neighboring star favorable for life (repeated directions are especially interesting);
-   The existence of a large space object unidentified or classified by NORAD as debris near the emission's origin point on the sky (the problem of unidentified Earth satellites has been considered by Bagby, 1981). A promising object must be investigated in detail for definitive conclusion.

As the "Obzor" system is not available, all-sky monitoring could be realized, perhaps, by members of the Society of Amateur Radio Astronomers (SARA) who are interested in SETI. Even a satellite TV reception system (with only a few modifications) and a home computer can make a low-cost SETI complex (Ayotte, 1988). In the case of an antennae deficit, it is necessary to eavesdrop, at least, on the Moon and in opposite directions from nearby stars favorable for life. For example, an intriguing radio emission has been received by amateur equipment (Gray, 1985) just in the opposite direction from ρ^1 Cnc star (its spectrum: G8V ; distance: 14 pc). The narrowband frequency channels are not necessary in the discussed case because the high amplitude signal is expected near the probe. Apparently the most promising frequency range for a probe search is the terrestrial microwave window (1-10 GHz; Oliver, 1977), especially its high-frequency part.

It is unlikely that the communicating radio emission of the relativistic($V_\perp \sim 0.1c$) probes can be found during the sky monitoring. Indeed, for a limit of t<20 yr and w = 0.9, it follows from Eqs. (1) and (2):$N \geqslant G_e r/1.5$ AU The average registration distance of probes (50-kW transmitter; $G_p = 10^6$) isotopically dispersed in space is $r=10^3$AU for the "Obzor" flux limit of 10^{-19} W/m^2. For $G_e=1$:$N \geqslant 667$. The NASA Targeted Search will have the flux limit $\sim 10^{-27}$ W/m^2 but $G_e \sim 10^6$ (Martin, 1983). For this case: $r=10^7$ AU and $N \geqslant 6.7 \times 10^{12}$. Such high numbers of alien probes do not appear to be acceptable. Nevertheless, the radar emission of probes could be found. That is why the reports about the unidentified narrowband radio flashes form space (Dixon, 1985; Amy, Large and Vaughan, 1989) could be of interest for SETI explorers.

To summarize, the radio search for nearby alien probes is a promising task not beyond the ability of amateur radio astronomers and "Obzor"-like systems.

References

Amy, S.W., Large, M.I., and Vaughan, A.E.: 1989, Proc. Astron. Soc. Aust., 8, 172.
Arkhipov, A.V.: 1989, 40th IAF Congress, Paper: IAA-89-649.
Ayotte, J.: 1988, Signals, The NAAPO Newsletter, 4, No. 4, 5.
Bagby, J.P.: 1981, J. Brit. Interplanet. Soc., 34, 289.
Bracewell, R.N.: Nature, 186, 670.
Deardorff, J.W.: 1986, Quart. J. Roy. Astron. Soc., 27, 94.
Dixon, R.S.: 1985, in Proc. IAU Symp. No.112, ed. M. Papagiannis, Reidel, Dordrecht, 305.
Freitas, R.A.: 1983a, Icarus, 55, 337.
Freitas, R.A.: 1983b, J. Brit. Interplanet. Soc., 36, 490.
Gray, R.H.: Sky and Telesc., 69, 354.
Krupenio, N.N.: 1971, Radar Research of the Moon, Nauka, Moscow (in Russian).
Ksanfomaliti, L.V.: 1981, in A Problem of Search for Extraterrestrial Civilizations, Nauka,
    Moscow 55 (in Russian).
Lawton, A.T., and Newton, S.J.: 1974, Spaceflight, 16, 181.
Lunan, D.: 1973, Spaceflight, 15, 122.
Martin, A.R.: 1983, J. Brit. Interplanet, Soc., 36, 239.
Oliver, B.M.: 1977, in SETI, eds., Morrison, P., Billingham, J., NASA SP-419, Washington, 63.
Papagiannis, M.D.: 1985, Nature, 318, 135.
Suchkin, G.L., Tokarev, U.V., Luk'anov, L.G., and Shirmin, G.I.: 1981, in A Problem of Search
    for Extraterrestrial Civilizations, Nauka, Moscow 138 (in Russian).
Troitskii, V.S.: 1986, in A Problem of Search for Life in the Universe, Nauka, Moscow 227 (in Russian).

KARHUNEN-LOÈVE VERSUS FOURIER TRANSFORM FOR SETI

Claudio Maccone
'Giuseppe Colombo' Center for Astrodynamics
c/o Istituto di Fisica Matematica 'Joseph Louis Lagrange'
Via Carlo Alberto, 10 - I-10123 Torino, Italy

1. INTRODUCTION

Recovering weak signals out of noise is the central problem of experimental SETI. Most searches to date have been for narrowband signals, and, consequently, only Fourier transform techniques have been used. However, from a theoretical point of view, this is not the only possible approach. A given stochastic process $X(t)$ can be either studied by the classical Fast Fourier Transform (FFT) or by another more general transform, named Karhunen-Loève (K-L) after its discoverers (refs. [1], [2], [3]), and reading

$$X(t) = \sum_{n=1}^{\infty} Z_n \, \phi_n(t) \qquad (0 \leq t \leq T). \qquad (1.1)$$

Assuming that the noise (auto)correlation $E\{X(t_1)X(t_2)\}$ is a known function of t_1 and t_2, it can be proved that the functions $\phi_n(t)$ ($n = 1,2,...$) are the eigenfunctions of the correlation, namely the solutions to the integral equation

$$\int_0^T E \, \{X(t_1)X(t_2)\} \, \phi_n(t_2) \, dt_2 = \lambda_n \, \phi_n(t_1) \, . \qquad (1.2)$$

These $\phi_n(t)$ form an orthonormal basis in the Hilbert space, and they actually are the best possible basis to describe the noisy signal, better than any classical Fourier basis. One can thus say that the K-L transform adapts itself to the shape of the signal, whatever it is.

The first SETI radioastronomer to point out various advantages of K-L over FFT for detecting wideband signals appears to have been François Biraud in 1983, ref. [4].

A further advantage of K-L is that the Z_n in (1.1) are orthogonal random variables, i.e., that $E \{Z_m Z_n\} = \lambda_n \, \delta_{mn}$. If $X(t)$ is a Gaussian process, this orthogonality amounts to statistical independence, meaning that the terms in the K-L expansion are uncorrelated. Since the constants λ_n are both the (all positive) eigenvalues and the variances of the random variables Z_n, any K-L expansion, when truncated to keep only the first few terms (corresponding to the largest eigenvalues ordered in decreasing magnitude), is the best approximation to the full K-L expansion in the mean square sense. This property immediately suggests a SETI application : use the first ('dominant') eigenvalue as the first natural threshold for rejecting false alarms.

Finally, the mathematical theory of K-L transform shows that the process X(t) need not be stationary. This too spells the difference against the classical Fourier techniques, which hold rigorously true for stationary processes only.

At this point one might well wonder what prevents radioastronomers from using K-L rather than FFT for SETI now. The answer is: the computational burden only. In fact, since the transform kernel is the correlation, this kernel is obviously not separable, and thus, in general, one cannot hope for the existence of a fast K-L algorithm. In other terms, the computer time required to calculate the eigenvalues and eigenvectors of a correlation matrix of order N is proportional to N^2, rather than to N log N as for FFT. Nevertheless, the steady improvements in computer hardware and parallelization techniques seem to give hope for a possible replacement of FFT by K-L in perhaps some years time.

This paper is devoted to explain the state-of-the-art in K-L research, with an aim to apply known results to SETI. In particular we consider:

(a)   The possibility, for stationary processes, that a fast K-L transform may exist.

(b)   The fast K-L algorithm already obtained by A. K. Jain in 1976 for a class of random processes used in digital picture compression.

(c)   The work by the author about the K-L transform of time-inhomogeneous Gaussian processes, and its relationship to special relativity.

2. K-L TRANSFORM FOR STATIONARY RANDOM PROCESSES

An interesting topic in K-L research, which might turn out useful for SETI applications, is the solution to the integral equation (1.2) that Srinivasan and Sukavanam (refs. [5], [6]) obtained for stationary random processes, namely for a correlation having the form

$$E\{X(t_1)X(t_2)\} = f(|t_1 - t_2|) \tag{2.1}$$

the right-hand side being an arbitrary real-valued function f(...) defined on the positive real axis. These authors assume that f(...) admits a Laplace Transform f*(...) and in practice confine themselves to the case where the latter is given by

$$f^{\star}(z) = \frac{g(z)}{h(z)} = \frac{\text{polynomial of degree not exceeding } (n-1)}{\text{polynomial of degree n}} \tag{2.2}$$

though they state that their arguments can be easily extended to the more general case where f*(...) admits a Mittag-Leffler expansion. For the case (2.2), they give explicit (complicated) formulas for computing the eigenvalues numerically, but apparently not for computing the eigenfunctions. This prevents further study to be carried on about the possibility that a fast K-L algorithm might exist for the stationary correlation (2.1). For SETI applications, it might be more useful to replace f(...) by some particular function of $|t_1 - t_2|$ and try to obtain explicit formulas for both the eigenvalues and the eigenfunctions. The topic is open for investigation.

3. THE FAST K-L ALGORITHM OF A. K. JAIN

The method of the K-L expansion was introduced into the realm of pattern recognition by Watanabe in 1965, ref. [7]. The goal of the approach is to represent a picture in terms of an optimal coordinate system. Among the optimality properties is the fact that the mean-square error introduced by truncating the expansion is a minimum. The set of basis vectors which make up this coordinate system is sometimes referred to as eigenpictures. They are simply the eigenfunctions of the covariance matrix of the ensemble of pictures. A recent paper discussing the state-of-the-art in this field (ref. [8]) is significant inasmuch as it proves that powerful enough computers already exist to apply K-L techniques to image processing, hence one is led to infer that it might be so for SETI too.

In 1976 A. K. Jain succeeded in finding the first fast K-L algorithm (ref. [9]), an excellent discussion of which is given in ref. [10]. The key idea is to make the correlation separable by resorting to exponential functions. For instance, let a picture, belonging to a given set of pictures (random field) be sampled on a NxN square sampling lattice, and let $f(m,n)$ denote the samples, where both m and n take integer values from 0 through N-1. Then the assumed correlation is of the type

$$E\{f(m,n)f(p,q)\} = \rho_1^{|m-p|} \rho_2^{|n-q|} = e^{-\alpha|m-p|} e^{-\beta|n-q|} \tag{3.1}$$

where ρ_1, ρ_2, α and β are constants, the former two being less than unity. For this (discrete) correlation both the K-L eigenvalues and eigenfunctions may be explicitly found, as in ref.[10]. Jain has shown that, if the image boundary pixels are known, they may be used to modify the rest of the image in such a way as to possess a K-L transform that can be implemented using FFT (or the more recently developed fast sine transform). Thus, Jain's result is essentially a reduction of K-L to FFT preserving the typical advantages of both.

4. K-L TRANSFORM OF SIGNALS EMITTED BY A RELATIVISTIC OBJECT

While studying the K-L transform of time-rescaled (time-inhomogeneous) Gaussian processes (refs. [11], [13]), this author found the relationship between K-L transform and the theory of relativity (ref. [12]). The key idea is that the ordinary Brownian motion $B(t)$ with variance t and initial condition $B(0) = 0$, can be rescaled in time by an arbitrary law $f(t)$ thus producing the new time-rescaled Gaussian process

$$X(t) = B\left(\int_o^t f^2(s)\, ds \right). \tag{4.1}$$

The K-L expansion of this new process was proved in ref. [11] to read

$$X(t) = \sum_{n=1}^{\infty} Z_n N_n \sqrt{f(s) \int_o^t f(s)\, ds} \ J_{v(t)} \left(\gamma_n \frac{\int_o^t f(s)\, ds}{\int_o^T f(s)\, ds} \right) \tag{4.2}$$

where the Z_n are Gaussian random variables with mean zero and variance λ_n, N_n are suitable normalization constants, $J_{v(t)}$ is the Bessel function of the first kind and (time-depending) order $v(t)$, and the constants γ_n are the zeros of certain linear combinations of the Bessel functions and their derivatives; see ref. [11] for details.

The relationship of these K-L results with the theory of relativity lies in the formula

$$f^2(t) = \sqrt{1 - \frac{v^2(t)}{c^2}}$$ (4.3)

where $f(t)$ is the time-rescaling function and $v(t)$ is the velocity of the moving body measured in the rest-frame, i.e., on the Earth.

Consider, for instance, an alien probe emitting signals and travelling close enough to the Solar System to be detected. The K-L expansion of the Gaussian noise (Brownian motion) in which the signal is embedded is given by (4.2) and (4.3) with the velocity $v(t)$ replaced by the constant probe velocity v_p. A few reductions show then that the Bessel function $J_v(...)$ in (4.2) reduces to a sine, so that

$$X(t) = \sum_{n=1}^{\infty} Z_n \sqrt{\frac{2}{T}} \; \sin\left[\left(n\pi - \frac{\pi}{2}\right) \frac{t}{T}\right] .$$ (4.4)

The important point here is that, in changing from the Earth reference frame to the inertial reference frame of the relativistic spacecraft, the random variables Z_n just change their variance according to the law

$$\text{New } Z_n = \left[1 - \frac{v_p^2}{c^2}\right]^{1/4} \times \text{ Old } Z_n$$ (4.5)

whereas the eigenfunctions of the K-L expansion keep just the same.

This K-L scenario is more complicated in case the motion of the relativistic spacecraft is not uniform, i.e., the changing of reference frame is not inertial. For instance, in the specific case of the proper-constant-acceleration "Ramjet" proposed in 1960 by R. W. Bussard (ref. [14]), the following K-L expansion was proved in ref. [15]

$$B\left[\frac{c}{g} \ln\left(2\frac{g}{c} t\right)\right] = \sum_{n=1}^{\infty} Z_n \frac{1}{\sqrt{T} \, | \, J_o\left(\gamma_n\right)|} \; J_o\left(\gamma_n \frac{\sqrt{t}}{\sqrt{T}}\right)$$ (4.6)

where g denotes the proper constant acceleration. By resorting to the asymptotic expression of the Bessel function of the first kind for large values of its argument, the approximation to (4.6) is obtained

$$B\left[\frac{c}{g} \ln\left(2\frac{g}{c} t\right)\right] \approx \sum_{n=1}^{\infty} Z_n \frac{1}{T^{1/4} t^{1/4}} \; \cos\left(\gamma_n \frac{\sqrt{t}}{\sqrt{T}} - \frac{\pi}{4}\right) .$$ (4.7)

It should be noticed that these K-L results may apply whenever the emitting 'object' moves at a relativistic speed with respect to the Earth. For instance, consider the Quasars, whose spectra exhibit such a huge redshift that, if they are regarded as cosmological objects, must be moving away from us at a relativistic speed. Now suppose we wish to detect possible 'intelligent' signals from Quasars (Kardashev Type III civilizations?). Then we only need to fix up a suitable mathematical expression for v(t), evaluate the corresponding f(t) from (4.3), and then use (4.2) accordingly.

Further results in ref. [12] that might be useful for SETI deal with the total energy of the stochastic process X(t) over the integration time ($0 \leq t \leq T$), i.e., the random variable

$$E = \int_0^T X^2(t) \, dt \tag{4.8}$$

Finally, the K-L expansion of the instantaneous energy $X^2(t)$ was also obtained in full detail (refs. [16],[17]).

4. CONCLUSIONS

Though the number of applied scientists using the K-L transform for their research is slowly increasing, one must admit that this transform is still outside the realm of most current scientific research. This seems to be due to a couple of reasons at least: the exceedingly heavy computational burden that K-L involves, and the many obscure points still plaguing the relevant mathematical theory. Whereas the first obstacle might be overcome relatively soon by the development of parallel computers, paving the mathematical way needs a considerable effort.

SETI is a research field where the K-L transform might reveal all its power against FFT : only K-L, in fact, would reveal wideband signals whatever the noise spectrum is, and whether or not the random process is stationary.

It seems to be high time for the K-L transform to be taken seriously by the SETI investigators all over the world.

References

[1] K. Karhunen, "Über lineare Methoden in der Wahrscheinlichkeitsrechnung," Ann. Acad. Sci. Fennicae, ser. A 1, Math. Phys., Vol. 37 (1946), 3-79.

[2] M. Loève, "Fonctions Aléatoires de Second Ordre," Rev. Sci., 84 (1946), No. 4, 195-206.

[3] M. Loève, Probability Theory, Princeton, NJ: Van Nostrand, 1955.

[4] F. Biraud, "SETI at the Nançay Radiotelescope," Acta Astronautica, Vol. 10 (1983), 759-760.

[5] S. K. Srinivasan and S. Sukavanam, "Photo-count Statistics of Gaussian Light of Arbitrary Spectral Profile," J. Phys. Vol. A-5 (1972), 682-694.

[6] S. K. Srinivasan, "Stochastic Point Processes and Their Applications," Griffin, London, 1974, in particular pages 83-87 and 95-98.

[7] S. Watanabe, "Karhunen-Loève Expansion and Factor Analysis Theoretical Remarks and Applications," Proc. 4th Prague Conf. Inform. Theory, 1965.

[8]   M. Kirby and L. Sirovich, "Application of the Karhunen-Loève Procedure for the Characterization of Human Faces," IEEE Trans. on Pattern Analysis and Machine Intelligence, Vol. 12 (1990), 103-108.

[9]   A. K. Jain, "A Fast Karhunen-Loève Transform for a Class of Random Processes," IEEE Trans. Commun. COM-24, 1976, 1023-1029.

[10]   A. Rosenfeld and A. C. Kak, "Digital Picture Processing," Second Edition, Volume 1, Academic Press, New York, 1982, in particular Chapter 5 and pages 132-133.

[11]   C. Maccone, "Eigenfunctions and Energy for Time-Rescaled Gaussian Processes," Bollettino dell'Unione Matematica Italiana, Series 6, Vol. 3-A (1984), 213-219.

[12]   C. Maccone, "Special Relativity and the Karhunen-Loève Expansion of Brownian Motion," Nuovo Cimento, Series B, Vol. 100 (1987), 329-342.

[13]   C. Maccone, "The Time-Rescaled Brownian Motion $B(t^{2H})$," Bollettino dell'Unione Matematica Italiana, Series 6, Vol. 4-C (1985), 363-378.

[14]   R. W. Bussard, "Galactic Matter and Interstellar Flight," Astronautica Acta, Vol. 6 (1960), 179-194.

[15]   C. Maccone, "Relativistic Interstellar Flight and Gaussian Noise," Acta Astronautica, Vol. 17, No. 9 (1988), 1019-1027.

[16]   C. Maccone, "The Karhunen-Loève Expansion of the Zero-Mean Square Process of a Time-Rescaled Gaussian Process," Bollettino dell'Unione Matematica Italiana, Series 7, Vol. 2-A (1988), 221-229.

[17]   C. Maccone, "Relativistic Interstellar Flight and Instantaneous Noise Energy," Acta Astronautica, Vol. 21, No. 3 (1990), 155-159.

Discussion

F. BIRAUD: The K-L transform is the way of being "prepared for the unexpected", as Sebastian's cookie says (see discussion following panel discussion on "Selection Criteria in Bioastronomy," chaired by I. Almar.)

C. MACCONE: Quite true. We actually know nothing about the nature of ETI signals: we have just made a set of "reasonable" (to us) hypotheses, and are trying to see whether they are able to put us in touch with the rest of the universe. Enlarging this set of technical hypotheses, such as shifting from FFT to K-L, can only be of help.

L. DOYLE: Does the reduction of the Karhunen-Loève technique to an FFT also require power of 2 data?

C. MACCONE: No, at least in the case considered by A. K. Jain. His main achievement is just to have shown a way to maintain the advantages of K-L while reducing the computational burden to $Nlog_2N$ for the special class of random processes he considered. Applying his results to SETI data should be a major goal to pursue.

S. LIKHACHEV: It is known, that for realizing of optimal results which are obtained by K-L transform (for example, root-mean-square error minimization) the zero-point mathematical expectation of your input data assumption is necessary. How can we obtain zero point mathematical expectation for SETI observations without loss of information?

C. MACCONE: Thank you for raising this important question. In the formal mathematical development of K-L techniques, such as those outlined in Section 3 of the present paper, the zero-mean requirement is of no harassment, but it _is_ for digital applications such as those needed for SETI. To my knowledge, no work has been done to limit the loss of information. Topics like this are still to be adequately studied for the K-L transform.

D. WERTHIMER: Is the K-L transform as good as the Fourier transform for detecting the "standard" sinusoidal wave buried in Gaussian noise?

C. MACCONE: Yes. In fact non-rescaled Gaussian noise is the same as Brownian motion. Thus, we must turn to the K-L eigenfunctions of Brownian motion, and these are just sines. This is the same as stating that the K-L expansion is the same thing as the Fourier expansion. Please notice that, in case the Brownian motion is rescaled in time, as outlined in Section 3 of the present paper, the K-L eigenfunctions are not sines any longer (they are Bessel functions), so K-L and Fourier differ, and K-L is the optimal transform, whereas Fourier is not.

A. LEGER: For those of us who are non-specialists of signal processing, could you give us examples of signals that can be detected this way but for nonspecialist of signal processing?

C. MACCONE: Only very few examples of signal detection by means of K-L have been developed because of the mathematical difficulties involved with finding the eigenvalues and eigenvectors of the (auto)correlation. One very simple case is that of white noise: since the autocorrelation is Dirac's delta function, any set of eigenfunctions is good. Another simple case is Brownian motion, for which the correlation is the minimum of the two time values, and, in this case, the eigenfunctions are sines. A further case is that of an exponential correlation, for which both eigenvalues and eigenfunctions can be determined. Not many more cases have been developed to my knowledge. Research work in this direction is needed.

D. CUDABACK: Has the K-L transform been applied to description of speech or music?

C. MACCONE: Not to my knowledge. As I mentioned, K-L is used in digital picture processing and data compression, but it seems clear that the majority of scientists and engineers may not be aware of the advantages of K-L over other transforms of non-statistical nature. Actually K-L is peculiar, just in that it is the only known _statistical_ procedure that emulates, and indeed surpasses, all the _deterministic_ transforms devised so far for technical purposes. When the computational burden involved by K-L is eventually overcome, this statistical procedure will likely develop into a standard investigation tool for a number of disciplines.

ON THE STRATEGY OF SETI

L. N. Filippova
"Orlyonok" Observatory, Krasnodarski krai, USSR

N. S. Kardashev
Astrospace Center, Moscow, USSR

S. F. Likhachev, and V. S. Strelnitskij
Astronomical Council, Academy of Science, Moscow, USSR

Paper Presented by V.S. Strelnitskij

ABSTRACT

Arguments in favor of narrow-beamed SETI signals transmitted by "civilization-senders" (CS) directly towards target stars are revealed. This consideration gives us, as a "civilization-receiver" (CR), the grounds to search near the ecliptic and around the moment when the candidate CS is in astronomical opposition with the Sun. A list of 29 candidate stars, with their dates of opposition, are presented for desirable international patrol observations.

CONVERGENT SEARCH STRATEGIES

During the 30 years following the pioneering work of Cocconi and Morrison (1959), the importance of rational strategies for SETI has become more and more clear. Without well-grounded strategies and without ideas of where, when, and how to search, SETI researchers may be doomed to turn the "SETI haystack" forever if one considers the huge the size of the "haystack".

Of course, we could abandon the passive search strategy completely and try to attract "their" attention. Such an active strategy was suggested two centuries ago here on French soil by Charle Gros, who proposed to construct a huge mirror and send the reflections of sunbeams to the presumed inhabitants of Mars. One of the last proposals of this kind was suggested by academician Andrej Sakharov in response to the questionnaire of the 1971 SETI symposium in Byurakan (L. M. Gindilis, private communication). Sakharov, the father of the Soviet H-bomb and the courageous fighter for democracy and human rights, explained that one could make good use of H-bombs by placing them in a geometrically regular pattern in circumterrestrial space where they could be exploded simultaneously, or in regular time intervals, to attract the attention of extraterrestrials (ETs).

However, these apparently more active strategies tend, in fact, to involve a passive wait for participation by a more clever partner. They seem to underestimate the readiness of humans to discover efficient, "convergent" strategies of mutual searches. P. Makovetski (1977) was the first to discuss in an obvious manner the possibility and the logic of convergent strategies. He also suggested some fine ideas about the possible convergence of the CS and the CR by choosing the appropriate moment of time for contact and using supernova explosions as a time reference.

But the very first convergent search strategy was, in fact, Cocconi and Morrison's clever idea that the 21-cm hydrogen line would be a plausible convergence wavelength for communication between CS and CR.

A NEW CONVERGENT STRATEGY

Proceeding from possible energetic and ecological limitation on the angular size of signal beams, we have recently come to a new convergent strategy for SETI that would lead any CR to a preferred zone on the sky and a preferred time schedule for SETI observations (Filippova and Strelnitskij, 1988; Filippova, Likhachev, and Strelnitskij, 1990). Here we shall briefly review the major points of this strategy, give the list of stars for cooperative patrol observations, and suggest a schedule of observations.

We first proceed from considerations by Townes and Schwartz (1961), the authors of the well-known hypothesis of laser interstellar connection. They have shown that, from energetic considerations, reasonable beamwidths for lasers used for interstellar communications should be very narrow–perhaps as small as $\theta \sim 10^{-7}$ radian.

Our new idea, contrary to traditional views, is that radio signals from ETs could also be this narrow. Our conclusion is based on our own ecological experience and on a belief that a developed CS will try to maximize, as much as possible, both the effective isotropic power of its transmitters and the sensitivity of its receiving systems. With this approach, the aim of CS is to establish a dialog with a potential CR without relying too much on the technical possibilities of CR. CS could simultaneously achieve maximum capability for transmitting or receiving by increasing the size of a directional antenna.

As one would expect, antennas placed in space can be made much larger than comparable ground-based ones. Following an analysis of natural limits for the size of a space radio telescope (SRT), where the quantity of material required for construction, the effects of tidal deformations of the dish, and the energy requirements to steer very large dishes were all considered, we came to the conclusion that SRTs as large as 1000 km in diameter could be constructed. These large SRTs could, in principle, be powered and steered using only the radiative energy from the parent star, which would make them "ecologically clean". Such an SRT would have a beamwidth $\theta \sim 10^{-7}$ radian at decimeter wavelengths. The similarity in beamwidth between the laser and SRT systems is useful for the following analysis.

Shklovsky (1987) pointed out a specific difficulty with the use of narrow beams for interstellar communications: if the cross section of the beam at the location of the target star is smaller than the "habitable zone", and if CS is unable to observe the planets orbiting the target star, then it will be forced to scan the circumstellar area with its narrow beam. Here we introduce the "Shklovsky factor", which is a measure of the increase in time required to scan the "habitable zone" of the target star. For a search program containing stars within distances of 100 light-years, the mean value of the Shklovsky factor could be ≥ 100. Developing a simple theory for a mutual stochastic search (Likhachev and Strelnitskij, 1990), we include estimates of the Shklovsky factor and show that the probable time for finding each other will be very long, perhaps more that 1000 years.

We suggest, therefore, that CS may prefer another strategy: to send narrow-beamed signals directly towards the target star. It is true that CS will miss candidate CRs with this strategy; e.g., if CS does not lie in or near the ecliptic plane of CR's planet orbiting its Sun, the latter will never pass through the narrow beam of CS directed toward that star. But, in return, if a favorably positioned CR does exist among the list of candidates, then the chance of finding each other with this "toward the star" strategy is better than using the "scanning" strategy. Indeed, if CR correctly guesses that CS is using the "toward the star" strategy, and if the frequency of the transmission is also correctly guessed, the signal might even be detected with CR's first attempt. In this scenario, CR knows that its planet will pass through CS's narrow beam around the times of astronomical opposition and conjunction, i.e., when the CR planet crosses the line between its own star and the candidate CS's star. CR can then make a concentrated effort to detect CS's signal during these special time intervals, which can be relatively short.

A CONVERGENT SETI PLAN FOR EARTH

Thus a new SETI strategy is available to us: first we select candidate stars near the ecliptic plane, and then we observe them in sequence as a patrol of quasi-continuous observations that coincide with the times of oppositions of each of the candidate stars. Note that the times around conjunction are less desirable because our antennas would be pointed near our own Sun and that would degrade receiver performance. The duration for each patrol observation can be estimated by calculating the time the Earth moves through the hypothetical cross section of the CS beam. For $\theta \sim 10^{-7}$ radian and for CS candidates within 100 light-years from the Sun, a typical duration might be several days. Quasi-continuous patrol observations lasting several days require, of course, international cooperation because the patrol must be done in a "relay-race" sequence of observations involving several observatories distributed in longitude around the Earth.

Table 1 contains a list of near-by ($D \leq 25$ parsecs), sun-like stars that are apparently single (no close companion stars), and that lie near the Earth's ecliptic. The list was prepared by one of us (Filippova, 1990) as an extension from a previous list, which was limited to the north ecliptic longitudes to the entire ecliptic. The low values of ecliptic latitude of the stars listed, indicated in the third column, guarantee that the Earth would intersect a hypothetical signal beam from CS if θ is greater than 10^{-7} radian. Since the patrol around opposition may continue for several days, it is sufficient to know the opposition time with an accuracy of about 1 day. With these assumptions, the dates of opposition listed in Table 1 are appropriate for at least the next 20 years.

Table 1.

Number in Wooley, et al. (1970)	Ecliptic Coordinates Epoch 2000		Date of Opposition	Distance, pc
	Longitude	Latitude		
9009	1°03'	−9°14'	24 Sept	19
9027	11 04	−2 48	4 Oct	23
33	13 11	0 06	6 Oct	7
44	12 29	−7 19	6 Oct	14
9056	26 31	8 54	20 Oct	17
9105	49 18	9 43	12 Nov	21
188	77 31	−4 14	10 Dec	17
209	84 38	−2 36	17 Dec	18
222	88 41	−3 09	20 Dec	10
9199	91 17	11 57	23 Dec	22
252	102 29	2 32	3 Jan	19
315	129 19	−6 37	29 Jan	20
9322	154 18	−7 16	23 Feb	23
449	177 09	0 41	18 Mar	10
9396	−175 54	−1 35	25 Mar	22
9408	−171 52	−0 06	29 Mar	20
9553	−112 12	−6 46	29 May	25
652	−102 14	−5 44	9 Jun	14
9604	−92 32	−10 36	19 Jun	20
9630	−81 37	−2 30	30 Jun	23
722	−80 56	2 05	1 Jul	15
9673	−64 09	−5 25	19 Jul	16
9678	−60 00	10 25	23 Jul	25
796	−53 46	−5 14	29 Jul	14
811	−50 47	−8 40	2 Aug	22
9745	−41 57	−12 32	11 Aug	22
9786	−22 57	2 33	31 Aug	22
9824	−12 44	−6 14	10 Sep	17
9829	−8 09	−0 55	15 Sep	23

References

Cocconi, G., and Morrison, P., 1959, *Nature*, 184, 844.
Filippova, L.N., 1990, *Astron. Tsirk.*, no. 1544, 39.
Filippova, L.N., and Strelnitskij, V.S., 1988, *Astron. Tsirk.*, 1531, 31.
Filippova, L.N., Likhachev, S.F., and Strelnitskij, V.S., 1990, in preparation.
Likhachev, S.F., and Strelnitskij, V.S., 1990, *Astron. Tsirk*, in press.

Makovetski, P.V., 1977, *Astron. Zh.*, 54, 449.

Shklovsky, I.S., 1987, "Vselennaja, Zhisn', Razum," 6th ed., Nauka, Moscow (in Russian).

Townes, C., and Schwartz, R.N., 1961, *Nature*, 190, 205.

Wooley, R., Epps, E.A., Penston, M.J., and Pocok, S.B., 1970, "Catalogue of Stars Within Twenty Five Parsecs of the Sun," *Royal Obs. Ann.*, 5.

Discussion

D. BRIN: First off, my congratulations for an ingenious concept. The problem of how to deal with tight beams is perplexing. Question number one: what sorts of frequencies do you plan to cover? And finally, do you conceive a satellite in solar polar orbit which might extend such a surveillance program beyond the ecliptic?

V.STRELNITSKIJ: (1) Observations following ZODIAC strategy can be made at any SETI frequencies. ZODIAC reduces the "SETI haystack" only in space and time. (2) No limitations to extend the ZODIAC strategy beyond the near-elliptical zone exist. ZODIAC only points out the space and time "attractors" - zones of a priori higher probability to detect the extraterrestrials signals.

L.DOYLE: Brent Sherwood at Goddard and Al Betz at Berkeley have suggested that interstellar communication could be reasonably achieved using the natural 10 μm maser activity evident in Mars' and Venus' atmosphere. Could you comment on the appropriateness of this wavelength for SETI?

V.STRELNITSKIJ: I think that for nearby communications it is not worse than the other "SETI wavelengths" because it is "prompted" to "us" and to "them" by Nature.

D.WERTHIMER: I am very pleased that someone is successfully competing with the Serendip acronym. Two questions: (1) What is the spectral region you are searching, and (2) Is there a Russian equivalent to your acronym?

V.STRELNITSKIJ: (1) We have made preliminary searches of two Zodiac stars at several cm wavelengths with the RATAN-600 radio telescope and obtained the high resolution visual spectra of these stars with the 6-m telescope at Zelenchuk. (2) In fact, ZODIAC is not an acronym, it is the natural name of the near-elliptical band on the sky. I deciphered it as "Zone Obviously Designed for Intelligent Aliens and Communications" for fun only yesterday, and did not yet think about a Russian version.

J. TARTER: I apologize for leaving the ZODIAC program out of my list of SETI projects on the telescope. I am very pleased to hear that it is operational. Could you tell me what the radio frequency search on the RATAN telescope is called? Also what is the "AILITA" program for SETI?

V.STRELNITSKIJ: ZODIAC is not the name of some concrete program, but rather the name of a strategy aimed at observations at different wavelengths. AILITA is the name of the complex of devices that are now being constructed for specialized SETI observations, at 21 cm, at the children camp astronomical observatory at Tuapse, on the Black Sea shore.

PAN-GALACTIC PULSE PERIODS AND THE PULSE WINDOW FOR SETI

Woodruff T. Sullivan, III
University of Washington, Seattle, WA 98195

ABSTRACT

It is argued that more attention should be paid in SETI programs to the possibility of finding rationalized, preferred pulse periodicities, in the same sense that many have argued for preferred frequencies. Within the range of detectable pulse periods, which is from 10^{-6} to 10^5 sec, the *Pulse Window*, from ~0.1 to ~3.0 sec and defined by the histogram of observed pulsar periods, is suggested as a natural galactic communications channel for pulse-like signals. Within this window, the best candidates for preferred or "Pan-Galactic" periods are the 0.122-sec period associated with the lifetime of hydrogen's 2s state, and the 0.17-sec period of the "Solar Pendulum". Outside the Pulse Window, the best candidates are periods associated with the lifetimes of the muon (2.19703 μsec) and of the neutron (896 sec), with the solar oscillations mode-splitting (124 min), and with the sidereal day (23 hr 56 min). Many other Pan-Galactic periods are suggested and discussed, some of which are based on the idea of "master" pulsars that serve to establish preferred periods for communications in our locale of the Galaxy.

1. INTRODUCTION

The dimension of time is certainly the most psychological and mysterious of the various parameters associated with the multidimensional "phase space" through which we must conduct SETI, yet it is also highly scientific. Rhythm and resonance, two important aspects of time, are fundamental to virtually every aspect of the arts and sciences. We well know the effects of time in our lives and we know how to measure its intervals with great accuracy, yet we do not know what time is. In the fifth century St. Augustine nicely summarized this with his famous dictum:

> What then is time? If no one asks me, I know. If I wish to explain
> it to someone who asks, I know it not.

He became so frustrated with trying to understand the nature of time that, when asked "What was God doing before the Beginning of Time?", he replied "Preparing proper punishment for people raising such deep questions!" Our concepts and perception of time have profoundly changed through history, and still today are remarkably varied amongst different cultures. This makes the prospect of ever being able to find a commonality in time for *interstellar* discourse, indeed discouraging. Nevertheless, we boldly press on, for in SETI we have a basic faith that despite probable chasms between the biology and culture of the two parties in question, there will be enough science, technology, and mathematics in common that one party can at least deduce the presence of the other, and perhaps even understand a message.

Over the modern history of SETI, much effort has been expended on rationalizing our choice of radio frequencies, but surprisingly little attention has been paid to frequency's "conjugate parameter", time. Yet a SETI beacon designer can in general, for a given mean rate of power generation, choose to squeeze the available transmitter power into a continuous narrow-band signal or into a pulsed, broader-band signal. Despite this dichotomy, only eight of the sixty SETI projects tabulated to date by Tarter have involved looking for pulses of any kind (and none of them for any special values of pulse period) — virtually all of the others were looking for a steady, monochromatic signal. Yet we know that even a monochromatic signal can carry no information unless its intensity (or some other property) varies with time, in effect turning it into a pulse-like signal. Moreover, a steady signal emitted into a sweeping beam, which is an efficient compromise for a beacon designer between signal strength and the desire to reach many targets, also creates a pulse-like signal for any fixed observer. Pulses also have many advantages for detection; for instance, Oliver (1986) has long made cogent arguments for the advantages of pulses from a signal-to-noise point of view. Current work at the Ames Research Center also indicates that within the guidelines of the planned NASA Microwave Observing Program (MOP) pulses seem to require fewer computational resources to detect than monochromatic signals. Indeed, the MOP includes an extensive search for pulses.

In light of the advantages of pulses and their likelihood in SETI, the goal of this paper is to answer the following question: Are there preferred "Pan-Galactic" periods or ranges of periods that we can identify to aid our search for a beacon? Such universal periods should arise from time scales in physics or astrophysics in a natural way, and be rationalized in the same sense that the 21-cm hyperfine line of hydrogen yields a preferred frequency. A Pan-Galactic period should also be well defined and amenable to SETI with our present technical capabilities. Some periods may require an ETI's knowledge of a specific aspect of our sun and solar system, while others are more general; the former would be applicable only in the case when the sun is being targeted.

Another aspect of time in SETI is to formulate preferred *epochs* when we should be searching for a signal from any given target star; that is, dates of observation that can be coordinated by some astronomical phenomenon observed by both parties (an event such as a supernova, the positioning of planets or binary stars, etc.). There is not the space here to discuss this, but I plan to do so elsewhere.

2. GENERAL CONSIDERATIONS

Our first task is to define the overall range of SETI pulse periods that we can detect; we will then focus on specific values. On the fast-pulse end, the limits are set by the effects of the interstellar medium on propagation of pulse-like signals.[1] For a given column density of electrons (dispersion measure), there exists a minimum detectable pulse periodicity because of the smearing effects of (a) varying arrival times due to multipath scattering, and (b) dispersion, that is, varying arrival times at different frequencies across one's bandwidth [see Backer (1988) for details]. Both effects are far worse at lower frequencies. For the 1 to 10 GHz range of the MOP and typical galactic distances and electron densities, the minimum detectable pulse periods range from 100 μsec in the worst cases (dispersion measures of 100 pc cm^{-3} [distances of ~3 kpc] at 1 GHz) to 1 nsec or less for the best cases (nearby 10 GHz transmitters observed with high spectral resolution). For the present study, I adopt

---

[1]Kardashev (1979) has also briefly discussed the limitations on usable SETI pulse periods, although his formulation of the problem leads to different conclusions from my own.

1 μsec as the minimum detectable pulse period. For transmtters as far away as the galactic center (9 kpc), the minimum detectable period is ~2 msec at 1 GHz and ~20 μsec at 3 GHz. The *maximum* detectable period, on the other hand, is difficult to determine. In practice it would seem to be set for us by social considerations (funding, careers, patience, problems of archiving data), in which case it is probably of the order of days or at most months. I adopt 10^5 sec, about one day, as the maximum detectable period. Within the wide range of 10^{-6} to 10^5 sec, however, the shorter periods would seem to be favored by: (a) the need to transfer information at a reasonable rate, and (b) the "psycho-biological" factor that in general the shorter the time scale of any stimulus of a given intensity, the more our attention can be captured (down to a limit of ~0.1 sec).

3. THE PULSE WINDOW

The range of 10^{11} (!) for detectable SETI periods is overwhelming and one would like somehow rationally to narrow the search range for Pan-Galactic periods. We should seek some astrophysical phenomenon that defines a time range, ideally one involving radio waves, our chosen means of communication. Along this line, the pulsars present themselves as *the* outstanding class of natural, pulsed, radio signals. Furthermore, it turns out that the dispersion in their periods is not large – 95% of known pulsars fall inside the period range of 0.1 to 3.0 sec (Fig. 1) – and that the present knowledge of this period distribution is thought not to have strong observational biases (Taylor and Stinebring, 1986).[2] Moreover, humankind's "pulse of life" felicitously occurs right in this range – the human heart beat averages 0.8 sec; this in turn means that music, one of the fundamental aspects of all cultures, tends to have a similar cadence. I am led to define the *Pulse Window* from ~0.1 to ~3.0 sec and to propose it as a natural communications channel, a temporal forum for interstellar intercourse. Its characteristics are nicely tied to the astrophysics of our galactic neighborhood, the biology of our species, and the fundamentals of our global culture. In fact, it matches not just our species, but virtually all of advanced life on earth. A wide variety of characteristic physiological periods for a species (including nerve reaction time, metabolism cycling time, cardiac period, gestation period, and even life span) vary as $M^{0.25}$ for birds and mammals, where M is the average mass of the species (Linstedt and Calder, 1981). Thus a 2-gram shrew "runs" about eight times faster than a human and a 7-ton elephant about three times slower in all these respects.[3] The 0.1 to 3.0 sec range of the Pulse Window thus encompasses not only the human pulse, but also fortuitously matches the range of heartbeats from the shrew to the elephant.

The Pulse Window is also superbly suited to the MOP; for instance, current plans for the Targeted Search of nearby solar-like stars call for seeking pulse periods in the approximate range 0.1 to 75 sec. During this meeting, Bill Calvin pointed out a further advantage of the Pulse Window: A good general strategy for any species signalling to get the attention of another species is to mimic some

---

[2]In late 1990, there are now about 500 pulsars known, with: (a) a very similar overall histogram of periods to the 398 pulsars of Fig. 1, and (b) about 30 at shorter periods than shown in Fig. 1. Don Backer and Dan Steinbring have pointed out to me, however, that theoretical and experimental reasons suggest that more sensitive searches may eventually reveal a secondary peak (window) for periods either closely around ~1.5 msec or in the 1.5 to 10 msec range.

[3]This scaling also provocatively implies that all we mammals get an "allotment" of about 2 to 3 billion heartbeats in a lifetime!

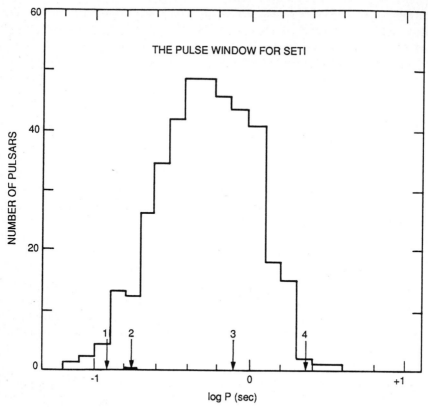

Figure 1. The Pulse Window for SETI, defined here by a histogram of periods for almost 400 pulsars known as of 1986 (Taylor and Stinebring, 1986); not shown here are about 30 known today of a very short period and about 100 more within the Pulse Window. The arrows indicate Pan-Galactic periods: (1) the 0.122-sec lifetime of the two-quantum $2s \rightarrow 1s$ transition in hydrogen, (2) the Solar Pendulum period of 0.17 sec (the bar indicates the range 0.15 to 0.18 sec for all FGK main sequence stars), (3) the ^{8}B lifetime of 0.772 sec, and (4) the 2.32-sec light-crossing time of the solar radius.

phenomenon that the first knows and cares about; the application to SETI is that an ETI might well send artificial pulses with periods in the range occupied by pulsars. Dixon (1973) has also argued that pulsar-like SETI periods are attractive to a sender in that they are more likely to be discovered by pulsar-hunting observers. His suggested range of SETI periods was 1 to 100 sec, on the basis of the slowest feasible rates of information transfer, that is, those that would yield the narrowest bandwidths. Even in their pioneer SETI paper, Cocconi and Morrison (1959) stated "we expect that the signal will be pulse-modulated with a speed not very fast or very slow compared to a second, on grounds of bandwidth and of rotations" [of planets].

Within the Pulse Window, we now seek exact values of Pan-Galactic pulse periods that should be closely scrutinized in SETI. There are several excellent candidates (see Fig. 1):

(a)   The mean lifetime[4] of 0.122 sec for the (forbidden) two-quantum 2s → 1s transition in the hydrogen atom (Spitzer and Greenstein, 1951). This transition contributes to a significant percentage of the continuum in ionized hydrogen regions and involves the hydrogen atom's fundamental Lyman-alpha energy states. It may well be as natural to think of this transition as defining the fundamental (and accessible) communications *time* associated with hydrogen as it is to think of the 21-cm hyperfine transition as defining the fundamental (and accessible) communications *frequency*.

(b)   The period of the "Solar Pendulum". A pendulum provides one of the most fundamental periodicities we know, its value depending on both its length and the local acceleration caused by gravity. For the length we naturally adopt the wavelength of the transmitted signal, say 21.1 cm. At the solar surface the period is then 0.17 sec, but remarkably the surface gravity of main-sequence FGK stars varies very little. The upshot is that the range of pendulum periods for "Good Suns" is only 0.15 to 0.18 sec. (The periods of planetary pendulums range from 0.6 to 1.5 sec, still within the Pulse Window. The Terrestrial Pendulum period is 0.92 sec.) For signals at other wavelengths λ, these periods should be multiplied by $(\lambda/21.1 \text{ cm})^{1/2}$.

(c)   The mean lifetime of 0.772 sec for decay of $^{8}\text{B} \rightarrow {}^{8}\text{Be} + \beta^{+} + \nu$. This is a link in the proton-proton fusion chains that power all low-mass stars. It is admittedly not one of the main links, but this particular decay provides the only well-defined time dealing with the nuclear astrophysics of the solar interior. This is a good marker if one is broadcasting for life made possible by the energy produced by main-sequence stars.

(d)   The light-crossing time of 2.32 sec across the solar radius. The radius would be inferred by an ETI from accurate observations of our sun's surface temperature and luminosity.

Note that some suggested periods could also be defined as a multiplicative constant $(2, \ln 2, 2\pi, \sqrt{2\pi},$ etc.) times the quoted value; in such cases I have chosen what seems the least ambiguous definition.

Besides defining the Pulse Window, pulsars can be used for SETI in other ways. Heidmann (1988) has suggested a scheme that uses pulsars to define a preferred SETI *frequency* for any specified target direction in the sky. It seems to me that it is far more satisfactory and straightforward to modify this scheme to assign natural SETI *periods*. Under this modification I suggest that for any target of interest one should search for SETI pulses whose period matches either that of the nearest pulsar on the sky or that of the nearest pulsar to the direction exactly opposite to the target. An ETI sends a signal bearing one of these periods because he too sees his target (the sun) lined up with one of these same two pulsars.

Another possible system employs a particular pulsar to establish a preferred period for all interstellar communications circuits in its locale of the Galaxy. Such a controlling, "master" pulsar could be defined, for example, in terms of its outstanding luminosity, its extremely short or long period, or simply its propinquity to the sender and/or transmitter. Excellent candidates for the greater solar neighborhood would seem to be the three known pulsars associated with supernova remnants, given their extreme luminosities, ages, and spin-down rates. The main trouble with any such scheme, however, is that pulsars do not radiate isotropically—their beaming is estimated to illuminate only 20-

---

[4]I define the mean lifetime of a spectral transition as the inverse of its Einstein A coefficient for spontaneous emission. I define the mean lifetime of decaying particles as the mean for an ensemble.

50% of the sky. This means that the entries in the pulsar catalog of any ETI will only partially match our own. Another potential problem is the slow increase of most pulsar periods as they spin down. Nevertheless, it would be worthwhile to pay attention, no matter which direction one is searching in a SETI program, to the periods in Table 1, many of which also fall in the Pulse Window. The pulsar 0529-66 in the Large Magellanic Cloud is especially nice because its great distance means that it does not suffer from a beaming problem: it should illuminate all observers within 5 kpc of the sun!

Table 1: Candidates for Master Pulsars

Period (msec)	Pulsar	Distance (kpc)	Characteristics
1.56	1937+21	2.4	shortest period known
1.61	1957+20	1	very short period, eclipsing binary
6.22	1257+12	0.3	nearest of millisecond period
33.1	Crab	2.0	large luminosity, supernova remanant
59.0	1913+16	5.2	best for general relativity tests
89.2	Vela	0.5	large luminosity, supernova remnant
150	1509-58	4.2	large luminosity, supernova remnant, largest $\dot{P}$
226	1929+10	0.08	nearest to sun
253	0950+08	0.09	second nearest to sun
957	0529-66	50	most luminous in LMC
4308	1845-19	0.5	longest period known

[Data mainly from Taylor and Strinebring (1986) and Manchester and Taylor (1981).]

4. OTHER PAN-GALACTIC PERIODS

Even though searches should be concentrated within the Pulse Window, we should not ignore any especially appealing candidates for Pan-Galactic periods that happen to fall outside, but still within the acceptable range of 10^{-6} to 10^5 sec. Here are the best ones I have found (see Table 2):

(a)   The mean lifetime of 896 ± 10 sec for the neutron, one of the three basic building blocks of the atoms and molecules that make up our universe.

(b)   The mean lifetime of 2.19703 μsec for the muon, the most abundant unstable particle in the interstellar medium, or in fact in the earth's atmosphere or any other atmosphere.

(c)   The period corresponding to the spacing of the innumerable modes of solar "five-minute" (actually 3 to 10 min) global oscillations. Such oscillations are now on the brink of being reliably detected by us from other stars. The 134 ± 2 μHz mode spacings correspond to a period of 124 ± 2 min (see Bahcall and Ulrich, [1988]).

Table 2: Pan-Galactic SETI periods in the 10^{-6} to 10^5 sec range

Period	Rationale	
2.20 μsec	muon	
18.3 μsec	2^{128} × Planck time	
1.6 - 4300 msec	master pulsars (Table 1)	
0.122 sec	H 2-quantum 2s $\rightarrow$ 1s	
0.138 sec	α^{-7} × Planck time (inverse of Bohr frequency)	
0.17 sec	solar pendulum	Pulse
0.15–0.18 sec	FGK V pendulums	Window
0.294 sec	α^{-20} × Planck time	
0.772 sec	^{8}B decay	
1.45 sec	e^{100} × Planck time	
2.32 sec	solar radius	
3.02 sec	2^{32} × Planck time (inverse of H hyperfine frequency)	
47.6 sec	[0 III] opical line	
896 sec	neutron	
43.3 min	Jupiter's orbit	
54.3 min	solar dynamical time	
2.07 hr	solar global oscillations	
3.12 hr	[0 I] far-infrared line	
1.54 hr	[0 II] optical doublet	
6.61 hr	[0 II] optical doublet	
23 hr 56 min	sidereal day	

(d)   The dynamical time scale for any massive system is $(G\,\rho)^{-1/2}$. The solar mean density ρ of 1.41 g/cm^3 then produces a value of 54.3 min.

(e)   The light-crossing time of 43.3 min across Jupiter's orbital semimajor axis. Jupiter is chosen because it is by far the dominant planet in our planetary system and would be detectable as such.

(f)   The mean lifetimes for the most luminous optical interstellar lines whose lifetimes fall in the acceptable range. These are perforce forbidden lines, and the lifetimes are 47.6 sec for the familiar 5007 Å [0 III] line and 1.54 hr and 6.61 hr for the 3727 Å [0 II] doublet. In addition, one of the two most luminous far-infrared interstellar lines, [0 I] at 63 μm, has a lifetime of 3.12 hr.

(g)   Earth's sidereal day of 23 hr 56 min. This is on the presumption that a neighboring ETI has detected our leakage radiation (which invariably will have our sidereal periodicity within it (Sullivan et al., 1978)) and is transponding signals back to us.

Finally, one should consider the Planck time $T_p = (G\,\hbar/c^5)^{1/2}$, a fundamental constant of nature, but so small (5 x 10^{-44} sec) that it is difficult to know how to use it! It is amazing that when Max Planck (1899) himself first set forward his fundamental units of time, mass, and length, he was so taken by their universality that he wrote:

> [These quantities are] independent of specific bodies or materials and
> necessarily maintain their meaning for all time and for all civilizations,
> even those which are extraterrestrial and nonhuman.

Fundamental periods such as the Planck time that fall outside the Pulse Window can of course always be brought within the desired range by a multiplication or division of powers of dimensionless constants such as the fine-structure constant $\alpha = e^2/\hbar c = 1/137.03599$, e, π, 2, or the ratio of the proton and electron masses. In general, however, I find this procedure too awkward and ambiguous to offer much hope that one can end up with a commonly reasoned Pan-Galactic period. Gruber and Pfleiderer (1989) have argued for such multiples based on 2 raised to powers that are themselves powers of 2. Starting with the time defined by the inverse of the 1420 MHz hydrogen line frequency ($t_H = 0.704$ nsec), they then define a basic time unit of $2^{32}\,t_H = 3.02$ sec as a preferred SETI period. This falls on the edge of the Pulse Window. Similarly, Frisch (1987) has suggested a preferred period defined by the inverse of the Bohr frequency of the hydrogen atom times a multiple of α^{-1}: $\alpha^{-7}\mu_B^{-1} = 0.138$ sec, which also falls in the Pulse Window. Adapting these ideas to the Planck time T_p, we find $T_p/\alpha^{20} = 0.294$ sec, $T_p e^{100} = 1.449$ sec, and $2^{128}\,T_p = 18.3\ \mu$sec.

Perhaps a more useful application of dimensionless constants is in conjunction with the primary Pan-Galactic periods. A good strategy for a transmitter sending a period P would be to interlace this period with other sub-periods or longer periods, perhaps of lower intensity and bearing more information. These derivative periods might have values such as $2 \times P$, $2^n \times P$, $\pi \times P$, $e \times P$, P/α, or their inverses. Along this line, Von Hoerner (1974) has argued that our musical scales may have some degree of universality because there only exist a few workable tempered, harmonic scales. He showed that only by dividing an octave into 5, 12, 31, 270 ... equal parts could one achieve accurate harmonic ratios of higher and higher order. Although these ideas can be more directly applied to SETI *frequencies*, they can also work for SETI pulses by using: (i) subpulses with periods 1/5, 1/12, ... times P; (ii) signals that split a period into ratios such as 5/4, (a musical third) or 4/3 (a fourth); or (iii) periods having ratios with P such as 5/4, 4/5, 4/3, or 3/4. This would indeed be the Music of the Spheres!

5. BIOLOGICAL PERIODS

If we assume that extraterrestrial life might be similar to ours, could biological periodicities be a source of exact Pan-Galactic periods for SETI? Timing mechanisms are fundamental to all organisms on earth, and many of the associated periods are driven by astronomical phenomena. For instance, the profound imprint on terrestrial life of 10^{12} day-night cycles is evidenced by ubiquitous circadian rhythms. But these biological clocks are never exact, as the prefix in their name indicates, having typically $\pm 10\%$ variations amongst different species or even within the same species under various conditions. Circalunar and circannual clocks behave similarly. In the end, despite searching for acceptable periods in a wide variety of biological and biochemical phenomena, I have not been able to find any viable candidates for exact SETI periods.

6. CONCLUSIONS

The time domain is critical for a rational SETI strategy. I suggest that special attention should be paid to the Pan-Galactic pulse periods suggested above (Table 2), just as we pay special attention to preferred frequencies. The Pulse Window seems a particularly fruitful resource to exploit, an ideal rendezvous in the time domain analogous to our use in the frequency domain of the Microwave Window or of the "Water Hole" between 18- and 21-cm wavelength. Several good pulse-period candidates lie within the Pulse Window, but the single best is the 0.122-sec lifetime of hydrogen's 2s state, which, like the 21-cm hyperfine line frequency, refers to a fundamental physical and astrophysical phenomenon associated with the most abundant element in the universe. The Solar Pendulum period of 0.17 sec (for 21-cm transmission) is also especially appealing as a fundamental time interval and therefore as a Pan-Galactic pulse period. Outside the Pulse Window, the strongest candidates are the lifetimes of the muon (2.19703 μsec) and of the neutron (896 sec), the solar oscillations mode-splitting (124 min), and the sidereal day (23 hr 56 min). But we have only just begun thinking about this problem, and undoubtedly further good candidates will emerge.

In thinking about SETI and time, I have profited from discussions with many people. In particular, I wish to acknowledge that it was Pat Thaddeus who suggested consideration of the lifetime of hydrogen's 2s state. I also thank George Lake, Adam Frank, Don Backer, Dan Stinebring, Craig Hogan, Wick Haxton, John Harvey, Jeff Wilkes, Bill Calvin, and Bruce Balick.

References

Backer, D.C., 1988, *Galactic and Extragalactic Radio Astronomy*, eds. G. Verschuur and K.I. Kellermann (New York: Springer), ~p. 480.
Bahcall, J.N., and Ulrich, R.K. 1988, Rev. Mod. Phys., 60, 297.
Cocconi, G., and Morrison, P., 1959, Nature, 184, 844.

Dixon, R.S., 1973, Icarus, 20, 187.

Frisch, D.S., 1987, unpublished manuscript.

Gruber, G.M., and Pfleiderer, J., 1989, Acta Astronautica, 19, 903.

Heidmann, J., 1988, Comptes Rend. Acad. Sci., 306, Ser. II, 1441.

Kardashev, N.S., 1979, Nature, 278, 28.

Linstedt, S.L., and Calder, W.A., III, 1981, Quart. Rev. Biology 56, 1.

Manchester, R.N., and Taylor, J.H., 1981, Astron. J., 86, 1953.

Oliver, B.M., 1986, *The Search for Extraterrestrial Intelligence*, eds. K.I. Kellermann and G.S. Seielstad (Green Bank: NRAO), p. 121.

Planck, M., 1899, Sitz. Ber. Preuss. Akad. Wiss., Fünfte Mitt., 440.

Spitzer, L., Jr., and Greenstein, J.L., 1951, Astrophys. J., 114, 407.

Sullivan, W.T., III, Brown, S., and Wetherill, C., 1978, Science, 199, 377.

Taylor, J.H., and Stinebring, D.R., 1986, Ann. Rev. Astron. Astrophys., 24, 285.

von Hoerner, S., 1974, Psychology of Music, 2, 18.

Discussion

E. OLSEN: Only a comment – the preferred periodicity range you identified (0.16 → 4.6 sec) will be nearly totally covered in the Targeted Search Pulse Detection of the NASA MOP. That detector will cover ≈1/2 sec → 300 sec. The sky survey will be able to search for pulses over range 0.1 sec to ≈1/2 sec.

W. SULLIVAN: I would hope that the Targeted Search could also be designed to look at the special pulse period corresponding to A^{-1} for the 2s → 1s two-quantum transaction in hydrogen, namely 0.122 sec.

F. BIRAUD: The period of a pulsar could be used to define both the carrier frequency and the pulse period. This opens interesting possibilities in phasing the modulation relative to the carrier.

W. SULLIVAN: I think the greatest value of Jean Heidmann's 1988 paper (*C.R.A.S.*) is the general idea of using pulsars as an aid for SETI. It seems to me, however, that it is much more direct to have pulsars imply the pulse *periods* that we should seek, rather than the radio *frequencies* we should use, since the latter logic requires several more inferential steps.

COSMIC BACKGROUND RADIATION LIMITS FOR SETI

S. Gulkis

Jet Propulsion Laboratory, California Institute of Technology, Pasadena, CA 91109 USA

Summary

Searches for extraterrestrial intelligence (SETI) depend ultimately on the noise introduced in the receiving system by the photon flux from cosmic background radiation. The major sources of the diffuse background radiation include galactic radiation at decimeter wavelengths and longer, 2.7 K radiation at millimeter wavelengths, interstellar dust at submillimeter wavelengths, and interplanetary dust and stellar emission at infrared wavelengths. Our knowledge of these has greatly improved due to recent measurements from the IRAS satellite. We give estimates for various background radiation components over wavelengths from 1 micrometer to 10 meters. Signal-to-noise estimates are presented for background limited linear and direct photon detection systems. Optimal wavelength regions for SETI are discussed. The discussion in this presentation follows a similar discussion given by C.H. Townes (Proc. Natl. Acad. Sci. USA, Vol. 80, pp. 1147-1151, February 1983).

Discussion

N. EVANS: Could you define more precisely what you mean by the microwave window?

S. GULKIS: The sum of three noise sources - 1) galactic synchrotron emission, 2) cosmic microwave background radiation, and 3) the spontaneous emission associated with coherent receivers - defines a broad spectral region from about 1 GHz to 60 GHz where background noise emission is at a minimum. This spectral region is known as the free-space microwave window. From the surface of the earth, the window is somewhat narrower due to the emission from the terrestrial atmosphere. The resulting minimum brightness region including the terrestrial atmosphere is known as the terrestrial microwave window. It extends from approximately 1 GHz to 10 GHz.

VLBI AND INTERSTELLAR SCATTERING TESTS FOR SETI SIGNALS

V. Slysh

Astro-Space Center, Lebedev Physical Institute, USSR

Summary

Assuming that a coherent extraterrestrial intelligent transmission can be considered as a point-like radio source, three tests are proposed to discriminate them against natural radio sources: The first test requires that strong scintillations be observed due to propagation through a turbulent interstellar medium. The second test assumes that the refractive interference should be present when the signals are refracted by larger scale interstellar turbulence. The third test involves very long baseline interferometry (VLBI) or space VLBI like the RADIOASTRON satellite to measure the angular size of the proposed SETI radio sources.

Discussion

I. ALMAR: Using space-VLBI looks very promising, in particular since there will be two dedicated satellites (Radioastron and VSOP) active in the second half of this decade. Nevertheless these satellites are for astrophysics and will observe extragalactic objects almost exclusively (quasar, AGN etc.). Do you think that appropriate objects for SETI search can be put on the observation program of these satellites?

V. SLYSH: Yes, there is a program committee which would accept reasonable proposals.

J. DREHER: If an ETI signal is narrowband, the group-delay methods of normal VLBI astrometry will not work. Will radio-astron allow phase-referenced VLBI measurements?

V. SLYSH: Yes, since there will be a phase-line between the ground-based and orbiting radio telescopes. The phase-referenced VLBI measurements will be employed for astrometric observations.

A TEST FOR THE INTERSTELLAR CONTACT CHANNEL HYPOTHESIS IN SETI

D.G. Blair*, R. Norris†, K.J. Wellington†,
A. Williams*, and A. Wright†

* Department of Physics, University of Western Australia, Nedlands, Western Australia 6009
† CSIRO Division of Radiophysics, Epping, New South Wales, Australia

Paper Presented by D. G. Blair

ABSTRACT

This paper describes a SETI research programme which commenced in 1990 in Australia. The project is intended to be a specific test for the existence of definable interstellar contact channels. The rationale for the choice of interstellar contact frequency is described. In the initial observations 100 targets were searched, down to a flux limit of 2 Jy.

1. INTRODUCTION

The Drake equation (1980) yields a relationship between the number of concurrent electromagnetically telecommunication civilizations (ETCs) in the galaxy, N, and their mean lifetime L. Optimistic scenarios for the number of suitable planets and for the inevitability of life, evolution and technology lead to several questions. If L is large, especially if indefinite survival of an ETC is possible, then N could be very large. For example, if L ~ 10^8 years we could expect the mean spacing between ETCs to be less than 50 light-years (Blair,1985). The question arises, where are they? Tipler has argued (1980) that since they have not shown themselves, we are alone: N=1.

An alternative hypothesis is that technological civilization is an extremely short-lived manifestation of the evolution process. Thus L is very small, and consequently N is very small. The nearest ETC could be a minimum of several thousand light-years away for L = 100 years.

A third possibility, on which this work is based, does not do away with the idea that L is large and consequently that N is very large. Instead it assumes that physical laws make interstellar travel impractical. They are out there, but they do not travel. They communicate by radio. Detectable signals may be expected on an interstellar beacon frequency. (Blair, 1986, Bracewell 1974). A search on an interstellar beacon frequency, of several hundred stars, will allow significant restrictions on the basic hypotheses.

The initial search described here is to our knowledge the first in the southern hemisphere (Papagiannis,1985). It covers 100 southern stars, and includes 40 nearby solar type stars south of declination 40°. The search tests specifically the interstellar contact channel hypothesis, and a null result will allow limits to be placed on at least one of the hypotheses, which are described below.

2. FIVE HYPOTHESES

To place the tests described here on a rigorous logical footing, we present five hypotheses which outline an optimistic and specific scenario for extraterrestrial intelligence in the galaxy. A modest observation program can significantly constrain the hypotheses, especially hypotheses 1, 4 and 5.

Hypothesis 1. Life and Technology : Inevitable and Abundant. That the evolution of life and, ultimately, technological civilization is an inevitable outcome of physical laws and occurs profusely throughout the galaxy.

Hypothesis 2. Temporal Mediocrity. That the Earth is mediocre in its time development and civilizations far advanced with respect to ours exist in the galaxy, it is not the first and the time range for formation is a moderate fraction of the age of the galaxy.

Hypothesis 3. Non-existence of Superscience and Hyperdrives. That the physical laws, energy sources and phenomena known to physics today represent a complete approximation to the laws of nature and that consequently interstellar travel is impractical. In particular the relationships between mass, energy, velocity and distance known today, can not be radically violated. "Completeness" implies the non-existence of radical new laws, while "approximation" refers to the fact that they are not necessarily exact, and that future science may refine our understanding. Thus while technological development may carry far beyond our own there is no radical new physics.

Hypothesis 4. Exploration by Data Exchange on Universal Contact Frequencies. That physical laws make interstellar exploration by spacecraft non-viable (Olivier,1981). That is, communication and exchange of information constitutes the one viable means of exploring life in the galaxy. Radio communication in the 1-20 GHz range is ordained by noise and energy considerations and interstellar contact channels can be universally defined.

Hypothesis 5. The Galactic Club Exists and Welcomes and Assists Members. That civilizations far ahead of us (say 500 million years ahead) have already developed communications and chosen beacon channels based on universality and logic. That target stars and planets are known through advanced astronomy and that beacons are directed at likely systems and are designed to attract ETCs at their most primitive stage. The goal is to spread and acquire knowledge and possibly even to assist emerging civilizations during dangerous phases of possible self-destruction.

3. CHOICE OF INTERSTELLAR CONTACT CHANNEL (ICC)

It has been frequently shown that the range 1-20 GHz represents the minimum energy band (MEB) for interstellar communication (Papagiannis, 1985). On the low side galactic synchrotron radiation creates a broadband background. On the high side the quantum limit and photon energy imposes an energy barrier, since each photon can convey a maximum of 1 bit of information. Water absorption limits the upper frequency to 10 GHz below the atmosphere. A logical choice of frequency, independent of anthropocentric bias can be defined as follows :

a) Choose a base frequency, f_B. It should be near or in the MEB, it should be universal, it should be the 'number 1' of our frequency scale. The natural choice is the 1420 MHz hydrogen line, F_H, since H is element number 1.

b) Choose a Civilization Signature Constant (C_{SC}). This should be non-integer so that no natural source can mimic the frequency through harmonic generation. The product $C_{SC} \times f_B$ defines the interstellar communication band, f_I modified only by possible Doppler shifts. The number π is the natural choice of C_{SC}, since it is fundamental to civilization, is universal, is irrational, and places the ICC in the middle of the MEB at 4.462336275 GHz. While π^{-1} could be equally universal, it translates the ICC to too low a value (0.45 GHz). Arguments could be made regarding other irrationals such as 'e' or the golden mean. They may be secondary choices but from our perspective π is the first choice.

c) It is worth considering the optimum choice for the type of beacon which might be used. An omni-directional beacon could be considered to reduce the numbers of antennas required, but this would be at an enormous energy cost. Such a beacon, however, would be unable to correct for local Doppler shifts, and hence would be less able to allow receiving civilizations to home-in on the ICC. It would also use very high powers. Secondly the beacon could be directed but time shared between targets. This would represent a huge loss of duty cycle, greatly reducing the probability of successful detection. The only efficient type of beacon, therefore is the directed beacon. While it requires multiple antennas, it minimizes energy usage and maximizes the probability of successful contact. For this reason we consider the directed beacon, corrected for local Doppler shifts from the source, to be the most likely choice.

d) We are unable to determine the correct choice of reference frame for Doppler corrections. In this present search we have chosen to observe in three frequency bands. The first assumes that the beacon is broadcast at f_I in the target barycentre reference frame. The second assumes that the beacon is broadcast in the solar barycentre reference frame. The third assumes sufficient knowledge of the earth's orbit, and the solar system distance to broadcast in the Earth geocentre reference frame. In future searches we intend to use also the reference frame defined by the relic cosmic microwave radiation. This universal frame could well be a much better choice as it is independent of both the source and receiver local conditions. Unfortunately the frame is not known to sufficient accuracy to be useful given the present bandwidth of our spectrum analyzers.

4. OBSERVATION

The Parkes radio telescope was used in a configuration as illustrated in Figure 1. The Parkes correlator was used to give two 512 channel, 100 Hz resolution spectrum analyzers. An additional spectrum analyzer HP 3582A provided another 256 channels with the same resolution. The latter was interfaced by GPIB to the Parkes VAX 750 to enable long averages, and the display of multiple spectra.

While all data was archived, the intention of this search was to undertake real time processing of all data, allowing immediate repeats for confirmation of possible spectral lines. Observations took place from 10.00 UT on 7 May 1990, to 2300 on 9 May 1990.

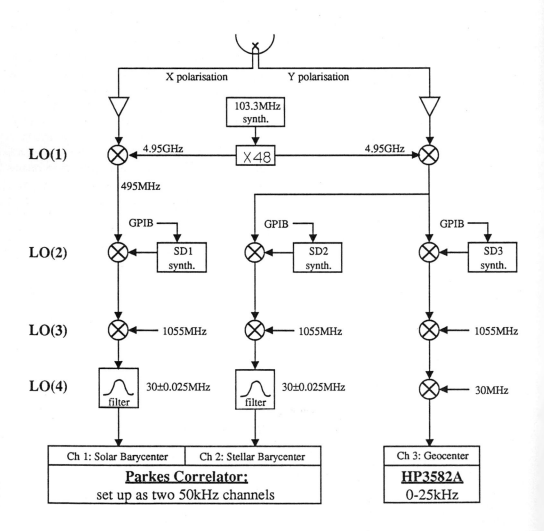

Figure 1: Set-up for SETI observations at 3 Doppler correction frequencies.

TARGET STARS

Target stars are chosen from the bright star catalog (Hoffleit) and are listed in Table 1. All were F,G, and K main sequence stars in the solar neighborhood. Reference spectra were taken several times in each 24 hour period, and these were used to normalize the observed spectra. The system noise temperature was consistently ~30 K, and the 3σ detection limit was less than 2 Jy for all sources examined. Table 1 lists the observational parameters of the 100 stars examined. Results are expressed as the threshold power detectable from a beamed narrow band beacon from a 1km dish. Threshold powers are within the range of terrestrial transmitters, as illustrated in figure 2. All stars were searched in 50kHz bands centered on the ICC in the target barycentre and solar barycentre reference frames. Most stars were also examined in a 25kHz band centered on the geocentre reference frame (see Table 1). Radial velocity data rounded to 1 km/s implied that the ICC should appear within 7 kHz of bandcentre.

The system sensitivity was tested by using standard calibrators, and the spectral analysis was tested using calibration signals injected at the feed. There was a surprising lack of interference and false alarms. Only one source yielded data worthy of repeat observation. A narrow line was seen while observing HR 6171 in the target barycentre reference frame. It was present for a period of about 1 hour, from UT1420 to UT1520 during 5 succesive observations on the 9th of May. In the following hour it was absent. It was initially thought to be interference from a geosynchronous satellite. However this is unlikely as it persisted while the RA of a satellite would have shifted 15°. It was absent when the telescope moved off source. Note also that while HR6171 has declination -2°19', a geosynchronous satellite by parallax appears well north of the equator. After an hour a new reference spectrum was taken, and thereafter the line was not seen. The line was not strong however, its amplitude was about 2Jy.

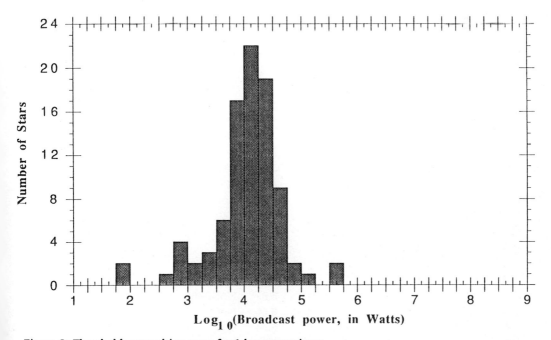

Figure 2: Threshold power histogram for 1 km transmitters.

Table 1: List of target stars, mostly taken from the Bright Star Catalog. Both the HD and HR catalog numbers are given, where known, as well as the radial velocities (RV) in km/sec. The distances (in parsecs), and the spectral type, where listed, are taken from 'Sky Catalogue 2000.0' (Hirshfeld). Asterisked stars were not searched in the Geocenter reference frame.

HR#	HD#	RA	DEC	VMAG	TYPE	Dist.	RV	IntTime	
33	693	0 11 15.8	-15 28 5	4.89	F7V	18	14	240	
77	1581	0 20 4.2	-64 52 30	4.23	F9V	7.1	9	240	
159	3443	0 37 20.6	-24 46 2	5.57	G8V	13	17	240	
176	3823	0 40 26.4	-59 27 16	5.89	G1V	20	2	360	
209	4391	0 45 45.5	-47 33 6	5.80	G1V	33	-11	240	
210	4398	0 46 11.7	-22 31 19	5.50	G3V	100	-15	360	
370	7570	1 15 11.1	-45 31 54	4.96	F8V	15	12	360	
486	10360	1 39 47.4	-56 11 53	5.86	K0V	-	23	240	
506	10647	1 42 29.3	-53 44 26	5.52	F8V	17	13	240	
509	10700	1 44 4.0	-15 56 15	3.50	G8V	3.6	-16	480	
512	10800	1 37 55.6	-82 58 30	5.87	G2V	17	-5	240	
650	13612	2 12 47.4	- 2 23 37	5.54	F8V	22	-3	240	
695	14802	2 22 32.5	-23 48 59	5.20	G1V	14	18	240	
810	17051	2 42 33.4	-50 48 1	5.41	G0V	17	16	240	
962	19994	3 12 46.4	- 1 11 46	5.06	F8V	18	18	240	
963	20010	3 12 4.2	-28 59 14	3.87	F8V	14	-21	240	
1006	20766	3 17 46.1	-62 34 32	5.54	G3-5V	13	12	240	
-	20794	3 17	-43 16	4.26	G5	6.2	87	240	
1010	20807	3 18 12.8	-62 30 23	5.24	G2V	12	12	240	
1084	22049	3 32 55.8	- 9 27 30	3.73	K2V	3.3	16	240	
1111	22713	3 39 1.0	- 5 37 34	5.96	K1V	21	40	240	
1232	25069	3 58 52.3	- 5 28 12	5.83	G9V	11	36	240	
1325	26965	4 15 16.3	- 7 39 10	4.43	K1V	4.3	-42	240	
1532	30495	4 47 36.2	-16 56 4	5.51	G1V	15	23	240	
1536	30562	4 48 36.3	- 5 40 26	5.78	F8V	22	79	240	
1538	30606	4 48 32.4	-16 19 47	5.77	F6V	23	35	480	
1608	32008	4 59 50.4	-10 15 48	5.38	G4V,K0III	110	-12	240	*
1665	33093	5 7 24.9	-12 29 26	5.97	F7V	-	50	240	
1674	33262	5 5 30.6	-57 28 22	4.72	F7V	13	-2	240	
1747	34721	5 18 50.4	-18 7 49	5.96	G0V	20	40	240	
2007	38858	5 48 34.9	- 4 5 40	5.97	G4V	24	29	480	
2022	39091	5 37 8.8	-80 28 9	5.65	G1V	15	9	240	
2086	40151	5 56 14.2	-22 50 25	5.96	K0V	10	34	240	
2186	42443	6 9 47.8	-22 46 27	5.71	F6V	25	22	480	
2261	43834	6 10 14.6	-74 45 11	5.09	G6V	8.7	35	240	
2313	45067	6 25 16.4	- 0 56 46	5.87	F8V,dF0V	33	44	240	
2400	46569	6 31 18.2	-51 49 34	5.60	F8V	19	16	240	
2667	53705	7 3 57.2	-43 36 29	5.54	G3V	13	86	240	
2866	59380	7 29 25.6	- 7 33 4	5.86	F8V	22	9	240	
-	61421	7 37	+5 21	0.37	F5V	3.5	-3	240	

HR#	HD#	RA	DEC	VMAG	TYPE	Dist.	RV	IntTime	
3018	63077	7 45 34.8	-34 10 23	5.37	G0V	17	103	240	
3064	64096	7 51 46.2	-13 53 53	5.17	G0V:	15	-18	240	
3259	69830	8 18 23.8	-12 37 55	5.98	G7.5V	11	30	480	
3677	79807	9 14 57.1	-37 36 9	5.86	G0V	-	-2	480	
3750	81809	9 27 46.7	- 6 4 16	5.38	G2V	15	54	480	
4158	91889	10 36 32.3	-12 13 49	5.70	F7V	23	-6	480	
4413	99453	11 25 43.2	-63 58 22	5.17	F7V,dF3V	22	-5	480	
4443	100286	11 32 16.0	-29 15 48	5.76	F8V	-	10	480	
4444	100287	11 32 16.3	-29 15 40	5.64	F8V	19	4	240	
4458	100623	11 34 29.4	-32 49 53	5.98	K0V	9.5	-23	480	
4488	101198	11 38 40.0	-13 12 7	5.48	F7V,dF5V	23	-24	480	
4523	102365	11 46 31.0	-40 30 1	4.91	G5V	10	15	240	
4548	103026	11 51 41.5	-30 50 5	5.85	F7V	26	33	240	
4788	109409	12 34 42.3	-44 40 23	5.77	G1V	33	18	480	
4903	112164	12 54 58.4	-44 9 7	5.89	G1V	38	32	480	
4979	114613	13 12 3.1	-37 48 11	4.85	G3V,dG5V	13	-15	480	
4995	114946	13 14 10.8	-19 55 51	5.33	G6V,dG5IV	30	-45	240	
5019	115617	13 18 24.2	-18 18 41	4.74	G6V	8.4	-9	240	
5257	122066	14 0 0.0	-25 0 37	5.77	F7V,F3V	14	-17	480	
5317	124425	14 13 40.7	- 0 50 44	5.91	F7Vw,F6IV	54	18	480	
5459	128620	14 39 36.2	-60 50 7	-0.01	G2V	1.3	-25	480	(α Cen)
5460	128621	14 39 36.2	-60 50 7	1.33	K1V	1.3	-21	240	
5698	136351	15 22 8.2	-47 55 40	5.00	F8V,dF5V	20	-11	480	
5699	136352	15 21 48.1	-48 19 3	5.65	G2-5V	16	-69	480	
5700	136359	15 23 10.3	-60 39 25	5.67	F7V,F5V	27	6	480	
5954	143333	16 0 19.5	-16 32 0	5.47	F8V	20	-25	240	
6060	146233	16 15 37.1	- 8 22 11	5.50	G2Va	17	11	240	
6094	147513	16 24 1.2	-39 11 35	5.40	G5V	15	10	480	
6098	147584	16 28 28.1	-70 5 4	4.91	F9V,G0V	12	9	480	
6171	149661	16 36 21.3	- 2 19 29	5.75	K2V,K0V	11	-15	2880	
6269	152311	16 53 25.1	-20 24 56	5.88	G3V,G5IV	22	-17	240	
6314	153580	17 3 8.6	-53 14 13	5.29	F6V,F8V	19	7	240	
6401	155885	17 15 20.9	-26 36 10	5.33	K1V	-	0	480	
6402	155886	17 15 20.7	-26 36 5	5.29	K0V	5.4	-1	480	
6748	165185	18 6 23.6	-36 1 11	5.95	G5V	14	13	480	*
6761	165499	18 10 26.1	-62 0 8	5.49	G0V,G3IV	23	30	240	
6907	169830	18 27 49.4	-29 48 59	5.92	F9V,G0V	19	-17	240	
6998	172051	18 38 53.3	-21 3 7	5.86	G4V,G5IV	34	36	240	
7161	175986	19 3 29.7	-68 45 19	5.88	F8V,dG0V	19	-	240	
7213	177171	19 6 19.8	-52 20 27	5.16	F7V,dF2V	21	2	240	
7226	177474	19 6 25.0	-37 3 48	5.01	F8V	12	-52	240	
7227	177475	19 6 25.0	-37 3 48	5.01	F8V	-	-52	240	
7597	188376	19 55 50.3	-26 17 58	4.70	G5V	11	-21	240	
7631	189245	20 0 20.2	-33 42 14	5.66	F7V	23	-8	240	
7637	189340	19 59 47.3	- 9 57 30	5.88	F8V,G0V	22	23	240	
7715	191862	20 12 25.8	-12 37 3	5.85	F7V	26	23	360	

HR#	HD#	RA	DEC	VMAG	TYPE	Dist.	RV	IntTime	
7722	192310	20 15 17.3	-27 1 58	5.73	K0Vv	8.3	-54	1440	
7776	193495	20 21 0.6	-14 46 53	3.08	F8V+A0	32	-19	360	
7875	196378	20 40 2.4	-60 32 56	5.12	F8V	21	-32	240	*
8013	199260	20 56 47.2	-26 17 47	5.70	F7V	22	-16	360	
8323	207129	21 48 15.6	-47 18 13	5.58	G0V,G2V	15	-7	240	*
8387	209100	22 3 21.3	-56 47 10	4.69	K4-5V	3.4	-40	480	*
8501	211415	22 18 15.4	-53 37 40	5.37	G3V,G1V	14	-14	240	*
8635	214953	22 42 36.8	-47 12 39	5.98	G0V	19	17	240	*
8843	219482	23 16 57.6	-62 0 5	5.66	F7V	22	-9	240	*
8883	220096	23 21 15.4	-26 59 13	5.64	G4V	11	13	240	*
8935	221420	23 33 19.5	-77 23 7	5.81	G2V,K0V	-	26	480	*
-	delta Pav	20 8 43.3	-66 10 56	3.56	G5IV	5.7	-22	240	*
-	Sag. A	17 43 00	-28 50 00	-	-	7000	-	240	
-	moon	-	-	-	-	-		480	

5. CONCLUSION

We have searched 100 target sources at a proposed interstellar communication channel frequency. Three of four possible Doppler corrections have been used. At the frequencies of observation no narrow band signals were detected above 2 Jy. A narrow spectral line was seen while observing HR6171, which is worthy of repeat observation. Future searches will concentrate on using improved spectral resolutions, additional southern stars, over a broader range of spectral class, and an ICC frequency defined in the cosmic microwave background reference frame. Present results do not disprove the hypotheses on which this work is based, but if the choice of ICC was correct, and the sensitivity was high enough, would imply a limit to the duration of ETCs $\sim 10^8$ years.

References

Blair, D.G.: 1985, The Australian Physicist 22, 98-104.
Blair, D.G.: 1986, Nature 319, 270.
Bracewell, R.N.: 1974, The Galactic Club, Norton, New York.
Drake, F.D.: 1980, in Strategies in the Search for Extraterrestrial Intelligence, ed. Papagiannis, Reidel, Dordrecht, p. 333.
Hirshfeld, A. and Sinnott, R.W.: Sky Catalogue 2000.0, Cambridge University Press.
Hoffleit, D.: The Bright Star Catalogue, Yale University Observatory.
Olivier, B.M.: 1981, in Life in the Universe, p. 351, ed. J. Billingham, MIT Press, Cambridge.
Papagiannis, M.D.: 1985, Nature 318, 135.
Tipler, F.: 1980, Quarterly Journal Roy. Astron. Soc. 21, 267-281.

Discussion

J. HEIDMANN: With respect to your comments about which frequencies to start first, I reached the same strategy for referenced frequencies based on remarkable pulsars (J. Heidmann, C. R. Acad. Sc. Paris, II, 306, 1988, 1441; Planetary Soc. 1988 International SETI Conf., Univ. of Arizona Press, in press; IAA 89-642, 40th Intern. Astronaut. Congress of the IAF, 1989).

D. BLAIR: The proposal of $\pi\, f_H$ was first made by me in a letter to nature in 1986, but it was possibly made by Bracewell and others much earlier. The fact that several people independently made similar suggestions should perhaps give us confidence that we may in fact be able to guess the extraterrestrial strategy.

B. CAMPBELL: Do you allow for the fact that Doppler velocity due to surface convective motions of dwarf stars differs from the center of mass velocity by typically 0.5 kms^{-1}?

D. BLAIR: No, we have only used catalog values. Our bandwidth of 50 kHz per reference frame is sufficient to encompass almost 2 km per second error.

F. DRAKE: Please define the golden mean.

D. BLAIR: The golden mean is obtained from the Fibonacci number. It is the limit of the ratio of F_{n+1}/F_n, where F is defined by the series 1, 1, 2, 3, 5, 8, 13..... i.e.: $F_{n+1} = F_{n-1} + F_n$. The Fibonacci series predicts the structure of numerous living structures, and the golden mean has been used to define the "most beautiful" shape for paintings and sculpture.

Editor's Note: The series converges to 1.61803399.....

THE SETI PROGRAM OF THE PLANETARY SOCIETY

Thomas R. McDonough

SETI Coordinator, The Planetary Society, Pasadena, CA 91106

In 1980, The Planetary Society was formed with two major goals: supporting planetary exploration and SETI. This paper surveys the accomplishments of the Society in the field of SETI.

In 1981, when the NASA SETI program was canceled by the US Congress, the Society reached an agreement with Paul Horowitz of Harvard University to build a portable SETI receiver, called Suitcase SETI. With assistance from NASA and Stanford University, this receiver was built. It was then used at the Arecibo radiotelescope for two weeks. (In 1982, the NASA SETI project was reinstated by Congress, in part due to the help of the Society's assistance in educating Congress about the importance of SETI.)

The Society's Suitcase SETI consisted of a 128k-channel receiver controlled by a microcomputer. It used ultra-narrowband channels of approximately 0.05-Hz bandwidth. The philosophy of this system was to focus on "magic frequencies" such as the 21 cm line of hydrogen. The minimum bandwidth was chosen from the Drake-Helou criterion, which calculates the narrowest bandwidth that can be transmitted without broadening by the interstellar plasma.

Although the instrument was originally designed to be portable, it was then discovered that a 26-m antenna at Harvard was available. The Society reached an agreement with Harvard University and the Smithsonian Astrophysical Observatory, the operators of the telescope, allowing us to refurbish the facility, and to install and operate Suitcase SETI permanently.

In 1983, the installation was completed and renamed Project Sentinel. This system then operated continuously, apart from occasional periods of maintenance. The technique used was to fix the antenna at one position for 24 hr, letting the Earth's rotation scan the sky. In this way, a strip of sky of approximately 0.5 deg could be observed each day.

One of the problems with this approach was knowing precisely which frequency to observe. Because of the 0.05-Hz bandwidth of each channel, even 128k channels encompasses a mere 6 kHz. The difficulty is that every object in the universe is moving, and hence the Doppler effect shifts the frequency of the presumed alien transmitter.

One possibility is that the other civilization may adjust its frequency so that the frequency received by us is the magic frequency as measured in the rest frame of the stars near the Sun. If they have not made this choice, they might choose to broadcast at the rest frame of the center of the Milky Way Galaxy. A third possibility was suggested by Dixon in 1973: the use of the rest frame of the universe, as deduced from observations of the 3K cosmic black-body radiation.

In order to encompass the uncertainties in these motions, it was proposed by Horowitz to expand the Sentinel system, increasing the number of channels by 64 times, to 8 million channels. The Planetary Society funded this proposal, and in 1985, the new, expanded system was completed at the Harvard site, and renamed Project META (Megachannel Extraterrestrial Assay). META has since operated continuously, with minor exceptions, observing at 21 cm, at the first harmonic of 21 cm, and in the OH (hydroxyl) band. The receiver is switched periodically between the three rest frames.

In 1989, The Planetary Society and the Argentine Institute of Radioastronomy agreed to build a duplicate of META. Hurrel and Olalde of the Institute constructed the duplicate at Harvard, and the instrument was then shipped to Argentina. In 1990, the instrument, called META II, began operating with a 30-m antenna. Argentine researchers use this 8-million channel system at least half of the time for SETI.

These have been among the most important contributions of The Planetary Society to SETI. We have also continued to support NASA SETI over the years in Congress and in the media. Additionally, we have published articles on SETI in our periodical, The Planetary Report; arranged support for the travel of American scientists to the Soviet Talinn SETI conference when NASA funding was not available; provided seed money for Gatewood's extrasolar planet search at Allegheny Observatory; supported brief SETI observations in Australia; supported the IAU Bioastronomy newsletter; provided partial support to the Ohio State University (Kraus and Dixon) and Canadian (Stephens) SETI programs; and organized an international SETI conference in Toronto.

Discussion

J. TARTER: I wish to make a slight clarification to the figures you just gave. "Suitcase SETI" operated with 0.015 Hz resolution, "Sentinel" used 0.03 Hz resolution and META SETI now operates with 0.05 Hz. The only reason to mention this is because it illustrates the general difficult problem of trying to explore as much of the multi-dimensional search space with limited resources. Each time Paul increased his bandwidth, he was making a conscious decision to sacrifice sensitivity in order to achieve greater frequency coverage.

T. McDONOUGH: This is correct. The basic criterion is that of Drake and Helou, who calculated that a purely monochromatic signal would be broadened to ~0.01 Hz bandwidth by galactic plasma. The current choice of 0.05 Hz allows META to encompass the uncertainties in the rest frames used.

L. DOYLE: Do you perform any cross correlation between channels to account, for example, for a signal coming uncorrected from a planet?

T. McDONOUGH: We do not use cross-correlation. We think it is likely that the frequency of a beacon transmitter would be shifted by their civilization to one of our three rest frames.

D. BLAIR: Do you have sufficient accuracy on the cosmic black body radiation reference frame that you can cover the uncertainties within your spectrometer bandwidth?

T. McDONOUGH: Yes. With 8 million 0.05 Hz channels, we have a total coverage of 400 kHz centered around our chosen frequency. The COBE spacecraft will allow substantial reduction in the uncertainties of this rest frame.

I. ALMAR: How all the observations are archived and how the interesting cases are selected? What is the procedure with the interesting cases?

T. McDONOUGH: All interesting observations are archived on videotape. There are typically 6-12 interesting cases each year. These observations are repeated, but the result is either RFI, or the signal is not found again.

W. SULLIVAN: Although I know that it was Phil Morrison who suggested to Paul Horowitz that the reference frame of the cosmic background radiation would be a natural one to use, it should be noted that Bob Dixon published this suggestion in 1973. [Dixon, R. S. (1973) Icarus 20, 187-199 (see p. 194)].

T. McDONOUGH: Thank you for this information on the origin of an ingenious and important concept.

TRYING TO DO SCIENCE USING HIGH SPECTRAL RESOLUTION SETI PROTOTYPES

Jill Tarter and Peter Backus
SETI Institute, 2035 Landings Dr., Mountain View, CA 94043

Kent Cullers, John Dreher, Chris Hlavka and Jane Jordan
NASA Ames Research Center
Moffett Field, CA 94035-1000

Paper Presented by J. Tarter

<u>Poster Paper</u>

Since 1968, a number of observers have attempted to detect a departure from Gaussian amplitude statistics in the radiation from saturated astrophysical masers. In the laboratory, this behavior is exemplified by the mode selection observed in saturated masers. The observed amplitude statistics from astrophysical sources have consistently shown no measurable deviation from Gaussian behavior. Several years ago, a preprint by DeNoyer and Dodd pointed out that all previous measurements were doomed to failure, having been made with frequency resolutions on the order of the Doppler-broadened line widths characteristic of the masers (kHz). They argued that the appropriate resolution for these observations is really that characteristic of the intrinsic coherence width of the maser line (sub-Hz), and they reported the results of some initial observations at Arecibo Observatory with 1/4-Hz resolution. The results presented in this preprint were inconclusive, but promising. We had the opportunity to make radio frequency interference observations at Arecibo in June, 1989 with two different high resolution SETI prototypes. At that time we used the MCSA 1.0 with 1/2-Hz resolution and the upgraded SERENDIP II+ with 1-Hz resolution to make long observations of a number of circumstellar and interstellar masers thought to be highly saturated. The initial reduction of data from the MCSA 1.0 has provided a tentatively positive result for one of the features in W51. This poster will discuss these results as well as the data reduction from the SERENDIP II+.

THE SUMMIT OF THE CBR AT MM WAVELENGTHS
AND OTHER QUERIES

J. P. Vallee

Herzberg Institute, NRCC, Ottawa, Ont., Canada K1A OR6

Summary

As a strategy for the Search for Extraterrestrial Intelligence (SETI), a renewed proposal is made to look at the summit of the Cosmic Background Radiation (CBR). The CBR peaks near 1.85 mm in the frequency spectrum of the Universe and near 1.05 mm in the wavelength spectrum of the Universe. I review the current status (best CBR temperature, best receivers, etc).

In addition, a modelling approach to existentialist questions and SETI is developed briefly, as a game. Existentialist "answers" (religious, scientific) are tabulated for questions on the Universe, Life, the Mind, according to their various choices.

THE POTENTIAL CONTRIBUTION
OF THE NORTHERN CROSS RADIOTELESCOPE TO THE SETI PROGRAM

C. Bortolotti,[1] A. Cattani,[1] N.D'Amico,[1,2] G. Grueff,[1,3]
A. Maccaferri,[1] S. Montebugnoli,[1] and M. Roma[1]

[1]Istituto di Radioastronomia, CNR, Bologna
[2]Istituto di Fisica, Univ. Palermo
[3]Dipartimento di Astronomia, Univ. Bologna

Paper Presented by S. Montebugnoli

Poster Paper

The large Northern Cross radiotelescope is located near Medicina (BOLOGNA) and is run by the Institute of Radioastronomy of the Italian National Council of Research. It operates at a frequency of 408 MHz with a bandwith of 2.7 MHz, with two arms aligned in the E-W and N-S directions. The E-W arm is a parabolic cylinder, 534 m long and 35 m wide, with 1536 equiphased dipoles in the focal line. To reduce cable losses, this arm is subdivided into 6 channels. Summing these 6 channels with proper gradient of phase, 3 beams (4×110 arcmin) are obtained with a spacing between each beam of 4 arcmin in hour angle. The receiver system temperature is 140 K and the antenna gain is 2.7 K/Jy. The N-S arm consists of 64 parabolic cylinders (10 m spacing) of 24×8 m each and with pointing range $0° + 90°$ dec., subdivided in 8 groups of 8, each of them phased via an original 'phase shifting' system using liquid dielectric. Summing these 8 channels with a proper phase, 5 (105×4 arcmin and 2 arcmin spaced) total power beams are implemented with an antenna gain of 1.6 K/Jy. Cross correlation of the 6 E-W single channels with the 8 N-S channels using 48 complex correlators allows the overall beam to be synthesized.

Different operational modes of the telescope are possible depending on the program in progress: with single channels, interferometers, 5 E-W total power, or 5 N-S total power beams alone or together (multibeam). A schematic view of the multibeam system appears as in Figure 1, where 15 pencil beams (about 4×4 arcmin) are obtained by correlating the 3 E-W and 5 N-S total power beams.

A data acquisition system for the SETI program may be connected to the radiotelescope in several ways, each of them with different characteristics, see Figure 2. Connecting the acquisition system to the output of the E-W total power multiplexing block allows the characteristics of this arm to be exploited: an antenna gain of about 2.7 K/Jy, (4×110) arcmin beamwidth, pointing range from $-30°$ to $+90°$, and integration time of ($96/\cos[\text{dec}]$) seconds, reached via a computer controlled total power beams switching system. Connecting the same data acquisition to the 5 (105×4) arcmin N-S total power beams, one obtains an antenna gain of about 1.6 K/Jy, a pointing range from $0°$ to $90°$, and an integration time of ($480/\cos[\text{dec}]$) seconds.

Another way to use the telescope with the SETI acquisition system is to connect the multibeam system. This exploits the global antenna capabilities with a pointing range from 0° to 90° and 15 pencil beams covering simultaneously an area of 12×12 arcmin. This is the more attractive mode in using the telescope for continuum sky survey at maximum instantaneous sensitivity.

Because of the high instantaneous sensitivity and the large portion of sky potential observable, the Northern Cross radiotelescope may be attractive for a complementary 408 MHz survey in the MOP-SETI observation program.

We thank V. Albertazzi for drawing the figures.[1]

Figure 1a. Northern Cross (E-W arm and part of N-S arm)

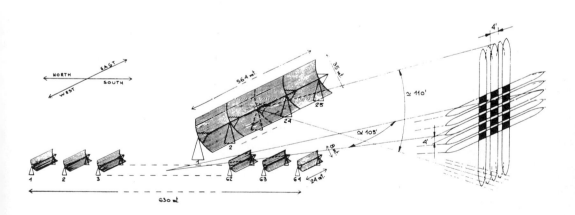

Figure 1b. Schematic view of multibeam system

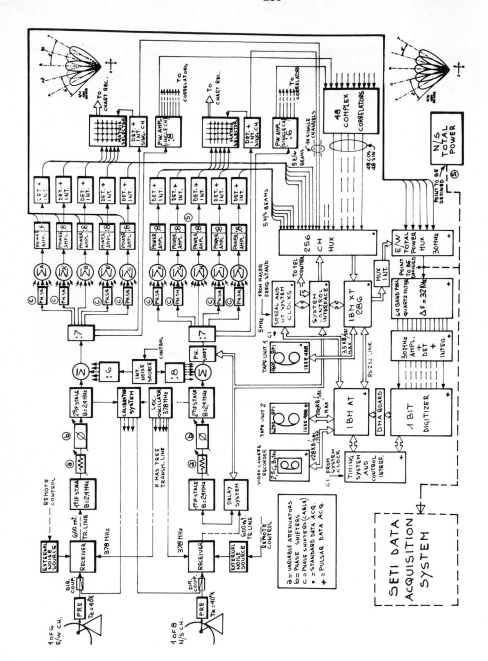

Figure 2. Northern Cross radiotelescope block diagram

A PROPOSAL FOR A SETI GLOBAL NETWORK

J. Heidmann
Observatoire de Paris
F-92195 Meudon, France

Poster Paper

After its generation-long gestation, SETI has become an internationally minded venture. In addition to still modest financial support, to recognition by academic bodies and to organization by international unions, it has numerous individual ties and collaborations between workers from different institutes in different disciplines. This multidisciplinary and international enterprise has also deep roots in the general spirit of its workers who profoundly think that in case of success, the most tangible result, an artificial extraterrestrial signal, should belong collectively to all of mankind as one of its newly added cultural conquests.

This point of view is expressed and supported by the document drafted by a number of lawyers, academics, and government officials: the "Declaration of Principles for Activities Following the Detection of Extraterrestrial Intelligence," recently endorsed by the International Academy of Astronautics and the Institute of Space Law.[1]

However, in spite of this wide spectrum of international connection, I thought there is still a practical missing link in this long chain at the particular level of the individuals effectively involved in the SETI radioastronomical field work. There is a need for rapid common action and exchange of information between them.

WHAT IS THE SITUATION?

(1) When a signal is discovered, the Declaration proposes the creation of an international committee for continuing analysis;

(2) Prior to that, it proposes that a real discoverer informs other observers for confirmation.

MY PROPOSAL BELONGS JUST BEFORE THESE STAGES:

A SETI Global Network would coordinate the activities of potential discoverers through close, easy and fast exchanges of relevant technical informations directly related to their field work, whether technical, instrumental, computational, organizational, scientific, and whether done by technicians, engineers, and scientists.

Figure 1. A Symbolic View of the Global SETI Network.

This proposal is in fact more at the technical level and more at the roots of the propositions made by the Declaration. At the present moment I would confine the organization of this SETI Global Network to SETI in the radio domain, as a start, because most of SETI was done in this radio domain (35 radio telescopes out of 40 instruments used and 49 radio searches out of 55 searches made[2]). However, in case of successful organization, it could be extended to all SETI searches.

Typical topics which could be included in the Network activities could cover:

SETI strategies
receivers and signal detectors research and development,
interference investigations and countermeasures,
logistics of searches,
field tests of fast cross checking of alarms,
field tests of continuous time monitoring of selected targets,
cross linking of the member workers through a fast communication network,
training for instant common action as soon as necessary.

In addition to being global this Network should be international. Results of SETI ratio observations have been published by workers from 10 different nations[2] and at recent SETI international meetings the desire for international collaboration has been clearly and frequently voiced.

My proposal for this SETI Global Network has already been endorsed by the SETI Committee of the International Academy of Astronautics at its October 8, 1989 Meeting in Torremolinos and I shall work for its endorsement by COMMISSION 51 (Bioastronomy) of the International Astronomical Union.*

If you agree, please express your support and if you wish to collaborate, please provide suggestions for a common effort.

A suitable frame for the Network creation might be sought at the occasion of the 1992 International Space Year.

References

1.   Billingham J., Michaud M.A.G., Tarter J.,1991, this volume
2.   Biraud F., l'Astronomie, December 1989, 566

---

*Editor's note: This has been done at the June 20, 1990 IAU Commission 51 Business Meeting in Val Cenis.

V. ADVANCED EVOLUTION - POSSIBILITIES

A SEARCH FOR DYSON SPHERES AROUND LATE-TYPE STARS IN THE IRAS CATALOG

Jun Jugaku
Research Institute of Civilization
Tokai University, Hiratskuka, Kanagawa 259-12, Japan

Shiro Nishimura
National Astronomical Observatory, Mitaka, Tokyo 181, Japan

Paper Presented by J. Jugaku

Thirty years ago Dyson (1960) suggested searches for artificial stellar sources of infrared radiation. Dyson offers a scheme in which planets are broken down and reconstructed into a swarm of individual fragments in the form of a spherical shell. A Dyson sphere effectively absorbs visible light from the central star and utilizes it as an energy source. The exterior of the Dyson sphere radiates infrared radiation into space as waste heat. Because of the large dimensions of the sphere, the infrared radiation from the Dyson sphere should be detectable over distances of several hundred light-years.

For a shell maintained near 300K, peak emission would be expected in the $10\mu m$ range, if the spectrum is approximately thermal. The Infrared Astronomical Satellite (IRAS) mapped the sky at 12, 25, 60 and $100\mu m$ in an unbiased way, covering 96% of the celestial sphere. About 65% of the sources detected by IRAS are stars, including stellar photospheres as well as stars with infrared excesses due to circumstellar shells of gas and dust (Aumann 1985, 1988; Backman and Paresce 1990). Therefore the IRAS data, particularly those of the Point Source Catalog (PSC) (1988), provide us with good materials to search for Dyson spheres.

We assume that Dyson spheres are associated with solar-type stars, namely luminosity class V stars of F, G, and K spectral types. (Luminosity class V indicates that the star is a main-sequence star like the sun.) The most comprehensive catalog for a selection of these stars is the Michigan Catalog of Two-dimensional Spectral Types for the HD Stars (Houk and Cowley 1975; Houk 1978, 1982; Houk and Smith-Moore 1988). (HD denotes a prefix used to designate a star as listed in the Henry Draper Catalog.) So far the Catalog has been published for all the HD stars south of declination -12°, so that coverage of the celestial sphere is ~44% The total number of stars is 130,375, of which only 594 F, G, and K stars of luminosity class V were detected with IRAS. Most of the stars in the IRAS catalog that have been identified with previously known objects are K and M giants. Even for those 594 stars, only 82 stars have been previously observed in the infrared region $\lambda \geq 1\mu m$ (Gezari, Schmitz, and Mead 1987), and 54 stars have data for $2.2\mu m$. These 54 stars are the subject of our analysis.

Papagiannis (1985, p. 268) demonstrated from a consideration of the amount of available solid matter that only a small fraction (~1%) of the central star's radiation could be intercepted by a Dyson sphere. On the other hand, if 10^{-8} of the radiative energy of the central star is converted as thermal

Table 1. List of Stars

HD No.	Sp.type	$m(2.2\mu m)$	$m(2.2\mu m) -m(12\mu m)$	$m(12\mu m) -m(25\mu m)$
693	F5	3.64	0.01	0.13
1581	G0	2.83	0.07	-0.01
1835	G3	4.91	0.24	0.74
4307	G2	4.64	0.03	0.82
10700	G8	1.67	0.10	-0.03
10800	G1/2	4.48	0.18	<0.19
11507	K5/M0	5.21	0.23	1.19
12311	F0	2.09	-0.07	-0.07
15798	F5	3.58	0.06	-0.05
17206	F5/6	3.31	-0.03	0.44
17925	K1	4.06	0.08	0.17
20010	F8	2.53	0.04	0.03
20766	G3/5	4.01	0.04	0.09
26491	G3	4.84	0.25	<0.65
26612	F2	4.06	0.06	<-0.09
29875	F1	3.67	0.17	-0.03
38393	F7	2.42	0.05	0.01
40136	F1	2.90	0.05	0.07
44594	G3	5.16	-0.08	<1.30
45184	G2	4.94	0.08	<0.92
53705	G3	4.15	0.53	0.13
59967	G3	5.15	-0.03	<1.24
65907	G0	4.23	-0.03	<0.32
76932	F7/8	4.38	-0.02	<0.46
79940	F3/5	3.57	0.15	0.03
90589	F2/3	3.12	0.01	0.13
100623	K0	3.0	-0.88	-0.02
102365	G3/5	3.32	0.04	0.11
104138	F7	5.32	-0.16	2.44
107439	G3/5	6.71	4.62	1.04
112164	G1	4.51	-0.11	<0.68
114613	G3	3.27	0.07	0.02
114837	F5	3.66	0.00	-0.04
128620	G2	-1.50	0.35	-0.01
130819	F3	4.14	0.04	0.72
131977	K4	3.02	0.55	-0.12
136352	G3/5	4.10	0.11	<-0.52
144585	G5	4.87	0.16	1.57
147513	G3/5	3.94	-0.10	1.04
152404	F5	5.70	2.71	2.31
155203	F2	2.35	0.10	0.30
156384	K4	6.04	2.97	0.50
156846	G1	5.17	0.18	2.28
165185	G3	4.57	-0.01	1.09
181484	F0	4.22	0.05	<0.23
189567	G3	4.57	0.15	<0.48
203608	F7	2.90	0.05	-0.05
206301	G2	3.65	0.18	0.18
209100	K4/5	2.18	0.04	-0.02
210918	G5	4.73	-0.03	0.84
215456	G0	5.11	0.11	<1.06
216435	G0	4.70	0.14	0.89
217357	K5/M0	4.46	0.00	0.70
219571	F3	2.97	0.09	0.20

emission from a Dyson sphere, it can be detected as an infrared excess at $12\mu m$ with the present capability of infrared photometry. We note that $2.2\ \mu m$ is a good measure of the photosphere of a star unless there is emission from circumstellar material at $T \approx 1500K$, which is very high for solid dust grains. Then the $2.2\mu m$ flux represents the stellar photosphere and the excesses at $12\mu m$ ($m_{2.2} - m_{12}$) and $25\ \mu m$ ($m_{2.2} - m_{25}$) above a blackbody normalized to the $2.2\mu m$ data represent the fluxes from circumstellar emission. Since the F, G, and K-type stars do not exhibit much difference in the shape of the energy distribution in the infrared region that we are concerned with, we will treat them together in the following.

Table 1 presents a list of our survey stars. The first column is the star's HD number; the second column is the spectral type given in the Michigan catalog. Columns 3, 4 and 5 are, respectively, the magnitude at $2.2\mu m$ taken from the compilation by Gezari et al. (1987), the magnitude difference between $2.2\mu m$ and $12\mu m$ and that between $12\mu m$ and $25\mu m$. To convert an IRAS flux into the magnitude system, we followed the *IRAS Catalog Explanatory Supplement*, 1988, p. VI-20; i.e., we assumed that zero magnitude corresponds to 28.3 Jy at $12\mu m$ and 6.73 Jy at $25\mu m$. When there is more than one $2.2\mu m$ flux given for an object star in the Gezari et al. compilation, we adopt the average value. We also applied color-correction factors 1.44, 1.43 and 1.42 at $12\mu m$ for F, G, and K stars, respectively, and 1.40 at $25\mu m$ for all stars. The scatter in the $m_{12} - m_{25}$ color is mostly due to the fact $25\mu m$ fluxes given by IRAS are upper limits for weak sources. These stars show apparent $m_{12} - m_{25}$ excesses but with no $m_{2.2} - m_{12}$ excesses. For most of the stars the value of $m_{2.2} - m_{12}$ is in the range from -0.2 to 0.2 magnitude. The scatter in this range can be explained by spectral line features of individual stars and by errors of observation. If a Dyson sphere exists, the value of $m_{2.2} - m_{12}$ should be larger than the above value.

We found three stars that have indeed large values of $m_{2.2} - m_{12}$. The star HD 107439 shows the largest color excess. It turned out that this star is actually one of the RV Tauri stars, a small group of very luminous pulsating variables. However, the optical spectrum of RV Tauri, stars at low resolution is similar to that of normal dwarf stars and impossible to differentiate in the Michigan survey. The second star HD 152404, turned out to be an RW Aurigae star. These are a kind of T Tauri stars which are young ($\sim 10^6$ years) stars associated with interstellar matter. Once again it is impossible to detect their characteristics with low-resolution classification. The third star that shows the $m_{2.2} - m_{12}$ excess is HD 156384. This is one of the nearest visual binaries with a second component of type K5V. Since IRAS has a large beam of 45 x 270 arcseconds, the value of m_{12} for this star is the sum of these two components. However, the contribution from the second component does not explain the large excess at $12\mu m$. The third component may be either a very cool star and/or nonstellar matter of $T < 500K$.

Soderblom (1986) presented a list of 63 SETI candidates, while Fracassini et al. (1988) selected 22 solar analogs, which do not overlap with Soderblom's stars. Twenty-five out of Soderblom's 63 stars have IRAS identifications, and 17 stars have been observed at $2.2\mu m$ (HD 3651, 4628, 4614A, 5015A, 10307, 10476, 48521, 75732A, 84737, 101177A, 102870, 126053, 159222, 166620, 197076A, 217014 and 222368). From the Fracassini et al. list, only two stars (HD 1835 and 20630) are available for both $2.2\ \mu m$ and IRAS data. All of these 19 stars are normal at $12\mu m$.

In conclusion, (1) no evidence of Dyson spheres was found in 54 stars selected from the Michigan catalog. An additional 19 stars in the lists of Soderblom (1986) and Fracassini et al. (1988) also do not show evidence of Dyson spheres. (2) More ground-based observations of F, G, and K dwarfs in the wavelength region between 1 and 4µm should be carried out to match the extensive data of IRAS. From the present study a list of 540 such stars has been prepared for 2.2µm observations.

We are grateful to J. Davidson and D. Backman for valuable comments on the original manuscript. One of us (J.J.) thanks T. Oshima for his moral support in studying Dyson spheres.

References

Aumann, H.H. 1985, *Publ. Astron. Soc. Pacific*, 97, 885.

Aumann, H.H. 1988, *Astron. J.*, 96, 1415.

Backman, D. and Paresce, F. 1990, in *Protostars and Planets III*, eds. Levy and Matthews (Tucson: University of Arizona Press), in press.

Dyson, F.J. 1960, *Science*, 131, 1967.

Fracassini, M., Pasinetti Fracassini, L.E., and Pasinetti, A.L. 1988, *Astrophys. Space Sci.*, 146, 321.

Gezari, D.Y., Schmitz, M., and Mead, J.M. 1987, *Catalog of Infrared Observations* 2nd ed., NASA Reference Publication 1196 (Washington, DC; National Aeronautics and Space Administration).

Houk, N. 1978, Michigan Catalogue of Two-dimensional Spectral Types for the HD Stars. Volume 2. Declinations -53° to -40° (Ann Arbor: Department of Astronomy, University of Michigan).

Houk, N. 1982, Michigan Catalogue of Two-dimensional Spectral Types for the HD Stars, Volume 3. Declinations -40° to -26° (Ann Arbor: Department of Astronomy. University of Michigan).

Houk, N., and Cowley, A. P. 1975, Michigan Catalogue of Two-dimensional Spectral Types of the HD Stars. Volume 1. Declinations -90° to -53° (Ann Arbor: Department of Astronomy, University of Michigan).

Houk, N., and Smith-Moore M. 1988, Michigan Catalogue of Two-dimensional Spectral Types for the HD Stars. Volume 4. Declinations -26° to -12° (Ann Arbor: Department of Astronomy, University of Michigan).

Infrared Astronomical Satellite (IRAS) Catalogs and Atlases, 1988, Volumes 1-7, NASA RP-1190 (Washington, DC: National Aeronautics and Space Administration).

Papagiannis, M. D. 1985, in *The Search for Extraterrestrial Life: Recent Developments*, IAU Symp., No. 112, ed. M.D. Papagiannis (Dordrecht: D. Reidel), p.263.

Soderblom, D. R. 1986, *Icarus*, 67, p. 184.

Discussion

D. WHITMIRE: How can you differentiate between Dyson spheres and dust shells at the same temperature?

J. JUGAKU: If there exist warm shells that are unlike cold ones usually observed around main-sequence stars, it may be difficult to differentiate them by broad-band photometry. Since I expect that artificial spheres have quite different geometrical structures, it should be possible to detect their signatures by means of high-resolution imaging and spectroscopy.

W. SULLIVAN: How can you distinguish a Dyson sphere from a circumstellar distribution of dust? Give the theorists a broadband spectrum, no matter how highly unusual, and they will eventually be able to explain it.

J. JUGAKU: I agree with you. In this case I would like to refer to Protocol Principle 1.

GRAVITATIONAL, PLASMA, AND BLACK-HOLE LENSES
FOR INTERSTELLAR COMMUNICATIONS

Von R. Eshleman
Stanford University, Stanford, CA 94305 USA

Summary

A star is a compound lens for electromagnetic waves. It has both a forward line focus due to gravitation and a forward conical focal surface due to surrounding plasma. The combination of the gravitational and plasma effects can produce a forward point focus of extreme magnification. A Schwarzschild black hole has an infinity of forward and backward foci, where there are two types of forward line foci and one kind of backward conical foci. Other advanced civilizations, if they exist, are assumed to be near stars. Thus, every society has at hand a potentially powerful lens to use as transmitting and receiving antennas for interstellar radio communications. Very advanced civilizations may be able to use black holes for this purpose. There could exist an essentially private but vast network of such links that could well escape detection by SETI. Its possible characteristics and a proposed search protocol are described. An important first step would involve a spacecraft mission to very deep space, on the order of 1000 astronomical units or six light days from the sun.

SETI THROUGH THE GAMMA-RAY WINDOW: A SEARCH FOR INTERSTELLAR SPACECRAFT

Michael J. Harris

SMSRC, Landover MD 20785, USA

ABSTRACT

Consideration of the Fermi Paradox leads to the conclusion that SETI strategies should focus on the detection of interstellar spacecraft. Although current concepts for such spacecraft are highly speculative, two very general distinctive features can be identified. First, autonomous propulsion on rocket principles is limited to a nuclear fusion or antimatter annihilation power source, both of which will emit γ-rays. Second, interstellar spacecraft are identified most readily by their very high proper motions. Our strategy is therefore to search for γ-ray sources with large proper motions. A variant of this method involves searching for transient sources which fall along a straight line in space. The limitations due to the poor measurement of γ-ray source positions are emphasized. Even the next generation (GRO-GRANAT) detectors will have useful resolutions of ~1°, limiting the radius of the search to ~100 pc.

The γ-ray spectra expected from spacecraft types are then reviewed, with particular attention to the antimatter annihilation ($p\bar{p}$ and e^-e^+) designs. The distinctive line emission from these objects is readily detectable if the spacecraft is sufficiently massive (a consumption of $\sim(1-10)R^2$ tons/s of antimatter is necessary for detection by GRO of a spacecraft at a distance R pc). To a first approximation, the spectra will resemble those of natural γ-ray sources. In particular, e^-e^+ annihilation spacecraft operating in a pulsed mode look very much like cosmic γ-ray bursts (GRBs).

No steady γ-ray source is known with a proper motion so large as ~1°/yr. Our search has therefore been concentrated on transient sources. Rigorous criteria have been developed by which the Doppler shifts of e^-e^+ annihilation features in GRB spectra, when combined with the poorly-known burst positions, identify GRBs lying along a straight-line trajectory. The GRBs in the 1978-80 Interplanetary Network catalog have been analyzed by this method. No suspicious alignments have been found.

1. INTRODUCTION

Let us consider the state of affairs which is implied if Fermi's Paradox is, in fact, valid. There will be a large number of spacecraft in transit from star to star, which might be detectable because of the very large energy consumption required for interstellar travel. The extreme implication of the Paradox, that the spacecraft ought to be "here, now", is interpreted as "within a distance short compared with the light travel time across the Galaxy", which I take to be ~100 pc, the limit for the search method which I will describe. In fact, if there are very few civilizations in the Galaxy (N ~1, for which there are compelling biological arguments), then the Fermi Paradox works in favor of SETI, implying that evidence of the civilization(s) will be brought within detectable range.

Spacecraft which are propelled by laser or particle beams sent out from a home base will not be considered here. I will focus on those which work on rocket principles, i.e., in which the fuel supply is carried on board the spacecraft. When account is taken of the necessity (implied by Fermi's Paradox) for the rocket to consist of at least two stages, it is apparent from Table 1 that specific impulses obtained from chemical and fission energy sources are too small for interstellar travel. The resulting mass ratios are prohibitive even for travel at a velocity 0.05c, which barely meets the criterion set by Fermi's Paradox. Nuclear fusion and antimatter annihilation are the only energy sources which need to be considered in this connection.

Table 1: Rocket capabilities - two stage, $\Delta v = 0.05c$[1]

Propulsion	I_{sp}	Mass ratio
Advanced chemical	10^3	$\sim 10^{1328}$
Fission (pulse)	2×10^4	2.5×10^{66}
Fusion (pulse)[a]	3×10^5	3×10^4
Annihilation (optimum)[b]	3×10^7	1.11

[a] Based on the DAEDALUS[2] design; the exhaust velocity is that of the reaction mass surrounding the nuclear fuel, which is ionized and expelled electromagnetically.

[b] Assumes photon exhaust with velocity c, from e^-e^+ annihilation (see section 3 below).

Fusion and annihilation both give rise to copious emission of γ-rays, which will be detectable by sufficiently sensitive detectors. In the absence of any detailed knowledge of aliens or of their spacecraft, it is impossible to specify the masses of the spacecraft, the absolute fluxes emitted or the sensitivity required for detection. In the next two sections I describe very general features of the γ-ray emission from interstellar rocket engines. Because their velocities will be of the order c, the most characteristic feature will be their very large proper motions, which will be 100 to 1000 times larger than those of any other bulk astronomical sources (with possible, easily recognizable, exceptions associated with AGNs). Unfortunately the angular resolutions achieved in γ-ray astronomy to date are very poor; characteristic position measurements of $\sim 1°$ over a ~ 10 yr period imply a limit of ~ 100 pc out to which spacecraft can be identified by their most distinctive feature.

Fusion-powered rockets may give rise to other detectable radiations, associated with the rocket exhaust, which are not discussed here, such as far-UV emission from interstellar matter excited by the exhaust, or low-frequency radio emission from propagating disturbances set up by the exhaust (Referee's communication). Antimatter rockets are less likely to be detected in this way because of the very small (or zero) exhaust masses necessary with such efficient engines.

2. NUCLEAR FUSION

The easiest fusion reactions to ignite are those involving deuterium and tritium:

$$d + d = t + p$$
$$(= {}^3He + n)$$
$$d + t = \alpha + n$$

Requiring higher temperatures or densities, but yielding more energy, is the reaction between deuterium and ^{3}He:

$$d + {}^3He = \alpha + p$$

This reaction is also favored by spacecraft designers[2] because it does not release neutrons, which are difficult to direct electromagnetically into thrust and are also a safety hazard. However it is impossible to avoid the d + d reaction, with its neutron exit channel, under the conditions in which d will fuse with ^{3}He. Therefore neutrons will be produced by any of the most plausible fusion reactions.

Neutron captures by the structural materials of the spacecraft or by the reaction mass give rise to the most distinctive γ-ray signature of fusion spacecraft. It is not possible a priori to specify what the structural materials are, but most of the obvious candidates emit characteristic γ-ray lines after neutron capture - for example ^{57}Fe emits γ-rays in two very closely spaced lines near 7.6 MeV following ^{56}Fe$(n,\gamma)^{57}$Fe. The reaction mass would probably be composed of hydrogen, in order to maximize exhaust velocity, in which case the 2.2 MeV line from $p(n,\gamma)d$ would be expected.

The fusion reactions may be ignited by either compression of fuel pellets by laser or particle beams ("cold" designs[2]) or direct heating to temperatures $\geq 10^8$ K ("hot" designs[3]). In "hot" designs optically-thin thermal emission at γ-ray energies would also be expected, from thermal bremsstrahlung or from thermal synchrotron processes if the thrust is directed electromagnetically. This radiation would peak in the tens or hundreds of keV, with significant emission up to ~1 MeV. If the ignition is accomplished by a nuclear fission device[3], a rich spectrum of γ-ray lines from the fission products would also be present. All of these possible components of the emission - narrow lines and continuum - will be strongly Doppler-shifted by the spacecraft's relativistic motion.

3. ANTIMATTER ANNIHILATION

An examination of the mass ratios in Table 1 reveals the superiority of antimatter annihilation as a power source. I have therefore focused attention on the detectability of this class of spacecraft[4]. However, the two practicable annihilation reactions - electron-positron (e^-e^+) and proton-antiproton ($p\bar{p}$) - both present serious problems. In the case of e^-e^+ designs, the products of the annihilation are two photons at energy 0.511 MeV, which are virtually impossible to focus so as to provide thrust for the rocket - dense electron gas mirrors have been suggested as a kind of "pusher plate" for them to react against[5]. The $p\bar{p}$ annihilation produces several charged pions, which decay to muons and electrons which can be focused by means of strong magnetic fields before they annihilate[6]. However, the production of $p\bar{p}$ in particle accelerators is extremely expensive. This form of propulsion is being considered by the U.S. Air Force for solar-system missions in the next century[7].

The e^-e^+ spacecraft will emit only one γ-ray line at 0.511 MeV. This line would also be expected, at some level, from the annihilation of the e^- and e^+ resulting from pion decay in $p\bar{p}$ spacecraft. The $p\bar{p}$ annihilation also produces 1-2 π^0 per event, which decay almost instantly into γ-rays distributed in a broad line centered at about $\sim$70 MeV. The operating temperatures of antimatter spacecraft are unknown. However it appears that $p\bar{p}$ designs must be fairly "hot", since the antiprotons must be injected into the annihilation medium at rather high energies of $\sim$10 MeV for most efficient annihilation[8], leading to heating of the medium. Thus there is reason to expect thermal radiation at energies up to $\sim$1 MeV.

Note also that "hot" ignition of nuclear fusion reactions may be accomplished by antimatter annihilation, in which case the fusion γ-ray spectrum described earlier would be superimposed on one or other of the two classes of annihilation spectrum.

4. KNOWN SOURCES

Several known astrophysical γ-ray sources display some or all of the spectral features just described, so that at low resolutions an interstellar spacecraft might be mistaken for such an object. Few steady sources of γ-radiation have yet been identified, and none is known to have any proper motion. Thus although "black hole" sources such as Cyg X-1 exhibit a continuum up to 1 MeV with a possible broadened, blue-shifted 0.511 MeV annihilation line superimposed, there is no reason to regard them as spacecraft candidates. A more intriguing source is 2CG195+04 (Geminga), which emits much more flux at energies around 100 MeV than in the X-ray range, and is not detectable at all at lower energies still. This energy distribution might be mimicked by a $p\bar{p}$ spacecraft.

Proper motions cannot be ruled out for most of the transient sources hitherto observed. In such cases, the motion would be revealed by successive outbursts at locations and times consistent with a straight-line trajectory in space (see next section). Several hundred γ-ray bursts (GRBs) have been observed, the origin of which is controversial, and whose spectra sometimes contain a red-shifted 0.511 MeV annihilation line. The continuum spectra from GRBs are consistent with thermal bremsstrahlung or thermal synchrotron radiation at temperatures $\geq 10^8$ K. The burst time-scales are generally $\sim$10s. These features are reminiscent of what might be expected from e^-e^+ spacecraft[10].

5. SEARCHES FOR LINEAR ALIGNMENTS OF GRBs

A search has been made for alignments of GRBs for which positions have been measured to within a few degrees[9]. The principle assumes that three GRBs lying upon a great circle will be alligned such that event times (corrected for light travel time to Earth) are consistent with travel along that line at constant velocity. In my recently published work[10], I use the Doppler shift of the 0.511 MeV line to determine the putative spacecraft velocity for cases where three GRBs are aligned. Seventy-seven GRBs were searched in groups of three, yeilding 135 alignments in which there was sufficient spectral data to apply the method. In no single case were consistant Doppler-shifted lines found in the second and third members of any trio; all alignments were rejected as spacecraft candidates.

At present, the measured positions of GRBs are limited by very poor quality. Error boxes on the order of several degrees imply that spacecraft with velocities of ~c can only be detected over distances of ~1 pc for measurements over a period of ~1 yr. This situation is expected to improve considerably when the Burst and Transient Source Experiment (BATSE) is launched in 1991 on NASA's Gamma Ray Observatory (GRO). This instrument will have the capability to measure burst positions to 1 deg or better (depending on the flux), and it is expected to be in operation for about 10 yr. The outer limit for spacecraft detection is then raised to ~100 pc, with at least 1000 GRBs expected to be observed.

Another experiment on board GRO, the Energetic Gamma Ray Telescope (EGRET), will detect γ-rays of energies above 30 MeV with a similar angular resolution. This energy range is appropriate for the detection of $p\bar{p}$ annihilation spacecraft, as described in section 3, out to similar distances of ~100 pc. The sensitivities of BATSE and EGRET imply that, at these limiting ranges, the spacecraft must consume 10^4-10^5 tons of antimatter per second.

Improvements in the measurement of γ-ray source positions may be expected from two directions. For steady sources and very bright transients, larger detectors will achieve improved signal-to-noise ratios and hence better source positions for a given angular resolution. However the positions of most GRBs are measured more accurately from the event delay times across interplanetary distances, so that progress is dependent on launch schedules in the various national planetary exploration programs. If UV emission can be detected from any GRBs, very good angular resolutions are obtainable; a High Energy Transient Explorer mission has been proposed for this purpose[11].

References

1. E.F. Mallove and G.L. Matloff 1989, The Starflight Handbook (Wiley: New York).
2. A. Martin and A. Bond eds. 1978, Project Daedalus (BIS: London).
3. F.J. Dyson 1968, Physics Today, 21, No. 10 p. 41.
4. M.J. Harris 1986, Ap. Space Sci., 123, 297.
5. E. Sanger 1953, Ingenieur-Archiv V., 21, 213.
6. D.L. Morgan Jr. 1982, JBIS, 35, 405.
7. R.L. Forward 1985, preprint.
8. D.L. Morgan Jr. 1989, in Annihilation in Gases and Galaxies, ed. R.J. Drachman (NASA Conf. Pub. 3058), p. 229.
9. J.L. Atteia et al. 1987, Ap. J. Suppl., 64, 305.
10. M.J. Harris 1990, JBIS, 43, 551.
11. G. Ricker et al. 1988, in Nuclear Spectroscopy of Astrophysical Sources, ed. N. Gehrels and G. H. Share (AIP: New York), p. 407.

Discussion

J. TARTER: You used data from the early 80's due to the interplanetary network - are there any other possibilities for more recent data sets?

M. HARRIS: There are possibilities from Soviet interplanetary missions since 1980 - Venera 13 and 14 in the mid-80's and the two recent Phobos probes all carried GRB detectors. We lost out when the "Challenger" exploded, which delayed GRO until it no longer overlapped with any of these.

D. BRIN: Mike, you are looking for very fast machines moving very near the solar neighborhood. In other words, they are "almost" here. This implies a coincidence of major magnitude. Not that you shouldn't look! But do you really rate this contact scenario very highly compared to others we've heard today?

M. HARRIS: I didn't sufficiently emphasize the poor quality of the γ-ray measurements. The locations I was using are only good to several degrees on average, implying a distance limit of order 1 pc for my method. You're right - I wouldn't have expected to find anything so close. However, the distance limit will expand by a factor of 100 or so when better limits are available from GRO/BATSE, and the method will be more promising.

V. BURDYUZHA: There are γ-ray bursters. Can they be spacecraft?

M. HARRIS: The spectra of GRBs qualitatively resemble electron-positron spacecraft at the poor energy resolutions which we have at present. However, the critical factor for determining artificiality is the proper motion, which is what I was looking for. I'm still looking, and I'll let you know when I find one!

CRITERIA OF ARTIFICIALITY IN SETI

V.V. Rubtsov

Ukrainian Extramural Polytech Institute, Kharkov, USSR

A search for astroengineering structures (AESs) in outer space presupposes a serious study of the "phenomenon of artificiality"; the more especially because the apparent lack of such structures may be interpreted as evidence of our cosmic solitude. It is quite insufficient to imagine AESs as "cosmic miracles" that can never be explained by natural causes.

What does the word "artificial" really mean? It is believed that it describes something that is manufactured. Yet there is another side of no less importance: an artificial object is involved in a sociocultural system; it performs certain functions in it, or, in other words, has a certain meaning for intelligent beings who created it. One may say that such an object is the unity of its meaning and its design.

A search is aimed at the discovery of an object or a phenomenon whose existence is predicted from some theoretical considerations. To find an object Q means to ascertain its existence and a close similarity between its theoretical (predicted) and empirical (real) characteristics.

If we know beforehand the most probable region of the object Q location, and if the region is not too large, we may study all the real objects which are situated there and compare them with our initial theoretical model. However, in practice the region of possible location of AESs is much too vast to be explored in such a way. Besides, our theoretical ideas of these structures may have little in common with reality. That is why the actual search for AESs should start from using some general criteria, not detailed models. These criteria would enable, when scanning the information on various objects, selection of supposedly artificial ones, almost irrespective of their true design.

At this stage of investigation, we try to single out only the objects and phenomena which deserve further exploration. In other words, we assume some (perhaps significant) likelihood of a mistake. It is rather difficult to unite the two opposite characteristics of these criteria: they must both be highly definite (so that we would get at the "output" an amount of phenomena, considerably reduced as against that at the "input") and at the same time "broad-sweeping" (in order to "intercept" as many "likely AESs" as possible). It is hardly feasible to create universal criteria of artificiality that would be suitable for every class of artificial phenomena.

In the real quest for AESs the most usual preliminary criterion of artificiality has been the "strangeness" of objects that do not have satisfactory "natural" explanations. Thus, V. Straijis (1986) paid attention to some peculiar cosmic objects (such as blue stragglers, carbonic and barium dwarfs, stars with abundance of 13_C, etc.) for which adequate "natural" explanations have yet to be found. Of course, nobody can prevent a researcher from regarding "strange" cosmic bodies as candidate AESs. But strictly speaking, "strangeness" as such is not and cannot be the selection criterion. At best it causes us to pay attention! The criteria must have a positive form, e.g., "If the object Q has a feature S, it is probably artificial". Unfortunately, at present we have no such criterion that would be suitable for the specific character of the SETI field. It is impossible to experiment with astronomical bodies; we can only observe them with more or less powerful instruments. So, there remains much room for various speculations and fruitless discussions. Naturally, under these circumstances an "artificial" explanation may seem artificial in another sense of the word.

Does it really mean that such an explanation is deficient in principle? I think not. The main cause of its secondary position is the contradictory nature of the existing methodological basis of SETI research. The late I.S. Shklovsky put forward in 1971 the well-known principle of "presumption of naturality". According to it, the assumption of artificial origin of an object may be considered only after all "natural" possibilities have been refuted. In reality a "normal" astronomical investigation will never need an "artificial" (A-) approach to its object of study. Any refuted hypothesis will be replaced only with a new "natural" (N-) one. On the contrary, searches for astroengineering structures require the equal status of A- and N-explanations from the very beginning of the investigation. When studying an object or phenomenon selected by some preliminary criteria, one should bear in mind both of these hypotheses.

An explanation from the artificial point of view (A-explanation) must contain the sociocultural component that reveals the function of the A-object, the technological component related to the object's design, as well as the "natural" component (those laws of nature, on which its existence and dynamics depend). It is obvious that a scientific explanation of a purely natural phenomenon (N-explanation) cannot contain the sociocultural or the technological components. If we cannot refuse in advance either "artificial", or "natural" hypotheses, the presumption of naturality should be replaced by joint evolution and competition of scientific research programs which are based on them. A- and N-programs should develop, interact and enrich each other, seeking, on the one hand, for the most complete representation of the object of phenomenon in its description and, on the other hand, for the best possible conformity between the description and a theoretical explanation of the phenomenon. During this process, one of the two explanations will be gradually superseded by the other, and a correct explanation will result.

Such a competition has not occurred as yet on a full scale in the real AES searches. But there are a number of cases which may be considered as models. The case of the Tunguska meteorite of 1908 seems to be especially significant. Let's regard it in some detail. I want to point out that I have no intention to defend or refute any viewpoint on this phenomenon; it is more important to trace the logic of the research.

For the first 40 years, there was one dominant hypothesis of the Tunguska cosmic body (TCB) origin: a usual meteorite, whose traces had been hidden by the marsh. All the investigations were aimed at finding these traces and remains of the meteorite (Krinov, 1949). In 1946 the Soviet engineer and science-fiction writer A.P. Kazantsev supposed that this body was in fact an extraterrestrial spaceship which met with disaster at the final stage of its flight. And so was initiated the alternative "artificial" research program. Its supporters tried at first to prove that the Tunguska explosion was nuclear and took place at an altitude of several kilometers.

The 1958 expedition of the Committee on Meteorites of the Academy of Sciences of the USSR established that the explosion of the TCB did occur in the air, and so it was premature to classify this body as a usual meteorite (Florensky et al, 1960). It was with considerable effort that the N-program assimilated this fact. Nonetheless, there soon appeared the hypothesis of a thermal explosion. According to this hypothesis, the TCB was a meteorite or the core of a small comet that exploded as a result of the rapid deceleration in the lower atmosphere (Stanyukovich & Shalimov, 1961).

At about the same time, A.V. Zolotov advanced a number of arguments in favor of a low speed for the body, which meant it could not explode due to its energy of motion (Zolotov, 1961). He thought a nuclear blast was rather likely. At present we can say, however, that the Tunguska explosion was not caused by any of the known nuclear reactions of fission, fusion or annihilation. All of them would have left detectable traces which are lacking (Kirichenko, 1975), Kolesnikov, Lavrukhina, and Fisenko, 1975). In other words, the initial hypothesis of A.P. Kazantsev is refuted by these results. At the same time, there are some peculiarities of the Tunguska explosion which liken this blast to nuclear ones, although without grounds for a more specific identification. The peculiarities include a slightly increased radioactivity of the soil and plants about the epicentrum, thermoluminescence of the rocks, and some biological anomalies (Vasiliev et al., 1976).

Studying the traces of the Tunguska explosion in the framework of the alternative scientific programs, the researchers ascertained its physical features, but could not choose those between the A- and N-hypotheses. Trying to find a way out of this deadlock, Soviet astronomer F. Yu. Zigel made a methodologically interesting step forward in 1966. Based on analysis of testimonies of the witnesses of the TCB in fall, he raised the question of a possible TCB maneuver at the final stage of its flight. It has been rather firmly established (from the structure of the area of the fallen trees and from some witnesses' testimonies which were collected in the 1960s) that immediately before the explosion the body was moving almost exactly east to west (Fast, Barannik, and Razin, 1976; Epiktetova, 1976). But the witnesses' testimonies, gathered in the 1920s, suggested with equal probability that the body might have arrived either from the south, or from the south-east (Sytinskaya, 1955). This evidence cannot be easily rejected because it was obtained shortly after the event. However, the procedure for gathering the surviving testimonies and determining the observed parameters of the body was much better developed in the 1960s. So, it would be wrong to discard the testimony of any of these two groups of witnesses. This contradiction may be due to a complex path followed by the TCB, which can not be approximated by a straight line. However, the eastern variant of the path has been traced to the Lena River. This casts doubt on the possibility of a maneuver at least for this body. It is not improbable that there were several objects flying on very different trajectories (because nobody seemed to see their joint flight), but in this case we should revise our understanding of the phenomenon as a whole.

Of course, the N-program may try to adapt itself to the maneuver possibility, assuming some peculiarities of the TCB shape, as well as those of the distribution of its density. But it is obvious that the A-program would explain it more naturally and simply.

Thus, while the N-program is directed mainly at proving the thermal model of the Tunguska explosion, as well as at looking for associated chemical anomalies in the soil and plants (Vasiliev, 1989), the A-program is orientated towards the search for "intelligent" features of the phenomenon. On the whole, we see that the equality of the alternative A and N programs has favored a more profound and impartial study of this event than a study under domination of one of them.

References

Epiktetova, L.E.: 1976, in Voprosy Meteoritiki, Izdatelstvo Tomskogo Universiteta, Tomsk, 20.
Fast, V.G., Barannik, A.P., Razin, S.A.: 1976, in Voprosy Meteoritiki, Izdatelstvo Tomskogo Universiteta, Tomsk, 39.
Florensky, K.P., et al.: 1960, Meteoritika, 19, 103.
Kazantsev, A.: 1946, Vokrug Sveta, 1, 39.
Kirichenko, L.V.: 1975, in Problemy Meteoritiki, Nauka Novosibirsk, 88.
Kolesnikov, E.M., Lavrukhina, A.K., Fisenko, A.V.: 1975, in Problemy Meteoritiki, Nauka Novosibirsk, 102.
Krinov, E.L.: 1949, Tungusskiy Meteorit, Izdatelstvo Akademii Nauk, Moscow.
Stanyukovich, K.P., Shalimov, V.P.: 1961 Meteoritika, 20, 54.
Straijis, V.: 1986, in Problema Poiska Zhizni vo Vselennoy, Nauka, Moscow, 47.
Sytinskaya, N.N.: 1955, Meteoritika, 13, 86.
Vasiliev, N.V., et al.: 1976, in Kosmicheskoye Veshchestvona Zemle, Nauka, Novosibirsk, 71.
Vasiliev, N.V., 1989, Zemlya i Vselennaya, 3, 29.
Zigel, F.Yu.: 1966, Zhizn v Kosmose, Nauka i Tekhnika, Minsk.
Zolotov, A.V.: 1961, Doklady Akademii Nauk SSSR, 136, 84.

Discussion

J. TARTER: In the past, when astronomers have discovered a new phenomenon and then understood how to best observe them, we always find other examples. However, there is one singular source for which our theorists have formulated natural explanations i.e. SS433. I wonder whether its uniqueness might not be a criterion for artificiality.

V. RUBTSOV: I would say a preliminary criterion. The story of the source SS433 is of interest indeed. Some time ago the Soviet radio-physicist V. M. Tsurileor offered an original method to attract attention of other civilizations: an imitation of phenomena which would violate the laws of the nature. As a matter of fact, he has predicted the object SS433, which displays the red and violet Doppler shifts at the same time. It is significant, however, that this prediction did not much influence the investigation of SS433; astrophysicists worked in the framework of a purely "natural" approach and arrived at last at a common agreement concerning its model. Perhaps it is just impossible to come to the "artificial" solution of a problem, unless we assume such a possibility from the very beginning of the investigation.

P. SCHENCKEL: In reference to Josip Shklorskii's principle on "Artificiality." Artificiality should suppose some sort of a message, such as a series of prime numbers, a mathematical formula or some other intelligent message.

If it does not transmit such a message, it could still be of natural origin, even though - today - it is not yet understood by astronomy's state-of-the-art. In this case, the premature declaration of artificiality would be misleading.

<u>V. RUBTSOV</u>: Certainly so, if you mean just a "declaration of artificiality." But I noted in my report, that "strangeness" as such cannot be a criterion of artificiality. There must be a competition between "natural" and "artificial" programs in order that we arrive after all at a correct solution ("natural" or "artificial"). If your viewpoint is fully correct, we may hope only for discovering an ETI message or a signal. But I think that searches for astroengineering structures are not a waste of time. We should not <u>assert</u> (or declare) the artificial origin of a "strange" cosmic phenomenon, but we can study it from this point of view.

THE ANTECEDENTS OF CONSCIOUSNESS:
EVOLVING THE "INTELLIGENT" ABILITY TO SIMULATE SITUATIONS AND CONTEMPLATE THE CONSEQUENCES OF NOVEL COURSES OF ACTION

William H. Calvin
University of Washington, Seattle, Washington 98195 USA

About 1915, Wolfgang Köhler did pioneering studies of problem-solving behaviors in chimpanzees, now found in most introductory psychology texts as the example of insightful behaviors, perhaps near the threshold of human-style intelligence from conscious contemplation. You probably remember the banana hung from the ceiling of a room occupied by a box and a frustrated chimpanzee. As a later observer described it:

"The matter gave him no peace, and he returned to it again. Then, suddenly -- and there is no other way to describe it -- his previously gloomy face "lit up." His eyes now moved from the banana to the empty space beneath it on the ground, from this to the box, and back to the space, and from there to the banana. The next moment he gave a cry of joy, and somersaulted over the box in sheer high spirits. Completely assured of his success, he pushed the box below the banana. No man watching him could doubt the existence of a genuine "Aha" experience in anthropoid apes."

In 1927, the philosopher Bertrand Russell wrote his own textbook of psychology. And he made a wry comment:

"Animals studied by Americans rush about frantically, with an incredible display of hustle and pep, and at last achieve the desired result by chance. Animals observed by Germans sit still and think, and at last evolve the solution out of their inner consciousness."

I'm not telling that story just because of the analogies to various approaches to SETI -- the people that emphasize random search of the sky vs. those who out of their inner consciousness evolve magic frequencies. Rather, I need to talk about random trial and error as a basis for intelligent behaviors: You don't always have to do your random trial and error in real time, running around and trying this and that, wasting energy and exposed to the hazards of your environment. You can sometimes just do all the trial and error inside your head: that chimp of Köhler's presumably simulated the necessary movements, pieced together the scenario of moving the box and standing atop it, and then acted. The only difference, I suggest, between what Russell characterized as the American and German approaches is whether the random trial-and-error is done on-line or off-line.

Offline trial and error is probably the basic way of solving novel problems and, where fancier procedures such as algorithms are used, trial-and-error was probably the main way of evolving them. We need to know how humans evolved such simulation abilities. And it would be nice to know some alternative paths that an extraterrestrial intelligence might have followed. I'm going to give an example of the neural machinery that intelligence might require, a brief description of how it might have evolved, then discuss hominid evolution more generally and speculate about what might happen in the next century as we better understand the machinery underlying our own higher intellectual functions.

SIMULATION TO SHAPE UP NOVELTIES

Offline simulation is a powerful technique for exploring complex scenarios of possible movements without risking life and limb. But it requires a lot of neural machinery:

1. You need to hold randomly-varied candidate sequences in a planning buffer.

2. You need to judge the candidate against a library of episodic memories (something analogous to "film clips" that retain the order of the actors and actions and incidents). Even if these memories are not identical to the novel plan, we can use these serial-ordered memories in a metaphorical manner, judging candidate scenarios on a poor-to-promising scale.

To do it rapidly, you need dozens to hundreds of planning buffers, a population of sequences, some of which will rate more highly than others. One can then randomly generate some variants of the top-rated candidate and conduct evaluations of this new generation of scenarios. In this manner, and often in a matter of seconds, we can shape up a plan (for a sentence to speak, or movements to make) that stands a better chance of fitting the situation.

This mental procedure seems quite analogous to Darwinian evolution, where the bodies produced by strings of DNA are evaluated by an environment of prey-predators-pathogens and only the survivors reproduce (with variations caused by a little gene-shuffling as sperm and ova are made). In a matter of millennia, the DNA strings are shaped up to roughly "match" the environment. In the immune response, the shaping up takes only a matter of days; the more successful of the initial array of antibodies reproduce with variations in their amino-acid sequence, some of the new generation are even better matched to the antigen-bearing cells, and so forth.

The mental strings are shaped up against a virtual environment of episodic memories, rather than the often noxious real-time environment of trial-and-error. They may be muscle activation commands, such as those needed to throw a rock or hammer a nut -- or perhaps to command something that even apes don't do, such as kicking a football or dancing at the discotheque. Indeed, we humans have greatly augmented serial-order propensities: we string together words into sentences, we string together musical notes into melodies, we try to create scenarios that explain the past and forecast the future.

Thanks to our mental Darwinism, each of us now has under our control a miniature world, evolving away, making constructs that are unique to our own head. Such Darwin Machines (totally unlike our usual deterministic von Neumann machines) may be a prerequisite to general-purpose intelligence, particularly the kind that might produce a technological civilization with a propensity to communicate.

EVOLVING A DARWIN MACHINE

What evolutionary situations might serve to augment such serial-order abilities, up to the point that many planning tracks are available and a Darwin Machine might emerge? Certainly the ballistic movements have the right hallmarks: they are the fraction-of-a-second actions such as throwing, clubbing, hammering, kicking (also, as I was reminded the other day by an orangutan, accurate spitting is also a ballistic task -- she was just trying to get my attention, wanting me to come back and groom her some more, rather than studying the pygmy chimp next door).

Because feedback is so slow (0.1 sec is about the fastest), you need a planning buffer for these faster-than-feedback movements. I should caution that this serial-order buffer is a lot fancier than the text buffer in your computer printer: it is more like a roll for a player piano, as there are at least 88 muscles to which one must send activation sequences to hammer or throw smoothly. Furthermore, you may need many such buffers, perhaps not for hammering but certainly for throwing. Tight launch windows demand precision timing that, given jittery neurons, can only be achieved by hitching many timers in tandem. To hit a target twice as far away, you need to reduce your timing jitter 8-fold. You usually get it only by assigning 64 times as many sequencing buffers to the task (imagine a roomful of player pianos, ganged into tandem, trying to sing as a chorus). Note that such massively serial machinery (dozens to hundreds of player piano equivalents) might be capable of secondary use, if not too hard-wired, e.g., a throwing sequencer might be used in the off-hours as a Darwin Machine for stringing words into sentences, spinning scenarios for future possibilities.

Many animals plan in a limited sense but they are standard behavioral sequences for their species, e.g., nest building when pregnant or nut gathering when the days shorten. For more novel actions, goal-and-feedback works pretty well; a detailed plan of action is seldom needed, because there is time to make corrections along the way. And most animals can modify standard behaviors by associative learning, in the manner of Pavlov's dogs. No player-piano-like serial buffer is really needed -- and certainly not hundreds, each with the means to judge a candidate sequence against serial-order-coded episodic memories. I simply do not know of any animal behavior that requires even a fraction of the sort of neural machinery needed by a Darwin Machine -- not even the throwing and hammering of chimpanzees seems precise enough.

Figure 1: Adapted from Calvin 1990.

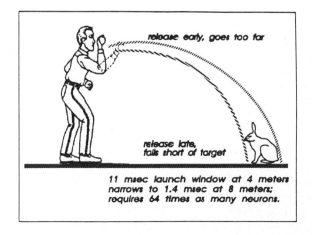

release early, goes too far

release late,
falls short of target

11 msec launch window at 4 meters
narrows to 1.4 msec at 8 meters;
requires 64 times as many neurons.

My argument is that our kind of intelligence arises from a secondary use of neural machinery evolved for its usefulness at the mundane task of throwing accurately and hammering skillfully. And that, as we have gotten better and better at them, there has been something of a jump in language and intelligence as the Darwin Machine secondary uses have emerged.

This is consistent with, though not particularly predicted by, the time course of prehuman evolution. The repeated climatic changes of the Ice Age might have been particularly important in avoiding the stabilities that slow evolution. Upright posture evolved long before the hominid brain started to enlarge beyond ape standards. Brain enlargement started between 2.5 and 2.0 million years ago (the previous major increment occurred more than 30 million years earlier, in the evolution of apes from monkeys). At about 2.5 million years ago is also when archaeological evidence of prolific toolmaking is first seen (my personal guess is that this is also when projectile predation got going too, but there is little agreement among anthropologists on the subject). And 2.5 million years ago is also when the Ice Ages started. One suspects that climate change sped up the evolution of those characteristically human skills and big brain (for climate change at 2.5 million years, see Shackleton et al, 1984, Vrba 1985; for arguments on species dates, see Grine 1989).

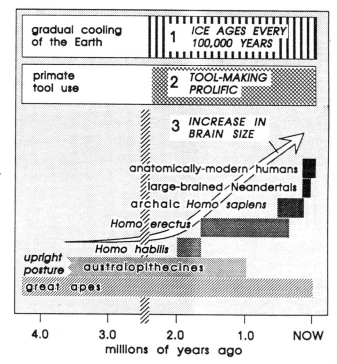

Figure 2: Adapted from Calvin 1990.

While the rapid melting of the ice sheets about every 100,000 years is the most obvious feature of the two-dozen Ice Ages, there are also abrupt changes in climate every so often, certainly within this last ice age and probably associated with switches in ocean currents. The North Atlantic Current, which is what keeps Europe warmer and wetter than Canada at the same latitudes, shuts down on occasion.

The last such episode was the Younger Dryas event of 11,500 years ago when Europe suddenly cooled and the forests died within decades. The North Atlantic Current apparently resumed about 10,720 years ago -- and over the course of just a few years, as you can see from the increase in rainfall, the decrease in severe storms (that's what dust measures), and warming. The cooling at 11,500 years ago was almost as abrupt.

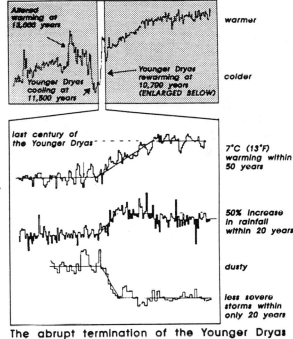

Figure 3: Data from Dansgaard et al. 1989.

The abrupt termination of the Younger Dryas

Evolutionary change can track climate change, so long as climate change happens slowly. But an abrupt change such as the Younger Dryas happens within the time of a single generation, as do the even more rapid changes we associate with droughts or El Niño. Rapid climate changes select, in effect, for versatility: having the body and brain to exist in either climate, able to find food and shelter, able to successfully raise offspring, able to endure the pathogens which come with each climate, etc.

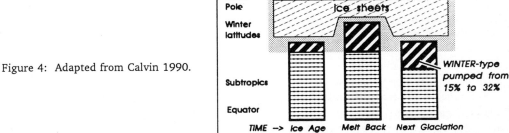

Figure 4: Adapted from Calvin 1990.

Climate change, via expanding and contracting the range of a species, also helps to make minorities into majorities, time and again. For example, living in the temperate zone year-around requires getting through the winter; in arctic aboriginal groups, we see such adaptations as larger bodies (reduced surface/volume ratio) and an emphasis on hunting grazing animals (grass remains edible year-around) or marine mammals (whose food supply is also not reduced in the manner faced by human gatherers in the dormant season). But arctic and temperate zone populations are often only a small fraction (say, 15 percent) of the world-wide population, and such genetic traits spread very slowly into regions where they are not under natural selection (hunting skills, while handy in the tropics, are hardly essential). Yet an ice sheet that melts back over the course of a few hundred generations serves to increase the temperate zone population; when the ice sheet again advances, the temperate-adapted subspecies is pushed into the subtropics, mixing genes with the main population. If the 15 percent minority had increased to 32 percent during the melt back, they might have remained at 32 percent of the smaller total population. Let this expand-and-compress cycle repeat a few times and the main population may start looking and acting remarkably like the temperate zone subtype.

PROSPECTS AND CAUTIONS

Consider the time scale of our own evolution: In only about the last 0.2 percent of the time since the hominid brain started enlarging have we had large-scale social organization; in only a few of the civilizations have technological innovations had any widespread consequences; in only the last dozen generations have we had much in the way of serious science. The radio-based technologies have influenced our thinking about SETI for less than four generations. Computers have been extending our simulation abilities for only several generations.

We have "grown up" very fast, with widening gaps between the abilities of the scientifically illiterate and the sophisticated. Yet I think that, in another century, we will remark on the leap we have made in understanding our own mental processes, and how this greatly expanded our abilities. Consider what happened to transportation in the wake of understanding Newton's physics of moving bodies -- in several centuries, we went from ox-carts to moon rockets. Consider what happened to communication in the wake of our nineteenth century understanding of electricity and magnetism -- we went from hand-carried letters to our ability to load up our memory telephones with a 15-digit sequence that dials a distant phone, connects you via communications satellites. At the mere touch of a single button right now, you can rudely awaken someone on the opposite side of the world. Consider what happened to medicine, once the circulation of the blood and the role of microscopic organisms were appreciated -- in several centuries, we went from purging and leeches to physiologically-based neurosurgery for epilepsy and Parkinson's disease, to persuading lowly bacteria to produce human growth hormone and insulin, merely from snippets of DNA. And computation: from pencil-and-paper to networked supercomputers in about a century. All of which would have seemed like magic to our ancestors, no matter how well-educated they were.

But most of the people in this world (including the 94 percent in the U.S. who lack basic scientific literacy) have no analogies with which to think about how a computer actually functions. Or an aspirin tablet. To many, our high technology is magic, just as a flashlight might appear to a remote tribesman encountering explorers. Because our thinking operates largely via analogies and metaphors, such people are often restricted to the analogy of the give-and-take of social relationships: tit for tat, gifts, deviousness, pleading, placating, flattery. And so they apply such analogies to everything incomprehensible. In the supernaturalistic approach to nature, everything -- say, the weather -- is said to have a spirit; personifying it explains its characteristics and provides humans with a way of influencing it via offerings and pleadings. Personification is still the most widespread "explanation" for mind (we imagine a "little person within" the brain, who acts as voyeur and puppeteer).

One of the key parameters to a Darwin Machine is the "good-enough" judgment. Premature closure -- making up your mind too soon by calling off the search for better scenarios -- is a major hazard, leading to erroneous associations and plans of action that won't work. Premature closure thrives on lack of appropriate mechanistic metaphors; scientists have a wider repertoire of mechanistic metaphors, but it might well be insufficient for interpreting the intelligence behind a message. Yet, once we establish a workable explanation for our thinking and language machinery, we will again see a great augmentation in our capabilities, just from more appropriate metaphors. Imagine what knowledge of the mental machinery will do to augment our sense of self, as we feel in better command of our destinies, better able to choose good courses of action -- able to move beyond the present limits in our abilities to appreciate complicated things, whether they be detailed logical arguments or the complexities of nonlinear systems exhibiting chaos and catastrophe. Rationality will take on an entirely new meaning, once we understand the neural machinery that we use to reason with. That's what another century might bring.

Understanding the mental machinery could make us less susceptible to being manipulated by the powerful. Or perhaps, were the knowledge of the pitfalls not widely known because of poor education, make us more easily manipulated: It would be sad if what our understanding of the mental machinery did was simply to make advertising techniques even more persuasive and pervasive, transferring even more power in our society to those who can afford to buy the best in media exposure. If you think that's a problem, just imagine an alien society that is even a century ahead of us in exploiting knowledge about Darwin Machine mental abilities and pitfalls. Trying to understand them might be difficult without advanced metaphors.

As enthusiastic as I am about the intellectual and technological benefits of the search itself, I remain concerned about how our scientifically illiterate society, with its limited metaphors, might react to exotic communications, should we succeed. Even though personification is understandably a staple of science fiction, social analogies are likely to be totally inappropriate when dealing with alien intelligences. And the more appropriate metaphors with which to imagine an exotic intelligence -- say, from sociobiology, the population thinking of evolutionary biology, neural circuitry, and Darwin Machines -- are still pretty much in their infancy.

References

Calvin, W. H., The River that Flows Uphill: A Journey from the Big Bang to the Big Brain (Macmillan, 1986).

Calvin, W. H., The brain as a Darwin Machine, Nature 330:33-34 (1987).

Calvin, W. H., Fast tracks to intelligence. In Bioastronomy -- The Next Steps, edited by G. Marx (Kluwer, 1988). pp. 237-245.

Calvin, W. H., The Cerebral Symphony: Seashore Reflections on the Structure of Consciousness (Bantam, 1989).

Calvin, W. H., The Ascent of Mind: Ice Age Climates and the Evolution of Intelligence (Bantam, 1990).

Dansgaard, W., J. W. C. White, and S. J. Johnsen, The abrupt termination of the Younger Dryas climate event. Nature 339:532-534 (1989).

318

Grine, F. E. (editor), Evolutionary History of the "Robust" Australopithecines. (Aldine de Gruyter, 1989).
Shackleton, N. J., et al, Oxygen isotope calibration of the onset of ice-rafting and history of glaciation in the North Atlantic region. Nature 307:620-623 (1984).
Vrba, E. S., Ecological and adaptive changes associated with early hominid evolution. In: Ancestors: The Hard Evidence, edited by Eric Delson, pp. 63-71 (Liss, 1985).

Discussion

D. BRIN: Bill, you list several successive circumstances contributing to our advancement which might be called fortuitous. Even most human cultures showed little technological ambition. Is it possible we may find that there are many marginally intelligent species out there in space, but only a few who benefitted from such a chain of lucky boosts to become so fiercely curious?

W. CALVIN: Mammalian cleverness seems far more likely than higher intelligence. Omnivorous diets require that animals accumulate a variety of strategies for finding food; fickle climates are likely the primary drive for spreading such versatility to the main population from small stressed subpopulations. We have the usual mammalian tendencies to poke our noses into things. But our "fierce" variety of curiosity seems secondary to scenario-spinning abilities; we construct (and mostly discard) scenarios that attempt to explain the past, and forecast the future. It is not an unmixed blessing: it is from scenario-spinning that we get our propensity to worry and suffer (which is quite different from pain sensation). I'd think twice about "uplifting" chimpanzees and dolphins to higher intelligence, as I discuss in the last chapter of The Cerebral Symphony. They might not thank us.

D. SCHWARTZMAN: I don't understand the role of ice ages in the last two million years in the evolution of human intelligence when all speciation of hominoids in that period apparently, at least according to present anthropological consensus, took place in East Africa, rather than in the rapidly changing climate of Europe.

W. CALVIN: 1) The ice sheets may have only been above 40° N, but climates were shifted even in Africa, e.g., glaciers developed on equatorial volcanos such as Mt. Kenya. 2) Subpopulations out of Africa may have spread their genes back into Africa when ice sheets pushed the temperate zone subpopulaton south.

L. KSANFOMALITY: Do you really believe that the human brain increased in size, say since Socrates' time?

W. CALVIN: The last size increase was about 120,000 years ago. Tooth size has decreased about 15% since then, but brain size has stayed about the same.

C. MATTHEWS: Could you comment on the Stephen Gould thesis (as in Wonderful Life, 1989) that biological evolution is so contingent that intelligent life is by no means to be expected on an Earth-like planet? Or anywhere.

W. CALVIN: I agree with Steve Gould. So far as I can see, there is nothing inevitable about higher intelligence evolving from mere cleverness, of the sort evolved by the ancestors of rat, raven, and octopus. Smarter may be better, everything else being equal, but everything else is seldom equal - and so species often settle into dead end stabilities.

I suspect that our uniquely human abilities such as plan-ahead consciousness, language, music, dance, accurate throwing and kicking, etc. are mostly due to a fortuitous consequence of getting better and better at projectile predation. Such styles of hunting are hardly inevitable; getting through the winter would seem to accelerate it, and the many climate changes would seem to spread these advantages around the world.

This is _not_ the usual narrow-minded argument that intelligence is unique to humans on Earth. I can imagine various other ways to evolve neural-like machinery that could be secondarily used as a Darwin machine. Fortuitous does not equate with improbable.

COGNITION IN AN AFRICAN GREY PARROT*

Irene M. Pepperberg

Department of Anthropology, Northwestern University, Evanston, Illinois 60208 USA†

Paper Read by M. J. Klein

Although parrots often reproduce the sounds of human speech,[1] the general perception has been that such vocalizations are never meaningful.[2] The unsuccessful use of standard psychological training paradigms by American researchers (e.g., Mowrer; Grosslight & Zaynor) to teach birds to engage in referential communication[3,4] further entrenched the perception of parrots as mindless mimics both in the public mind and also in the scientific literature.

A few studies, however, challenged this idea. During the 1940s and 1950s, European researchers showed that Grey parrots could learn symbolic and conceptual tasks often considered as pre- or co-requisites for complex cognitive and communicative skills (reviews in Pepperberg[5,6]). Data from subsequent field and laboratory studies also indicated that natural psittacine vocalizations might mediate social interactions and are likely learned from other flock members.[7] These two lines of research suggested that the American psychologists' failures might be due to inappropriate training techniques, rather than to an inherent lack of ability in the psittacine subjects.[5]

I decided to test this premise. I developed techniques to integrate the experimental rigor of the laboratory with what little was known about psittacine communication in nature and with data from studies on human social learning (reviews in Pepperberg[6,7,8]). These techniques have been successful, and the following sections provide details of the procedures and a short summary of the results.

THE SUBJECT OF THE STUDY

The subject, Alex, an African Grey parrot, has been the focus of a study on interspecies communication and avian cognition since June 1977. At the start of the project he was about 13 months old and had received no prior formal vocal instruction. He has free access to the laboratory room while trainers are present (~8 hrs/day), but is confined at other times to a cage (~ 62 x 62 x 73 cm) and the desk upon which it rests. Water and a standard psittacine seed mix are continuously available; fresh fruits, vegetables, specialty nuts (cashews, pecans, etc.), and toys are used in training and are provided at the bird's vocal requests.

---

*Material in this review has been drawn primarily from publications by the author; see references 5, 6, 7, and 8.

†Author's present address: Department of Ecology and Evolutionary Biology, University of Arizona, Tucson, Arizona 85721 USA.

TRAINING TECHNIQUES

Use of Intrinsic Rewards (Pepperberg 1981, 1990a,b)

Several training procedures are used in this project, but one feature common to all is the consistent, exclusive use of intrinsic reinforcers — i.e., the reward for a correct response is the object to which the targeted question refers, rather than any single, extrinsic item. Thus, if Alex correctly identifies a key, that is what he receives. This procedure insures, at all times and at every interaction, the closest possible association of the label or concept to be learned and the object or task to which it refers.[5,9]

Programs such as Mowrer's relied on extrinsic rewards[3]: Identifications of food or nonfood items or any appropriate responses to various specific commands were rewarded with a single, preferred food that neither directly related to, nor varied with, the specific task being targeted. But extrinsic rewards can delay label or concept acquisition by confounding the label of the exemplar or concept to be learned with that of the food reward.[9,10,11] Alex therefore never receives extrinsic rewards.

If Alex is not interested in obtaining the objects that are being used to train a particular concept, his reward will be the right to request vocally ("I want X") a more desirable item than the one he has identified. Such a protocol provides some flexibility but maintains the referentiality of the reward: Alex will never, for example, automatically receive a nut for identifying a cork; the nut must specifically be requested ("I wanna nut"), and trainers will not respond to such a request until the appropriate prior task is completed.[12]

The Model/Rival (M/R) Technique (Pepperberg 1981, 1990a,b)

The primary training system, the model/rival (M/R) technique, involves three-way interactions between two competent human speakers and Alex. The technique is based on a protocol developed by Todt, an ethologist who studied social learning in parrots,[13] and on Bandura's studies on the effects of social modeling on learning in humans.[14] In Alex's presence, two humans handle objects that are related to the task. One human then acts as a trainer of the second human. The trainer presents an object, asks questions about the object (e.g., "What's here?", "What color?", "What shape?"), and gives praise and the object itself as a reward for a correct answer. Disapproval for incorrect responses (erroneous answers that are similar to those being made by Alex at the time: unclear vocalizations, partial identifications, etc.) is demonstrated by scolding and temporarily removing the object(s) from sight. Thus the second human not only acts as a model for the bird's responses and as a rival for the trainer's attention, but also allows Alex to observe the effects of an error: The model is asked to try again or to talk more clearly if the response was (deliberately) incorrect or garbled.

Unlike Todt's modeling procedure (and that of some other researchers[7]), our protocol also repeats the interaction while reversing the roles of the human trainer and model, and includes Alex in the interactions. We thus demonstrate that interaction is indeed a "two-way street": that one person is not always the questioner and the other always the respondent, and that the procedure can be used to change the environment. Inclusion of role reversal thus counteracts what would be, for our project, the drawbacks associated with Todt's method: Todt's birds were exposed only to pairs of individuals maintaining their respective roles, and his birds did not transfer their responses to anyone other than the human who posed the questions. In contrast, Alex responds to, interacts with, and learns from all of the trainers with whom he comes in contact.

Sentence Frames (Pepperberg, 1990b)

After Alex begins to produce a new label in the presence of a new exemplar, we use an additional procedure to clarify his pronunciation. We present the new exemplar to him with a string of sentence frames — e.g., "Here's your paper!", "Such a big piece of paper!". These sentences allow us to produce a target word, such as "paper", frequently and with consistent stress, without presenting it as a single, repetitive utterance. This combination of a particular form of vocal repetition and the physical action of presenting the object resembles the behavior parents sometimes use when introducing labels for new items to very young children,[15,16] and appears to have two effects: (1) Alex hears the label employed in normal, productive speech so that he experiences the label in the way in which it is to be used; and (2) he learns to reproduce the emphasized, targeted label without associating simple word-for-word imitation of the trainers with reward.

Referential Mapping (Pepperberg, 1990a)

We use another technique, called referential mapping, to assign meaning to novel vocalizations that Alex may produce spontaneously. Such vocalizations are generally combinations and phonetic variations of English labels already in his repertoire. Thus, after learning "grey", he produced "grape", "grate", "grain", "chain" and "cane". Unlike vocalizations that are trained by the M/R procedure, these recombinations are not necessarily initially used intentionally to describe or request novel objects or circumstances. The procedures for dealing with these spontaneous modifications, however, neither attempt nor need to evaluate Alex's intentionality. Rather, the procedures instruct his trainers to respond to the novel speech acts as though he were intentionally commenting about or requesting objects, actions, or information.

Three procedures constitute the technique of referential mapping:

(1) Because other studies on humans and birds suggest that experiencing the appropriate consequences of an utterance may assist learning, trainers respond to Alex's vocalizations with an appropriate object or action — i.e., as if he does indeed understand the significance of what he is saying. Whether Alex intended to produce the combination is not important; we simply demonstrate that these phrases can be meaningful and that they can be used to control, or at least influence, his environment and the actions of his caretakers. Analogous mechanisms may occur in animal systems in nature: Certain young birds, through their interactions with adult conspecifics, learn not only what to sing but also how song is to be used (see reviews in Pepperberg[12,17,18]).

(2) We also use a variation of the M/R technique to demonstrate further the possible relevance of Alex's spontaneous recombinations. Here two humans model a communicative interaction corresponding to the novel vocalization: One human produces this vocalization while the other produces an object, an exemplar of the term (e.g., for colors), or demonstrates the action to which it refers. The roles of the humans are then reversed, so that Alex observes that the vocalization is neither specific to nor controls only a particular individual's actions. If Alex emits the vocalization during this demonstration, he is shown (and occasionally receives) the object or action (e.g., a "chain" of paper clips). Humans not only model identification of the object or action (by responding to each other's queries of "What's this?"), but, when possible,

use objects that demonstrate various applications of the vocalization; e.g., produce "box" exemplars of two colors.

(3)   Finally, trainers use sentence frames to provide further cues about the appropriate context in which the object or action label should be used. While the targeted object is manipulated or the action demonstrated, either by a human or Alex, humans produce sentences such as "You're eating a green nut!", "I'm holding a green nut!", "Do you want another green nut?", in which only the label for the action or object remains fixed and stressed. As before, Alex hears numerous repetitions of the label, but in a way that demonstrates the connection between label and referent.

RESULTS AND IMPLICATIONS OF TRAINING

Alex's Accomplishments (adapted and updated from Pepperberg, 1988)

Using these techniques, my students and I have taught Alex tasks that were once thought beyond the capability of all but humans or, possibly, certain nonhuman primates.[19] Alex has learned labels for more than 35 different objects: paper, key, wood, hide (rawhide chips), grain, peg wood (clothes pins), cork, corn, nut, walnut, showah (shower), wheat, banana, box, pasta, gym, cracker, scraper (a nail file), chain, shoulder, block, rock (lava stone beak conditioner), carrot, gravel, back, chair, chalk, water, nail, grape, cup, grate, treat, cherry, wool, popcorn, citrus, green bean, and banerry (apple). We have tentative evidence for labels such as bread and jacks. He has functional use of "no", phrases such as "come here", "I want X" and "Wanna go Y" where X and Y are appropriate labels for objects or locations. Incorrect responses to his requests by a trainer (e.g., substitution of something other than what he requested) generally results (~75% of the time) in his saying "No" and repeating the initial request.[12,20] He has acquired labels for 7 colors: rose (red), blue, green, yellow, orange, grey, and purple. He identifies five different shapes by labeling them as 2-, 3-, 4-, 5-, or 6-cornered objects. He uses the labels "two", "three", "four", "five", and "sih" (six) to distinguish quantities of objects up to 6, including collections made up of novel objects, heterogeneous sets of objects, and sets in which the objects are placed in random arrays.[21,22] He combines all these vocal labels to identify proficiently, request, refuse, categorize, and quantify more than 100 different objects, including those that vary somewhat from training exemplars. His accuracy has averaged ~80% when tested on these abilities.[5,20,21,22,23]

We have also examined Alex's comprehension of the concept of "category". He has learned not only to label a number of different hues or shapes, but also that "green", for example, is a particular instance of the category "color". He has learned that, for a particularly colored and shaped object, "green" and "three-corner" represent different categories of markable attributes of this single exemplar. Thus he can categorize objects having both color and shape with respect to either category based on our vocal query of "What color?" or "What shape?"[24] Because the same exemplar is often categorized with respect to shape at one time and color at another, the task involves flexibility in changing the basis for classification. Such flexibility, or capacity for "reclassification", is thought to indicate the presence of "abstract aptitude".[25]

Alex has also learned abstract concepts of "same", "different", and to respond (with the vocalization "none") to the absence of information about these concepts. Such faculties were once thought beyond the capacity of an avian subject (note Premack[19,26]; but see Zentall, Hogan, & Edwards[27]). Thus, when presented with two objects that are identical or that vary with respect to some or all of the attributes of color, shape, and material, Alex responds with the appropriate <u>category</u> label as to which attribute is "same" or "different" for any combination.[23] If, however, nothing is same or different, he has learned to reply "none".[18] He responds equally accurately to objects, colors, shapes, and materials not used in training, including those for which he has no labels. Furthermore, Alex is indeed responding to the specific questions, and not merely responding on the basis of his training and the physical attributes of the objects: His responses were still above chance levels when, for example, the question "What's same?" was posed with respect to a green wooden triangle and a blue wooden triangle. If he were ignoring the question and responding on the basis of his prior training, he would have determined, and responded with the label for, the one anomalous attribute (in this case, "color"). Instead, he responded with one of the two appropriate answers [in this case, "shape" or "mah-mah" (matter)].[18,23]

In a further demonstration of his comprehension skills, Alex showed that he could view different collections of 7 physical exemplars (each collection chosen from among 100 objects of various combinations of shapes, colors, and materials), be asked 1 of 4 possible vocal questions, each of which requested a different type of information (e.g., "What color is object-X?") about a single object in the collection, and reply vocally to each question. A correct response indicated that he understood all the elements of the question and used these elements to guide the search for the one object in the collection that provided the requested information. Alex responded with an accuracy of 81.3%.[28]

Implications of the Results

The import of these findings is two-fold. First, note that a successful research design was possible only after data had been collected on the behavior of these birds in the wild. Second, the data indicate the existence of complex cognitive capacities in a species whose brain organization is considerably different from that of terrestrial and aquatic mammals and that may be specialized for behaviors we have yet to discover. That his abilities are comparable to those of mammals, and that such capacities have been found in so unexpected a source, suggests that our search for other forms of intelligence is not likely to be in vain.

References and Notes

1. See, for example, Baldwin J.M.: Deferred imitation in West African Grey parrots. IXth Int'l Cong. Zool. 536, 1914.
2. Lenneberg E.H.: Biological Foundations of Language. Wiley & Sons, NY, 1967.
3. Mowrer O.H.: Learning Theory and Personality Dynamics. Ronald Press, NY, 1950.
4. Grosslight J.H., Zaynor W.C.: Vocal behavior of the mynah bird. In Salzinger K., Salzinger S. (eds): Research in Verbal Behavior and Neurophysiological Implications. Academic Press, NY, 1967, pp 5-9.
5. Pepperberg I.M.: Functional vocalizations by an African Grey parrot (Psittacus erithacus). Z Tierpsychol 55:139-160, 1981.
6. Pepperberg I.M.: An investigation into the cognitive capacities of an African Grey parrot (Psittacus erithacus). In Rosenblatt J.R., Beer C., and Slater P.J.B. (eds): Advances in the Study of Behavior, Academic Press, 1990b, pp 357-409.

325

7. For details and original references, see Pepperberg I.M.: The importance of social interaction and observation in the acquisition of communicative competence: Possible parallels between avian and human learning. In Zentall, T.R., Galef, B.G. (eds): Social Learning: Psychological and Biological Perspectives. Erlbaum, Hillsdale, NJ, 1988, pp 279-299.
8. Pepperberg I.M.: Referential mapping: attaching functional significance to the innovative utterances of an African Grey parrot (Psittacus erithacus). Appl Psycholing 11:23-44, 1990a.
9. Pepperberg I.M.: Object identification by an African Grey parrot (Psittacus erithacus). Abstr, Midwest Animal Behavior Meeting, W. Lafayette, IN 1978.
10. Greenfield P.M.: Developmental processes in the language learning of child and chimp. Behav Brain Sci 4:573-574, 1978.
11. Miles H.L.: Apes and language. In de Luce, J., Wilder, H.L. (eds): Language in Primates. Springer-Verlag, NY, 1983, pp 43-61.
12. Pepperberg I.M.: An interactive modeling technique for acquisition of communication skills: Separation of 'labeling' and 'requesting' in a psittacine subject. Appl Psycholing 9:31-56, 1988.
13. Todt D.: Social learning of vocal patterns and modes of their applications in Grey parrots. Z Tierpsychol 39:178-188, 1975.
14. Bandura A.: Analysis of social modeling processes. In Bandura, A. (ed): Psychological Modeling. Aldine-Atherton, Chicago, 1971, pp 1-62.
15. Berko-Gleason J.: Talking to children: Some notes on feedback. In Snow, C.E., Ferguson, C.A. (eds): Talking to Children. Cambridge University Press, Cambridge, 1977, pp 199-205.
16. De Villiers J.G., De Villiers P.A.: Language Acquisition. Harvard University Press, Cambridge MA, 1978.
17. Pepperberg I.M.: Social modeling theory: A possible framework for understanding avian vocal learning. Auk 102: 854-864, 1985.
18. Pepperberg I.M.: Comprehension of "absence" by an African Grey parrot: Learning with respect to questions of same/different. JEAB 50:553-654, 1988.
19. Premack D.: On the abstractness of human concepts: Why it would be difficult to talk to a pigeon. In Hulse, S.H., Fowler, H., Honig, W.K. (eds): Cognitive Processes in Animal Behavior, Erlbaum, Hillsdale, NJ, 1978, pp 421-451.
20. Pepperberg I.M.,: Interspecies communication: A tool for assessing conceptual abilities in the African Grey parrot (Psittacus erithacus). In Greenberg, G., Tobach, E. (eds): Language, Cognition, and Consciousness: Integrative Levels. Erlbaum, Hillsdale, NJ, 1987, pp. 31-56.
21. Pepperberg I.M.: Vocal identification of numerical quantity by an African Grey parrot. Abstr. Psychonomic Society Meeting, San Antonio, TX, 1984.
22. Pepperberg I.M.: Evidence for conceptual quantitative abilities in the African Grey parrot: Labeling of cardinal sets. Ethology 75:37-61, 1987.
23. Pepperberg I.M.: Acquisition of the same/different concept by an African Grey parrot (Psittacus erithacus): Learning with respect to color, shape, and material. Anim Learn Behav 15:423-432, 1983.
24. Pepperberg I.M.: Cognition in the African Grey parrot: Preliminary evidence for auditory/vocal comprehension of the class concept. Anim Learn Behav 11: 179-185, 1983.
25. Hayes K.J., Nissen C.H.: Higher mental functions of a home-raised chimpanzee. In Schrier, A., Stollnitz, F. (eds): Behavior of Nonhuman Primates, Vol. 4, Academic Press, NY, 1956/1971, pp. 57-115.
26. Premack D.: The codes of man and beasts. Behav Brain Sci 6:125-167, 1983.
27. Zentall T.R., Hogan D.E., Edwards C.A.: Cognitive factors in conditional learning by pigeons. In Roitblat, H.L., Bever, T.G., Terrace, H.S. (eds): Animal Cognition, Erlbaum, Hillsdale, NJ, 1984, pp 389-405.
28. Pepperberg, I.M.: Cognition in an African Grey parrot (Psittacus erithacus): Further evidence for comprehension of categories and labels. J Comp Psych 104: 41-52, 1990a.

Discussion

G. MARX: Hungarians - like myself - having a different grammar and history, etc., "enjoy" the difficulties of communication abroad. If two intelligent races (e.g., Homo sapiens and Australopithecus) meet, would they be able to communicate to the top of their intelligence, or would they be restricted to concrete objects that they show each other? (This would mean restriction to "kiddy talk"!) Scientists like to say that Newton's and Maxwell's equations on quantum mechanics are the same everywhere, and there will be the "common objects" for SETI, but even humans have learned so many different formulations of the same laws. Are you optimistic about SETI beyond $1 + 1 = 2$?

W. CALVIN (for I.M.PEPPERBERG): Yes, the chimpanzee work shows that a substantial common vocabulary and sentence usage is possible. We'll have lots of physics to talk about, even if they consider ours to be at "Aristotelian" levels.

ARCHAEOLOGY IN SPACE: ENCOUNTERING ALIEN TRASH AND OTHER REMAINS

D. L. Holmes

Institute of Archaeology, University College, London, U.K.

1. INTRODUCTION

As of September 1959, when Cocconi and Morrison showed that it had become technologically feasible to detect a signal beamed in our direction by an alien intelligence, we human beings had only taken the first few tentative steps into space. We had managed to orbit a few small satellites, and to crash land a probe (Luna 2) on the Moon. But such early efforts in spaceflight were hardly relevant to the Search for Extraterrestrial Intelligence (SETI). Inspired by Cocconi and Morrison, those wishing to look for Extraterrestrial Intelligence (ETI) turned to ground-based radio telescopes.

However, as someone who studies the only technological intelligence that we know of, i.e. ourselves, I find the continued mainstream SETI preoccupation with searching for ETI-generated radio signals unnecessarily narrow. Radio signals are but one kind of artifact, and a signal reaching us from within our own galaxy could only have been transmitted less than 100,000 years ago--an extremely short time in astronomical or geological terms. An artifact may be broadly defined as anything, be it matter or energy, made or modified by a human being or other intelligence, or any kind of trace resulting from the technological activities of an intelligent species, although in this paper, the term "artifact" is used to denote material objects and traces rather than electromagnetic signals, except where specifically referred to.

If we are prepared to admit the possibility of ETI capable of interstellar radio communication, we should also presuppose alien intelligence capable of some form of interstellar travel. Even we are about to have our own unmanned probe in interstellar space, Pioneer 10, which left the Solar System seven years ago and should soon cross the heliopause. Numerous studies have been undertaken which indicate the feasibility of interstellar travel (see e.g. Forward 1986; Mallove & Matloff 1989), and many authors have concluded that if there are advanced technological ETs, they should have colonized the galaxy by now (e.g. Hart 1975; Papagiannis 1978). Thus, if there is ETI capable of either "occupied" or "unoccupied" interstellar flight, then there is the possibility of ETI presence, past or present, within the Solar System.

2. EXTRATERRESTRIAL ARTIFACTS

It has been suggested that extraterrestrial probes or starships could conceivably be already present in the Solar System (e.g. Freitas 1983; Papagiannis 1978), and that starship exhaust trails (e.g. Martin & Bond 1980), and astroengineering activities (e.g. Dyson 1960) beyond the Solar System could be detected. These few examples illustrate the tendency of other researchers to consider only macro-scale extraterrestrial artifacts (ETAs) existing at broadly the present time. Yet with ourselves, the most likely kind of artifact to be found of any past or present society is a piece of trash: a potsherd or a plastic bag. Alien trash could range from dust-sized particles to derelict spaceships.

We, ourselves, are rapidly littering the Solar System. Right above us, in low earth orbit, is a veritable midden heap comprising trillions of pieces of trash consisting of dead satellites, spent rocket stages and a whole range of smaller debris including minuscule objects like paint flakes and aluminum oxide dust particles (Furniss 1988). There are many space vehicles and numerous bits of equipment and debris on the Moon, several spacecraft both on and orbiting Mars and Venus, and various other spacecraft and related debris in interplanetary space.

The generation of trash is one of the hallmarks of human beings, and it will continue to be so for the foreseeable future. Based on our own example, why should we expect all alien species to behave any differently?

Unlike radio signals which may be viewed as transient artifacts, material ETAs could be of great antiquity if their preservation allowed. I shall not speculate here how different sorts of artifacts might survive in various environments, but it is certainly an area which will need addressing in any more detailed consideration of ETAs. Nevertheless, the majority of possible objects are likely to survive for at least a few million, or tens of millions of years in most space environments, and on small rocky bodies such as the Moon and asteroids.

3. DETECTION OF ETAs

A few searches for alien artifacts, both within the Solar System and beyond, have already been made (Freitas & Valdes 1980; Valdes & Freitas 1983; Papagiannis 1985). However, the detection capabilities of the instruments used in these searches have been limited, although Valdes and Freitas (1983) estimated that they could have detected objects with lunar albedo in some Earth-Moon libration orbits as small as 1-3 meters.

Except for the Moon, the planets and various small bodies of the Solar System generally have not been observed at better than 20 meters spatial resolution, and mostly considerably less. However, if an alien intelligence has ever visited the Solar System, then it is likely that with further exploration using unmanned space probes, any objects or traces left by it will be discovered within the next few decades.

There have already been suggestions that some features seen on one planet of the Solar System may not be natural--the so-called "face" and other surface features in the Cydonia region of Mars (Hoagland 1987; O'Leary 1990; Carlotto & Stein 1990). Even if they are natural, which I personally consider to be the case, they are perhaps at least worth studying from a SETI perspective. The Cydonia features are large (the "face," for example, is about 2 km wide), and were imaged by the Viking Orbiter at a resolution of about 50 meters. The Mars Observer, due to be launched in 1992, should provide a much better resolution capability with selected areas being imaged with a resolution of 1.4 m.

For seeing beyond the Solar System we have a variety of astronomical spacecraft, in particular the Hubble Space Telescope with its ability to detect objects 25-50 times fainter than Earth-based telescopes can view.

4. SOME THOUGHTS ON HANDLING AND ANALYZING ETAs

Since our ability to detect extraterrestrial artifacts is still somewhat limited, it is perhaps not yet worth undertaking a massive dedicated search effort within the Solar System. Nevertheless, we are in the process of exploring the Solar System in increasing detail, so that there is the possibility that we may, by chance, encounter an object of alien manufacture, and it is the contention of the writer that we should be properly prepared for such a discovery. With the detection of a suspected ETA, there are three basic issues of concern; verification, study, and conservation.

4.1 Verification

When a suspected ETA is found, its artificial nature may not be unequivocal, so it is necessary to consider ways of verifying whether an object is truly artificial and not natural. Carlotto and Stein (1990), for example, have recently applied fractal modelling to assess the "face" and other features in the martian Cydonia region. While the approach has yielded some interesting results, it has not provided a definitive answer.

4.2 Study

The study of an alien artifact is a rather large topic which is difficult to give justice to in this short paper. It naturally overlaps with both verification and conservation, and how one goes about investigating an artifact depends whether the object is something discovered by remote sensing, or whether it is something that we can physically examine. If it is something detected by the sensors of a spacecraft, then we do our best to analyze the data returned. If it is an object we have the option of picking up and carrying elsewhere for study, then that is a much more complex situation. Some broad guide-lines for such a case may be stated as follows:

(a)   Study the object in context as far as possible. Even if, at first, we can recognize an object as a definite ETA, we will know nothing else about it. It could be dangerous in some way, or it might be fragile and disintegrate if touched. It will also be extremely important to study the context of the find for clues as to how and when it got to where it is found. Various remote sensing techniques are advocated at this stage.

(b)   If it is considered safe to move the object and take it elsewhere for further study, it is probably advisable to take it to a laboratory on a space station rather than to bring it directly to Earth. Unless the object was completely understood, full Earth gravity might be damaging and there might still be a possibility that the artifact was in some way harmful.

(c)   Non-destructive methods of analysis should be used as far as possible. Obviously, we shall want to learn as much as we can about the object, but there may be a limit and with an artifact of great historic value there is no sense in needlessly destroying it.

Table 1 lists some possible general classes of artifacts. "Trash" may be defined as artifacts that have been used, but are now discarded and probably no longer functional. Trash also includes the traces of past activities which are no longer being continued. "Operational" artifacts can be regarded as items that are still functional and being used. Moreover, there are many objects produced by human beings which do not have straightforward logical functions. They reflect our sentiments and emotions, or more formally, our traditions and religions. If the output of trash is a hallmark of human beings, so is culture which at its simplest can be considered a learned social programming. Consequently, I have allowed for "symbolic" artifacts. However, the listing in Table 1 is a theoretical one. Archaeologists frequently have problems determining even the general function of objects and sites of past human cultures, so if we have difficulties with artifacts from our own less-technologically-advanced past, we should not expect to fully comprehend an object of advanced alien manufacture, unless of course the aliens deliberately left the object for us to find and interpret. We may have to content ourselves with knowing only that it is an artifact of alien origin.

Table 1: Possible classes of extraterrestrial artifacts.

A. ARTIFACTS WITHIN THE SOLAR SYSTEM

Trash:	Operational artifacts:	Symbolic artifacts:
rubbish, debris	sensors and other equipment	monuments
defunct equipment	unoccupied spacecraft	ritual artifacts
defunct spacecraft	occupied space vehicles	burials
activity traces		

B. ADDITIONAL ARTIFACT CLASSES BEYOND THE SOLAR SYSTEM

starship exhaust trails
astroengineering activities
radio signals

4.3 Conservation

If an object has survived several millions of years in space, it is possible that it is not so fragile as to disintegrate as soon as it is touched. Careful consideration will be needed if the object is to be moved to a laboratory for further study as it may start to deteriorate in a different environment. After study, the environmental requirements for the object's long-term storage need assessing. It might be safest to return it to its find locality, although as a discovery of tremendous historic significance it would be desirable to display the artifact in some museum situation, perhaps with a very carefully controlled environment, for all to see.

5. SUMMARY

As we continue to explore the Solar System with our spacecraft, and beyond the Solar System with increasingly sophisticated space-based observatories, the possibility of encountering a material alien artifact is as conceivable as the detection of a transient radio signal. We should be well prepared for such a discovery, particularly in the case where we have the option of collecting an ETA. Verification, study (including study of the context of an ETA find), and conservation are the key aspects to the scientific investigation of an alien object.

References

Carlotto, M.J., Stein, M.C.: 1990, J. Brit. Interplan. Soc. 43, 209.
Cocconi, G., Morrison, P.: 1959, Nature 184, 844.
Dyson, F.J.: 1960, Science 131, 1667.
Forward, R.L.: 1986, J. Brit. Interplan. Soc. 39, 379.
Freitas, R.A. Jr.: 1983, J. Brit. Interplan. Soc. 36, 501.
Freitas, R. A. Jr., Valdes, F.: 1980, Icarus 42, 442.
Furniss, T.: 1988, Flight International (July 30), 28.
Hart, M.H.: 1975, Quart. J. Roy Astr. Soc. 16, 128.
Hoagland, R.C.: 1987, The Monuments of Mars, North Atlantic Books, Berkeley.
Mallove, E.F., Matloff, G.L.: 1989, The Starflight Handbook, Wiley, New York.
Martin, A.R., Bond, A.: 1980, in Strategies for the Search for Life in the Universe; ed. M.D.
    Papagiannis, D. Reidel Publishing Company, Dordrecht, p. 197.
O'Leary, B.: 1990, J. Brit. Interplan. Soc. 43, 203.
Papagiannis, M.D.: 1978, Quart. J. Roy. Astr. Soc. 19, 277.
Papagiannis, M.D.: 1985, Nature 318, 135.
Valdes, F., Freitas, R.A., Jr.: 1983, Icarus 53, 453.

Discussion

M. HARRIS: One comment and one question. 1) If any suspect artifact is ever found, a guaranteed way of showing that it is alien is to determine that it has non-solar-system isotope ratios. 2) Do you think aliens will produce any trash at all? They will need to exist for years at a time in a completely closed environment while traveling between stars. They will therefore need to recycle everything very efficiently.

D. HOLMES: I agree with your comment because I think the material composition of a suspected ETA would be a probable indication of its non-natural origin, but I didn't have time to go into this in my talk. However, though I take your point about aliens being efficient recyclers during their interstellar voyage, there are abundant raw materials within the Solar System so if they stopped here they could afford to be wasteful. Wherever we go, even space, we leave litter and as I indicated in my presentation, we have no a priori reason to suppose that all alien intelligence should behave any differently. We should at least keep our minds open to the possibility. Of course, the recycling argument doesn't apply to, say, an "unoccupied" space probe which became inoperable within the Solar System.

B. FINNEY: One of the archaeologists in my department (Dept. of Anthropology, University of Hawaii) likes to tell the story about how, one day in the mid-1960s, when he was a doctoral student at the University of Arizona, he was approached by his professor who asked him if he would be interested in applying to be a scientist astronaut. It seems that NASA had contacted the professor and told him that they might want an archaeologist to go to Mars to investigate the possibility that intelligent (i.e., artifact producing) life might have once existed there. Nothing, of course, came of this, but do you think archaeological training might be part of the curriculum of future inter-planetary astronauts/cosmonauts?

D. HOLMES: Unless something suspicious had been seen on say, the Moon or Mars, by an unmanned space probe in advance of a manned mission, then a significant amount of archaeological training is probably unnecessary. The astronauts should perhaps be briefed and given guidelines such as I've outlined. Otherwise they could depend on instructions from appropriate Earth-based archaeologist-specialists if a suspected ETA were encountered.

D. BLAIR: I wonder if present activities on Earth including mining of coal (once swamps and forests on the surface) and limestone (sediments based on shell fish, etc.), and searches for meteorites and alluvial gold, do not presently constitute very extensive search. Modern high-strength materials, from glasses, other ceramics, plastics, titanium, aluminum and stainless steel would all survive much better than dinosaur bones. Most would be very conspicuous in a mining process in much the way that peat marsh bodies are often discovered. The absence of such discoveries perhaps sets strong limits on the frequency of extraterrestrial trash-generating visits.

D. HOLMES: The Earth is one of the worst places in the Solar System for preservation with its thick atmosphere, hydrosphere, etc., so I have my doubts that any ETAs would necessarily survive for millions of years. Also, I do not think we have explored all that much of the Earth's surface and its rock deposits built up over geological ages. Even if the odd piece of metal or whatever did survive, people like miners are not especially likely to think the object significant and bring it to the attention of some knowledgeable person. Recognizable human archaeological finds are frequently ignored or overlooked and only a few are eventually brought to the notice of, say, a local museum curator.

BIOLOGICAL CONSTRAINTS ON INTERSTELLAR TRAVEL

Jörg Pfleiderer

Institut fuer Astronomie, Leopold-Franzens-Universität

A-6020 Innsbruck, Austria

1. INTRODUCTION

Interstellar space travel necessarily includes small and closed ecological systems, multi-generation life cycles, and complete recycling of material. All three items are extremely difficult if not impossible to sustain, as soon as higher life, especially man, is included. While the list of relevant items is certainly not complete (M. Pfleiderer 1987), those mentioned will demonstrate some constraints which may play an unpleasant role for overly optimistic plans.

2. CLOSED ECOLOGICAL SYSTEMS

An ecological system is the combination of live forms, different species and different individuals of one species. This system is within a non-living supporting structure, such as soil, water, and climate that is sufficiently separated from the environment (for example, other systems) that it is recognizable or defineable as an entity.

All ecological systems on Earth are open in many respects. That is, the environment cannot be rigorously distinguished from the system. The biological and physical entities are closely intertwined. The systems are more or less stationary but the temporal constancy is not a result of physical equilibrium but rather a result of exchange with the environment. The system is in the state of a flowing equilibrium.

It is well known that definitions of life generally include the concept of flowing equilibrium because such equilibrium is necessary to maintain a state of decreased entropy in the system, as compared to the environment. No ecological system can be completely closed, with no exchange whatsoever with the environment. The entropy condition is, as a thermodynamic condition, related to an energy flow. Every system gets energy from outside and finally releases the same amount of energy into the environment.

The system also needs material resources. Air, water, and most elements are more or less ubiquitous and exchanged, in the course of time, within the whole Earth. No natural system is perfectly closed in respect to these resources. Some systems are near to being closed in respect to a particular resource. For example, a fish population in a lake with no spring and no sink: It depends on water recycling with non-deteriorating water quality. Most systems are open as far as resources are concerned.

Many systems exist where the food chain seems to be closed. That is, there is little exchange of organic matter with the environment, and the exchange does not directly contribute to the food situation. However, a closer look reveals that in most cases even a small exchange can contribute to the stability of the system. For instance, seeds are exchanged by wind or water. Even a plant population restricted to one system (endemic species) may profit from the stabilizing effect of exchange for another species. Other systems are open as far as food is concerned. Most deep-sea populations ultimately get their food from the surface, at least the first member of a food chain.

Nearly all life relies on microorganismic symbioses. Bacteria, algae, and protozoa are ubiquitous. In this respect, nearly all ecological systems are open. That is, the symbioses are stabilized by an open exchange with a large environment. It is not necessary that the exchange is permanent and quantitatively large. It is only necessary that abundance fluctuations of the microorganisms that may be large locally can be damped and reversed from outside.

We know of only one ecological system on Earth that is truly closed in respect to food and microorganisms. The Pacific Rift System is, however, extremely specialized. It has probably developed from a much larger stock of species in which only a few survived. It has not, however, become totally independent of other life depending on dissolved oxygen that was created photosynthetically by surface organisms. Studying the system has revealed, and will reveal, a lot of biological information but it probably has little relevance to space travel. The same is probably true for most other small ecological systems. Insofar as they are truly closed (endemic populations), they are also specially adapted, and rely on openness in other respects.

It cannot be excluded that microorganisms can indeed form truly stable small closed systems. Completely sealed aquatic systems containing algae, bacteria, and other small organisms, have been prepared by C. Folsome at the University of Hawaii. These continue to survive, now after many years. Again, the relevance to space travel is questionable because these systems do not contain higher life forms.

All ecological systems contain, for every species, a reasonably large stock size and gene pool in order to reduce vulnerability to small perturbations. Small stock sizes may occur but are generally a sign of intrinsic vulnerability. It is well known how easy it is to exterminate a species of small stock size. Species may become extinct from natural or man-made reasons, but it is always a change of the living conditions to which the species cannot adapt in a sufficiently short time. The adaptability decreases sharply with decreasing stock size.

Some species are at the verge of extermination even if the present stock size is large, if they had previously gone through a phase of very small stock size (bottleneck). This is because the gene pool was reduced too much. Such species can survive only if living conditions happen to stay fairly constant.

It is possible to prevent extinction of some species by keeping a stock in a zoo-like environment. A zoo, or a similar institution, is not a closed system but rather a very open one in which living conditions are adapted to the needs of certain species.

In space travel, we would certainly try to create a zoo-like environment with specially adapted living conditions for people. However, such a system would be small and ecologically closed. Any system not specifically adapted to such conditions is endangered by a breakdown, causing irreversible changes of one of the many conditional chains. The largest danger probably comes from the microorganismic symbioses. A well-known example of a breakdown in an individual is diarrhea caused by changes in the intestinal flora. Because there are many such symbioses in man and in man's food, some of them not yet sufficiently understood, the probability of one single total breakdown in a small system not adapted by evolution to smallness is significant. It is certainly possible that such breakdowns may be overcome, but the probability is not high. As a matter of fact, we lack any specific long-term experience.

We know that for specially selected people and conditions, the probability of symbiotic breakdown even in a very small system (manned space ship) is sufficiently low that it did not yet occur. However, the probability increases at least linearly with time, and decreases less than linearly with the system size: a larger size buffers more effectively, but the danger of infection increases.

In short: work on stabilization of small-sized systems is required.

3. RECYCLING

Any space ship which cannot rely on earthly resources must have complete recycling of material. There is not one known single separated ecological system doing that on Earth, except the artificial ones mentioned above. For instance, the Pacific Rift System does not recycle at all, being supplied with all chemical needs by the environment. More strictly speaking, the system is only biologically closed.

One might argue that some trace elements are needed in such small amounts that complete recycling is not necessary. Indeed, recycling of the common elements may be a more difficult task. However, the problem of recycling is not only one of not loosing elements or chemicals from the system, but also to avoid contamination within the system.

We normally consider recycling to be a task well done by nature. It may take its time but, except for special poisonous (for example, radioactive) refuse, its return into the life cycle or, in the case of contamination, its removal from it, is no problem. There is enough space somewhere to get rid of any refuse that cannot be reasonably recycled.

The situation in a space ship is quite different. Resources are not only limited but actually barely existing, as compared to terrestrial resources. Recycling must not only be complete but also very fast. This will create a rather inhuman situation. Since use of resources for any purpose not absolutely necessary has to be extremely limited, there will be a strong tendency to suppress those cultural activities that need material. For example, can a violin be built? Can lyrical poems be printed? When must a volume of Chaucer be recycled? The chemical recycling of bodies before their direct relatives are dead must have a profound influence on religious feelings. These are only some examples illustrating the necessarily large impact of recycling on the cultural and social conditions. I consider this impact to be more important than the unsolved questions of how to avoid unusable waste, how to recycle contaminated waste, and how to completely recycle any waste.

Energy is not recycled. One may save energy, or use energy several times, but a flowing energy equilibrium is in any case necessary. This does not mean that the energy problem is simple. Nuclear power is long-lived but needs recycling of nuclear material. Other energy sources, as solar power, can be used only near a star. Technical devices tend to have a short lifetime, as compared to interstellar travel times. A solar energy device would be unused and untested for generations before it is needed again. Our experience of how to keep technical devices in working condition under such unfavourable circumstances is practically nil.

4. MULTI-GENERATION PROBLEMS

The cultural and social life of a space-ship population is necessarily quite different from our terrestrial life. A strong loss of tradition must occur. A careful selection of the crew will help only for the first generation. Extensive training of following generations is possible only if the population is convinced of the necessity and can appreciate its merits. Nobody wants to learn something for the sole purpose of passing it on to the next, still more unwilling generation. There is no way out of the conclusion that a new tradition will develop that facilitates adaptation to the new situation at least for some people. However, no prediction is possible on how this will comply with the technical necessities. Knowledge will atrophy if it is not needed fairly regularly. That is, it is well possible that technical knowledge is lost to a degree that makes operation of the vehicle inefficient and finally impossible. The space ship, the size of which is necessarily smaller than that of a modern large city, will leave little room for a variety of cultures. Such uniformity to which man is not adapted will create a strong selectional pressure. Whoever does not adapt will less probably reproduce. Strong precautions must be taken to avoid a shrinking of the gene pool. As a matter of fact, this applies a fortiori to the animal population. However, it is impossible to prevent the tendency of health and reproductivity to be strongly influenced by psychological effects.

In summary, such a space ship has barely a fair chance of surviving, and a still smaller chance to re-adapt to the reverse situation of colonization of a new planet.

Lacking, at present, the experience necessary to even plan interstellar human travel, most people (but certainly with exceptions) are not anticipating it in the foreseeable future. However, such experience would, in my opinion, reveal that not only mankind but life in general, at least all higher life forms, are evolutionarily unsufficiently adapted to conditions of long-term travel. Space travel thus would encounter borderlines which are as effective as those absolute ones detected in physics in our century even if they certainly are not as absolutely defineable.

If ETIs are of similar biology, they will meet similar unsurmountable biological limits. Thus, the probability is very high that galactic colonization has never occurred. Only those that stayed at home survived.

Reference

Pfleiderer, M., 1987: Proc. IAU Coll., 99, 279.

Discussion

J. BILLINGHAM: Some work has been carried out in NASA over the last twenty years on the problems you have outlined for small closed natural ecosystems for space travel. One of the conclusions is that it may be necessary to actively control the ecosystem. While the development of such controlled ecosystems is still in its earliest stages, it might offer, if successful, higher reliabilities for spacecraft ecosystems. This is something you did not mention.

J. PFLEIDERER: Sorry I did not mention it. Active control of ecosystems, if done or - as at present - regulated from outside, resembles the zoo conditions I spoke of. If done completely within the system, it will certainly stabilize it for some time but will not avoid ultimate difficulties, only postpone them. Our control parameters generally originate from the - often unexpected - experiences of local breakdown. Adaptation of parameters to gradual changes of the system needs continuous research. Active control will only marginally respond to very gradual changes that possibly lead in the end to an uncontrollable instability.

G. MARX: A short illustration of Pfleiderer's theses: People are not especially interested to sit in prison, in spite of the fact that the chance of traffic accidents could be avoided.

J. PFLEIDERER: That is true for the first generation only, perhaps for the second. People having been born in prison and never having left it, perhaps not even having evidence from their grandparents that one could live outside, tend to be frightened when leaving it. Also, long-term prisoners have been frightened by the many changes in the modern world to which they felt they could not adapt. Zoologists experience similar problems in releasing zoo animals into the wilderness.

A PHILOSOPHICAL APPROACH TO THE EXTRATERRESTRIAL LIFE ISSUE

A. D. Fokker

Astronomical Institute of Utrecht University (Retired), 3721 AL Bilthoven, Netherlands

ABSTRACT

If, right from the epoch of the Big Bang, life was bound to occur with necessity sometime in the Universe, the niches where life emerged are the veritable interesting spots. In contrast, inanimate bodies and processes that are the object of astrophysical research are intrinsically uninteresting. An anthropic principle of the second kind is introduced:

"Since an uninteresting Universe is unacceptable to our mind, we require that the Universe was interesting enough, right from the start, as to be prone to the generation of (intelligent) life sometime and at several places. For that reason (intelligent) life is wide-spread throughout the Universe."

So far, we know of only one spot in the Universe where life, particularly intelligent life, has originated. That is our Earth. The constraints on the circumstances under which (intelligent) life could emerge on Earth were severe. Under slightly different circumstances terrestrial life would never have had a chance to develop. So, the course of events might as well have failed to give rise to (intelligent) life in our solar system.

1. It is conceivable that our Earth is the only spot in the entire Universe where life originated. If that were so, the origin of (intelligent) life apparently would not have the character of necessity. The appearance of intelligent life in only one niche in the Universe would then be entirely accidental. The Universe might as well have been devoid of life.

2. On the other hand, it is conceivable that life has emerged in several spots throughout the Universe. Actually, we are more inclined to this view than to the former one.

If life is wide-spread throughout the Universe, the niches where life flourishes are the veritable interesting spots. In our imagination we may explore these spots, and these would deserve our vivid interest, far more so than the inanimate astrophysical processes we are accustomed to.

As a matter of fact, we may consider a Universe, in which life is a phenomenon bound to occur, as a truly interesting Universe. On the other hand, a Universe that lacks the potentiality to give rise to forms of life is a totally uninteresting kind of Universe. Such a Universe is merely the scene of processes that occur blindly. As a matter of fact, these are the kind of processes that Astrophysics reveals to us.

In virtue of these considerations I dare to introduce the following assertion as an anthropic principle of the second kind:

"Since we cannot help imposing on the Universe the property of interestingness, we require that its constitution was and is necessarily of such a kind as to be prone to the generation of life sometime. A strong version of this principle is the statement that an interesting Universe should of necessity give rise to intelligent forms of life."

It may seem presumptuous to impose "interestingness" onto the Universe. This claim may be defended by the following argument. Our mind has the faculty of creating mathematics. It is by the very help of mathematics that we understand correctly several (most?) of the (astro)physical processes and phenomena.

Similarly, it is in virtue of our faculty of judging the interestingness of things, that we require that the Universe was bound to accommodate life right from the epoch of the Big Bang. We judge that the Universe should be a meaningful and an interesting Universe. And so the anthropic principle of the second kind enunciates that the Universe was interesting enough indeed, right from the beginning, to generate life as a matter of course, sometime and at several places.

In conclusion, (intelligent) life should be wide-spread throughout the Universe!

Discussion

L. DOYLE: A universe without life would be uninteresting because such a thing as interest would not exist in such a universe. Correct?

FOKKER: "Interestingness" is a concept of the human mind and so as a property of the Universe, it has no sense without the very existence of ourselves. It is for this reason that I introduced the postulate "The Universe should be interesting" as an anthropic principle. This would seem a rather presumptuous claim. But since we are intelligent enough to ponder about the Universe as a whole (a very astonishing fact!) we rightfully feel ourselves entitled to impose on the Universe the necessity of being "interesting," i.e., to give rise to the development of (intelligent) life phenomena. I admit that the argument is a bit queer, but the excuse is that we are confronted with mystery indeed.

THE INTRINSIC LIMIT TO THE SPEED OF INNOVATION
AND ITS RELEVANCE FOR THE QUESTION "WHERE ARE THEY?"

Peter Kafka

Max-Planck-Institut für Astrophysik, D-8046 Garching, F.R.G.

Why has the earth not been colonized from outside, although any developed technological civilisation should be able to diffuse through the galaxy within a few million years? Mainly two answers have been proposed:

1. We are alone (or nearly so) because some narrows along the path towards our level of complexity make the appearance of intelligence or civilisation extremely unlikely events. A whole universe (or even multiverse?) is then needed to let this possibility become realized on just one planet.

2. Technological civilisation itself is the narrows. Either it becomes self-destructive through global ecological (or social) disaster, or it succeeds in self-organizing technological restriction. Then, mind might be a long-lived phenomenon, but renounce the spatial expansion of its own physical structure.

The first answer has been favoured by Brandon Carter, Frank Tipler and many others. Carter's probabilistic argument [1, 2] is impressive at the first glance: A crossing of the narrows must be extremely unlikely, because otherwise it would have happened much earlier; that it happened only "near the end" (at about half the life-time of the sun) appears then as a natural implication of the fact "that we are here".

The weakness of Carter's argument lies in the fact that many known and unknown processes on earth happen on time-scales similar to that of solar evolution. Present knowledge may not even be sufficient to exclude the possibility that the decay of some abundant radioactive nuclides was necessary before life or nervous systems could reach their present level of complexity. Similarly, the decreasing frequency of large volcanic eruptions and of collisions with interplanetary bodies might have played a role, as well as the slow shaping of Gaia's crust and atmosphere as parts of the biosphere. Therefore, the idea that "intelligence" is likely to appear on "habitable planets" after a few billion years is still compatible with Carter's argument.

Tipler argued [3] that "the most solid experimental fact" in this whole discussion is the absence of foreign explorers or conquerors throughout the earth's history. In his opinion this makes the SETI project comparable to ESP-research: *"Virtually any motivation we can imagine that would lead extraterrestrial intelligences to engage in interstellar radio communication with us would also motivate them to engage in interstellar travel. In particular, radio communication is colonization of other inhabited star systems by memes (idea complexes) from alien star systems. If one opposed on moral grounds colonization by genes (via interstellar travel), one would also oppose colonization by memes (via radio). Interstellar colonization either by genes or by memes necessarily implies biological evolution on an interstellar scale: The first intelligent species to originate will occupy all ecological niches available to it, a behavior pattern adopted by all species that ever existed on earth. ..."* [3] Obviously, Tipler assumes that mind's ecological niches would have to be found in physical space. *"What have they been doing these billions of years?"*, he asks - but isn't this a childish question? Even human mind has already discovered quite different spaces for inward instead of outward expansion. And even some human minds do communicate with others without wishing to "colonize" them. Mind is a new front of evolution in the space of possibilities, not "property" of some individuals or species or cultures. From arguments like Tipler's we can certainly not exclude the possibility that there are intelligences around and communicate with each other.

Concerning the present and future attempts of search for extraterrestrial intelligence (SETI), Jill Tarter said [4]: "It's technology which we are trying to detect – not intelligence." This very relevant remark leads us on the right track to answer the questions "Where are they?", even if we don't think we are necessarily alone.

I have often argued [5] that technology itself is the narrows along the way to further mental evolution, because there is a purely logical upper limit to the speed of growth of complexity, and that so-called technological civilization surpasses that limit, thus destroying the conditions for further "creation of values". Even worse, from the theory of creation, i.e., self organization, there follows what I called the "devil-theorem": In a spatially finite system with unbounded evolution the speed of innovation must increase until a global instability sets in.

Why is that so? We do have a "solid experimental fact" (just look at the present situation of the earth), but we can gain more general insight by thinking about time and complexity. To remember what complexity is, consider the number of possible "relation structures" for a set of points with one line or no line between any two of them. How many points are needed to let the number of such possible structures surpass the number of baryons in our observed universe? The answer is: 24 points! How, then, have viable structures at all been found and kept any stability for some time? How is the history of our universe and all its details being selected? If one visualizes each momentary state of the universe as a point in a practically infinite-dimensional space, there is just this single line that is realized in the "space of possibilities". It started from an extremely special global state ("big bang") which offered immense "fossil" resources and sinks for later self-organization. (The two main sources of free energy are "fossils of the first few minutes": Because things were thrown apart, they stored gravitational potential energy with respect to each other, which can be re-gained in the formation of lumps, and because expansion was initially so fast, there wasn't time to go beyond Hydrogen and Helium, the fossil fuels in stars.) Ever since this unlikely beginning, the unavoidable fluctuations have been exploring neighbouring possibilities. Since there are so many of them, there are probably "better" ones found, more viable ones - if there is time enough to test the relevant relations between the new and the old. More viable possibilities survive by definition. A hierarchy of dissipative structures emerges, with more and more mutual adaptation, which also includes relative isolation as far as possible. The tautological principle of this Darwinian co-evolution is: "Probably, something more likely is going to happen". This is the meaning of time, the drive behind the growth of complexity in our universe - up to the speed-limit.

Does this mean that the "better" (the more complex, which we find more valuable) arises without any value-judgement? No, the selection process is the value-judgement, and its principle is the same on the levels of physical, chemical, biological and mental evolution. One can easily see, why the emerging world is hierarchical. Structures on lower levels, the viability of which has long been tested, will be used on higher levels with little modification because attempts to "improve" them must introduce many untested interactions and, therefore, probably lead to break-down. With too many new relations (remember the 24 points!) time is not sufficient to try them out, and no viable new structures will be found, even if they might be possible. Building upon time-tested feed-back loops is more successful. Still, a crisis is unavoidable.

At any moment, there is a "front of evolution" in the "space of possibilities", where innovation proceeds fastest. Speed itself is an "evolutionary success" and is likely to grow until feedback with the whole becomes insufficient. Then, this front collapses, but evolution goes on with whatever diversity is left. Of course, we cannot formulate a general systems-theoretical argument which would allow us to call certain developments "safe" in the sense that they will not destroy their own roots. However, even with the absurd assumption that the front might succeed in complete emancipation from its roots and the whole, a logical limit to the speed of creation of values is self-evident: The level of complexity reached has to be "re-learned by each generation". Thus, the critical speed is roughly defined by "essential change within the life-time of the individual structures at the front". If the (r)evolutionary process of fluctuation and selection gropes its way into the space of possibilities faster than that, the leading sub-systems cannot even properly take into account their own complex value. Self-organization of global simplicity sets in and increases the speed of "wrong" value-judgements further. Within a few generations of the leading sub-structures they start destroying themselves and the viability of the whole system from which they evolved.

As I wrote elsewhere [6]: "...*Evolution itself defines and creates a critical time-scale, which it then necessarily tries to surpass. But thereby it must destroy its own logical pre-conditions. The leading figures at the front of evolution don't give themselves enough time to judge values in the process of exploring the neighbourhood in the space of possibilities. Of course, the tautology remains valid that "more likely things will probably be realized" via the accidental fluctuations (including their more recent form of appearance, called planning) - but with a lack of time for selective adaptation, i.e. adaptive selection, the more likely is no longer a growth of complexity but rather its decomposition. In a very sophisticated way the entropy law seems to have conquered the Earth, an open dissipative system in which we thought it wouldn't be valid. While everybody was still worrying and quarreling about the resources, we have been filling up and blocking the sinks...*"

This kind of instability is quite similar to the "success" of a fast-growing water-lily on a pond, or of a cancer-cell in an individual organism. The characteristic difference, however, lies in the "globality". If the system is "isolated" or "spatially finite" in the sense that the time-scale for communication with the outside is long compared to the time-scale of the instability, no revival from "outside ponds" and no survival of "outside individuals" will stop or heal the local disaster. A "black hole" will remain, or "scorched earth".

If evolution doesn't stop due to external influences, this onset of global instability is probably unavoidable. Growth of evolutionary speed itself seems to be an evolutionary success as long as the errors can be pushed to the "borders" - i.e. until the global scale has been reached. This acceleration must certainly take place when evolution on a planet reaches the level of mental structures. The "discovery" (i.e. "detection", i.e. "apo-kalypse") of the "laws of nature" will start technological progress because this provides more power. Of course, like in our own history, many individual minds will understand the "devil-theorem" quite early, since the laws of logic are more fundamental than the "laws of nature". But in the fight between "God and Devil", dia-bolos (i.e. "he who throws things in disorder") will prevail because he is always quicker than the creator of true complexity.

Thus, any planet with intelligence is likely to run into our kind of technological crisis and to approach global ecological or social disaster. Still, I call it a crisis, and not the end. When deadly consequences of this "progress" are felt on the critical time-scale (the own life-time) by a majority, insight in the logical pre-conditions of creation may become dominant in the global society of minds. It may then still be possible to self-organize the restriction of power and of the speed of innovation and to shift the front of evolution to the mind - where creation of new complexity is possible without destruction of its whole basis.

Conclusions concerning SETI are obvious. If there are others in our universe, they will not be interested in simple material structures, except during a few generations before that crisis. Mind will recognize itself as infinitely more complex, i.e. valuable. Topics like astronomy would play a negligible role in an "Encyclopedia Galactica". If civilizations transmit signals, they will probably not use "variations of something expected", as William Calvin proposed here "because radio-astronomers are interested in pulsars" [4]. For mind the only interesting thing in the universe will be other mind. Even the "acquisition signals" (though probably on "magic frequencies" - e.g. as favoured by David Blair [7]) might not be perceptible on the human time-scale - another relevant remark by Jill Tarter [4]. Civilizations beyond the acceleration-crisis would not try and help others to overcome it, too. Not because they are selfish, but because such help is obviously impossible. They must know that many of us have understood the origin of the crisis, but that we can stop only (if at all) at the very edge of the abyss. The time scale of interstellar communication is longer than that of our instability. After the development of radio-technology there is no time left for help. Earlier interference, however, before the onset of the instability, would not mean help, but colonization - which is probably excluded by further mental evolution (or even by a fundamental incompatibility between long-distance space-travel and a mastering of the Devil). Hence, there is nothing important which we could learn from aliens on the time-scale of the crisis, on the human time-scale. It's all in our minds! Still, the discussion about whether we should listen or not, and why we don't hear anything, may contribute a little to the understanding of the devil-theorem...

References

1. Brandon Carter, "The anthropic principle and its implications for biological evolution", *Phil. Trans. R. Soc. Lond.* A310, 347-363 (1983).
2. John D. Barrow and Frank J. Tipler, *The Anthropic Cosmological Principle*, Oxford University Press, New York 1986.
3. Frank J. Tipler, *Letter to Physics Today*, Sept. 1988, pp. 88 and 146.
4. Panel discussion, Proc. 3rd Internat. Symp. Bioastronomy, Val Cenis, 1990 (this volume).
5. First in P. Kafka "On the world's Ends" preprint 1976, (publ. in *Munich Social Science Review* 1978/2, 91-99), last in *Das Grundgesetz vom Aufstieg*, Carl Hanser Verlag, München 1989.
6. P. Kafka, "Time and Complexity" in *Proceedings of the Workshop on Gravitation, Magneto-Convection and Accretion* at the Ringberg Castle, Tegernsee, May 28 - 31, 1989, (Schmidt, B., Schmidt H.U., Thomas H. - C. eds.) Max-Planck-Inst. Proc. MPA/P2 Sept. 1989.
7. David Blair, Proc. 3rd Internat. Symp. Bioastronomy, Val Cenis, 1990 (this volume).

SETI SEARCHES WITH THE 70m SUFFA RADIO TELESCOPE

N. S. Kardashev
Astrospace Center, Moscow, USSR

Paper Read by V. S. Strelnitskij

Summary

The 70-meter radio telescope, which is now under construction, will allow observations down to 1 millimeter wavelength and will provide the capability for SETI observations in the optimum wavelength range (Kardashev 1979) around 1.5 millimeters.

If our hypothetical correspondents possess instruments similar to ours, it would be possible to establish bilateral communications up to distances as great as 100 parsecs. For this calculation we assume a one megawatt transmitter power with 1-Hz bandwidth and an integration time of 100 seconds.

The detectability of aliens with highly advanced technology is also considered. If highly advanced aliens have engineered structures that surround their energy sources, e.g., stars or even the nuclei of galaxies, and if they have designed these structures to exist at low temperatures, then one can show that these constructions will have very large dimensions which could be detected with the 70-meter SUFFA radio telescope operating at millimeter wavelengths. Furthermore, it can be shown that a radio telescope operating at 1.5 millimeters will be capable of spatially resolving circumstellar constructions within the Galaxy and circumgalactic constructions at great distances throughout the Universe.

Reference

Kardashev, N.S., 1979, Nature, 278, 28-30.

Discussion

W. SULLIVAN: My great reservation in searching for broadband sources, such as infrared radiation from Dyson spheres, is that I cannot imagine how the observation of, say, an unusual infrared or radio spectrum will ever supply sufficient evidence to establish the presence of extraterrestrial intelligence. Such an extraordinary conclusion requires widespread consensus regarding the artificial nature of the spectrum. But theorists will always be able satisfactorily to explain broadband spectra (and perhaps even narrowband ones, too!) Please comment.

(ANSWER BY V.SLYSH): The long wavelength part of the spectrum of radiation from Dyson spheres must follow the Raleigh-Jean law while that of a dust envelope will be depressed due to diffraction on grains with sizes less than the wavelength. But this difference will be considered only as evidence, not as proof of the presence of an astro-engineering activity.

T. WILSON: **The 1.5 mm wavelength would not be optimal in the discussion given by S. Gulkis. Would this be a serious drawback?**

(ANSWER BY V.SLYSH): No, according to Kardashev's estimates it is 1.5 mm wavelength that is optimal for the SETI communication.

J. TARTER: **Is Sufa an acronym and will you decipher it?**

(ANSWER BY V. STRELNITSKIJ): "Sufa" is a proper name - name of the mountain, where the "70-m" is being constructed. But if you want to decipher it as an acronym, I could propose: "Search for Unidentified Flying Aliens!"

(COMMENT BY V. SLYSH): "Sufa" is "Table" in Uzbek.

STRATEGY OF THE MUTUAL SEARCH FOR CIVILIZATIONS BY MEANS OF PROBES

U. N. Zakirov

Physical-Technical Institute of the USSR Academy of
Sciences, 10/7, Sibirski Tract, Kazan 420029, USSR

Poster Paper

The strategy of mutual search for civilizations by means of probes is based on the considerations of the technological level required to utilize internal (rocket-borne) and external (cosmic) energy sources. The possibility of detecting internal energy sources is most developed because well-known physical laws are involved, e.g., the process of energy release from hydrogen isotopes at the expense of microtarget supercompression. It is possible to suggest several tests to search for evidence of internal energy sources in the vicinity of the 50 nearest stars (16 light-years). For example, it might be possible to detect power changes in a probe engine from the behavior of its kinematic characteristics as it moves to the nearest stars or from these stars toward the Sun. On the other hand, it might be useful to modify the operating plan for the infrared cameras and Multi-Object Spectrometer of NICMOS-type, which will, for several years, be parts of the Hubble telescope. These instruments could be programmed to search not only for extra solar planets, but for "self-luminous" probes as well.

The possibilities of detecting cosmic energy sources from interstellar probes are speculative. We can only speculate that hypothetical civilizations have developed and are using such technologies as probe accelerations by Ramjets, or the utilization of electric and magnetic fields, or strong gravitational fields, or antigravitation and vacuum energy.

VI. WIDER INTERDISCIPLINARY CONNECTIONS

CONNECTIONS: LIFE ON EARTH AND ATOMS IN THE UNIVERSE

R. E. Davies and R. H. Koch
University of Pennsylvania, Philadelphia, PA, U.S.A.

Paper Presented by R. E. Davies

ABSTRACT

Virtually all electrons and nuclei of the atoms that are or have been part of living matter on Earth have reached us from almost all stars in our and nearby galaxies and even from all other galaxies in the Universe that have produced observed high-energy gamma rays. However, a standard 70 kg human is always making about 7 ^{3}He, 600 ^{40}Ca, and 3000 ^{14}N nuclei every second by radioactive decay of ^{3}H, ^{40}K and ^{14}C, respectively.

1. INTRODUCTION

The history and evolution of the chemical elements comprising living systems continues to be of great interest. From nucleosynthetic theory and many observational analyses (e.g., Grevesse and Anders, 1989) it is accepted that 29.3% of the mass of the primordial H has been transmuted into elements heavier than H. The cosmic mass fractions of H, He, and Li through U are now, respectively, $X = 70.7\%$, $Y = 27.4\%$, and $Z = 1.9\%$.

At different times some of our H atoms have resided serially in many different stars. Weaver et al. (1978) show that about half of all H in a large-mass star survives the subsequent supernova explosion to be incorporated eventually into many other stars yet to be born. In addition, all stars have significant stellar winds and many other stellar mass-loss episodes (e.g. Boothroyd and Sackmann, 1988) operate even in single red giants and mildly-interacting close binaries. In all of these stars much more than half of their ejecta survives as H and subsequently becomes mixed into the interstellar medium (ISM).

If it is assumed that the Big Bang occurred 20x10^9 yr ago and formation of the Solar System 5x10^9 yr ago, then about 98% of the H remaining 15x10^9 yr after the Big Bang formation of He had not yet been synthesized into other elements. However, most of this H had already been part of at least one star and some of the presently non-stellar H atoms had already been incorporated into very large numbers of stars. So, virtually all galactic stars existing earlier than about 8x10^9 yr ago have actually contributed to the mass of every human.

On the basis of depletion of deuterium in the intergalactic medium (IGM), it has been noted by Chuvenkov and Vainer (1989) that exchange between the Milky Way Galaxy and the primordial IGM occurs at a rate of 3% of the galactic mass per 10^9 yr. In addition, some of the supernova ejecta from occurs at a rate of 3% of the galactic mass per 10^9 yr. In addition, some of the supernova ejecta from M31 and other Local Group members exceeds the escape velocity from that galaxy. Thus, it is possible that as much as 10% of the H in every average 70 kg human are nuclei not originally in the Milky Way disk.

In addition, gamma photons have been observed from more distant objects out to the quasar horizon of the Universe (Ramaty and Lingenfelter, 1986). These highly energetic photons, created originally by nuclear processes in quasars, themselves create matter and anti-matter in Earth's atmosphere. The anti-matter then annihilates and some of the matter (mainly electrons and protons) must be incorporated into humans. Clearly and remarkably, because of the ubiquity of energetic gamma rays, at least parts of atoms have actually been contributed to the totality of the biomass on Earth from all parts of the observable Universe.

After the formation of the Solar System and the appearance of Life on Earth, the history of our H atoms became even more complicated as they became incorporated repeatedly into many individual molecules, such as water, glucose, etc. Most of the approximately 6×10^{27} H atoms in a 70 kg lean human are combined with oxygen as water. With a mixing time of about 10^3 yr for virtually all of the hydrosphere, as cited by Stuiver et al. (1983), each human contains about 1 H atom from every average milligram from all species of organisms that existed over 10^3 years ago.

2. CONCEPTS OF STELLAR AND BIOLOGICAL GENERATIONS

Even as a given stellar Population endures, its composition must change because X continually decreases and Z increases. Even though "Population" terminology is of limited quantitative resolution, it continues to be convenient, but the term "stellar generation" is not satisfactory. Low-mass stars are not "generated" from older low mass stars, even though they do contain some ejecta from the stellar winds or flares of most of the older stars. The heavy-element content of, e.g., Sun and Earth was condensed largely from the ejecta of over 10^9 short-lived (e.g., 10^6 yr) high mass stars. This very large number of stars follows from the value of Z, which was 1.9% for the epoch of formation of the Solar System and from the total mass of galactic stars. However, this does not mean that the Solar System required 5000 serial stellar "generations", as the term is used biologically.

Biological generation is generally characterized by transmission of unique genetic information from parents to offspring. This lineage is unbroken back to the origin of present Life. In the case of an adult multi-cellular organism, only a trifling fraction of the mass is actually derived from the parent(s). This is akin to the situation in stars. An important difference does occur in that, for the biological case, the original fertilized cell came entirely from the parents whereas, for recent stars, there were more than 10^9 stellar progenitors. Thus, stellar and biological generations and evolution are similar but not identical concepts.

3. THE CHEMICAL COMPOSITIONS OF THE UNIVERSE AND BIOMASS

Among many others, Brock (1985) has summarized thinking and experiments concerning the intrinsic nature of matter from Hellenic times into the 17th century. Natural philosophers considered that the four alchemical elements - Earth, Air, Fire, and Water - were derived from a more fundamental substance, protyle. By the end of the 18th century, many real chemical elements had already been discovered and investigated quantitatively. Dalton (1810) produced compelling evidence for the existence of atomic elements but rejected the concept of protyle as originally it had been put forward by Empedocles and Democritus. Prout (1816) was the first to suggest that protyle, the building block of all the elements composing matter, is actually H. The discovery of isotopes meant that Prout's Hypothesis was valid: all elements have been and continue to be synthesized by processes starting with H, identifiable as the protyle of the ancients.

The cosmic abundances of the chemical elements updated by Grevesse and Anders (1989) are given in Table 1. Normalized to different regimes of Earth (Weast, 1988) Table 1 also gives the abundance-ratios of the 38 elements measured in humans, 27 of which are essential (e.g., Underwood, 1981). Since every human contains about 6×10^{27} atoms, it is expected that some atoms of every stable element (and most unstable ones) will actually be present in each person.

Additionally, each human contains 10^{14} atoms of ^{14}C and 10^9 atoms of ^3H made by cosmic ray interactions. The decays of these species then produce about 3000 ^{14}N and 7 ^3He atoms each second in the body. The very long-lived isotope ^{40}K, although made in stars, remains as 0.00118% of total K in the human body. By its decay path it makes about 600 ^{40}Ca per second in the body.

The entries in Table 1 for, e.g., P and K emphasize the impressive selectivity of human life with concentration factors of more than 800 compared to the cosmic values. On the other hand, most of the trace elements used in only tiny amounts are readily available cosmically and in Earth's crust and have not been evolutionary bottlenecks. However, it is also clear that the formation of Earth's crust and oceans led to selective concentrations and exclusions of many elements individually. For example, the trace element Mo is quite rare in sea water and would have to be concentrated about 500 times if this were its only source for Life. This element is, in fact, readily available in Earth's crust (Gualtieri, 1977) and in reality the human body selects against it by about a factor of 15.

4. THE COSMIC SOURCES OF SOME ELEMENTS SELECTED BY ORGANISMS

As a result of experimental and theoretical astrophysics, the processes and sites for synthesizing nuclides are now well-known (Grevesse and Anders, 1989; Aller, 1989). The modes of nucleosynthesis for biologically-important elements are summarized in Table 2.

In the galactic plane near the location of Sun, the average stellar plus interstellar density is of the order of 5×10^{-24} g cm^{-3} (cf., Allen, 1973). The mean densities of Sun, Earth, sea water, the terrestrial atmosphere, and the human body are, respectively, about 1.4, 5.5, 1.0, 1.3×10^{-3} and 1.0 g cm^{-3}. Selections by living cells continue to be many and various. For instance, the primary reservoirs of N, Cl, and Mo available for living matter are the atmosphere, sea water, and the crust, respectively. Since no one single reservoir can make all elements readily available and because of the known catalytic properties of many mineral surfaces, it seems certain that Life originated with access to the interfaces among the individual phases of Earth.

Table 1: Cosmic abundances by mass and human concentration factors by mass for elements with measured concentrations in humans.

Rank	Symbol	Cosmic Abundance Percent by Mass	Rank	Symbol	Human / Cosmic Abundance	Rank	Symbol	Human / Earth's Crust	Rank	Symbol	Human / Sea Water	Rank	Symbol	Human / Dry atmosphere
1	H	70.7	1	P	1000	1	N	1000	1=	Zr*	very large ratios	1	H	2500000
2	O	9.58×10^{-1}	2	K	800	2	C	600	1=	Ge*		2	C	1500
3	C	3.04×10^{-1}	3	Rb	650	3	H	65	1=	Cd*		3	O	3
4	N	1.10×10^{-1}	4	Hg*	600	4	S	10	1=	Ti*		4	N	0.065
5	Mg	6.56×10^{-2}	5	Cl	300	5	P	7	1=	Cr				
6	Si	7.06×10^{-2}	6	Ca	250	6	Cl	5	6	P	150000			
7=	Fe	1.26×10^{-1}	7	Br*	150	7	Cd*	2.5	7	N	70000			
7=	S	4.14×10^{-2}	8=	Mo	150	8=	Br*	2	8	Hg*	35000			
9=	Al	5.76×10^{-3}	8=	Zr*	100	8=	Hg*	2	9	C	6500			
9=	Ca	6.16×10^{-3}	8=	Cd*	100	8=	Se	2	10	Cu	6000			
11=	Na	3.32×10^{-3}	11	O	65	11	O	1.5	11	Fe	4000			
11=	Ni	7.26×10^{-3}	12	C	56	12	Cu	0.45	12	Zn	3000			
13	Cr	1.76×10^{-3}	13	Pb*	60	13	Ca	0.4	13	Ni	1000			
14=	P	8.09×10^{-4}	14	N	45	14	I	0.35	14	Mo	500			
14=	Mn	1.32×10^{-3}	15	Na	45	15	Zn	0.25	15	V	350			
16=	Cl	4.68×10^{-4}	16	Cu	35	16	Ge*	0.15	16	Pb*	250			
16=	K	3.70×10^{-4}	17	I	35	17	K	0.1	17=	Sn	100			
18	Ti	2.89×10^{-4}	18=	Sn	25	18=	Mo	0.065	17=	Co	100			
19=	Co	3.34×10^{-4}	18=	S	15	18=	Pb*	0.065	19=	Rb	50			
19=	Zn	2.70×10^{-4}	20	Zn	15	20	Na	0.05	19=	Se	50			
19=	F	4.02×10^{-5}	21	B*	15	21	B*	0.035	21	Ca	40			
22	Cu	8.35×10^{-5}	22	Ge*	4.5	22	Rb	0.03	22	Mn	20			
23	V	3.75×10^{-5}	23	As	4	23	Mg	0.02	23	K	8			
24	Ge	2.17×10^{-5}	24	Sr*	2	24	Zr*	0.015	24	S	7			
25	Se	1.23×10^{-5}	25	Se	1.525	25	As	0.01	25	As	2.5			
26	Sr	5.18×10^{-6}	26	F	1	26	Sn	0.0075	26	I	2			
27	B	5.76×10^{-7}	27	Mg	0.7	27	Ni	0.0065	27	Si	1.5			
28	Br	2.37×10^{-6}	28	Ba*	0.65	28	Fe	0.006	28	H	0.85			
29	Zr	2.61×10^{-6}	29	V	0.25	29	F	0.0015	29	O	0.75			
30	Rb	1.52×10^{-6}	30	Ti*	0.15	30	V	0.00065	30	Br*	0.45			
31	As	1.23×10^{-6}	31=	H	0.13	31=	Cr	0.00045	31	Mg	0.35			
32	Ba	1.55×10^{-6}	31=	Fe	0.065	31=	Co	0.00045	32	F	0.3			
33	Pb	1.14×10^{-6}	33	Al*	0.01	33	Mn	0.0004	33	Al*	0.25			
34	Sn	1.64×10^{-6}	34	Mn	0.0075	34	Sr*	0.00035	34	Ba*	0.2			
35	Mo	6.16×10^{-7}	35	Cr	0.005	35	Ti*	0.00009	35	Na	0.15			
36=	Cd	4.55×10^{-7}	36	Si	0.007	36	Ba*	0.00004	36	Cl	0.08			
36=	I	2.87×10^{-7}	37	Ni	0.007	37	Si	0.00002	37	B*	0.02			
38	Hg	1.71×10^{-7}	38	Co	0.003	38	Al*	0.000006	38	Sr*	0.0075			

*Not known to be an essential element for humans.

5. PHOSPHOROUS AND POTASSIUM

The odd-numbered elements P and K are among the 20 most common ones cosmically, in Earth's crust, in sea water, and in humans. Remarkably, P and K themselves determine the minimum cosmic volume necessary to make a human being.

As a nuclear species, P is synthesized mainly by Ne burning in high- and intermediate-mass single and binary stars. A smaller amount of P is also made in the subsequent supernova explosions. For the example of a $25M_O$ (Woosley and Weaver, 1982) star (where M_O is the mass of Sun), some 3×10^{30} g of P are mobilized into the ISM. The average density of P in the ISM is of the order of 10^{-29} g cm^{-3}. As a consequence of repeated shockings of a cloud sample by successive supernovae, density wave passages, and low-mass star evolutions, the primitive Solar System nebula condensed some 2×10^{28} g of P. A large portion of this survives in Sun and the gaseous planets. At about 1 A.U. from Sun, condensation processes have preserved some 2×10^{22} g of it in Earth's crust. Geophysical processes concentrated the element into assorted phosphate minerals. Humans obtain their necessary quantities of it from the plants and animals in their diets. A smoothed average density for P characteristic of the Milky Way disk (including stars, gas, dust, etc.) is of the order of 10^{-28} g cm^{-3}. For a 70 kg human with 560 g of P, we have calculated that about 2×10^6 present solar volumes of average galactic matter are required in order to make available the quantity of P essential for that human body.

Table 2. Contributions from individual nucleosynthetic processes to the 38 elements found in measured amounts in humans (cf. Table 1).

Process	Element
Bb = Big bang	H
H = hydrogen burning	N
He = helium burning	O, C
C = carbon burning	Na#, Cl#, Cu#, Co#
N = hot or explosive hydrogen burning	Mg#, F
O = oxygen burning	S#, Si#
Ne = neon burning	P, Na#, Al*#
S = slow neutron-capture onto Fe-peak "seeds"	Cl#, Rb#, Zr*, Ge*, Pb*#, Hg*, Cd*, Sn#, Sr*#, Ba*
R = rapid neutron-capture onto Fe-peak "seeds"	Rb#, Br*, Mo, Pb*#, Cd*#, Sn#, Se, I, As
Ex = explosive burning in supernovae	Ca, P, S#, K, Na#, Cl#, Mg#, Fe#, Zn#, Cu#, Si#, Ni#, Al*#, Ti*, V, Mn#, Cr
E = nuclear statistical equilibrium	Fe#, Zn#, Ni#, Mn#, Co#
X = spallation in ISM	B*

The names for the processes are taken from Grevesse and Anders (1989).
* Elements not known to be essential for humans.
# Elements for which more than one isotope is cosmically abundant (>20%).

However, about 90% of the observable galactic matter is now in the form of low-mass stars which are presumed to have the same initial composition as the ISM. These stars do not process nuclear matter quickly and do conserve their matter efficiently for very long intervals ($>10^{10}$ yr) so most of their P is really sequestered and unavailable for Life. Hence, the interstellar volume really necessary to form a human becomes about $2x10^7$ equivalent solar volumes.

As indicated in Table 2, the other limiting element, K, is made only explosively. The same example as formerly, a $25M_O$ model, disperses about $3x10^{29}$ g of K (about 10% the mass of P) into the ISM. Great varieties of concentrating processes transport some P and K from the ISM gas into interstellar dust, where they are temporarily poorly detectable, and then to both Sun and Earth, where they can be measured (cf. Table 1). This results in the conclusion that, for K as well as for P, the volume needed to form a human is about $2x10^7$ solar volumes.

6. THE NUMBER OF GALACTIC SUPERNOVAE NECESSARY FOR HUMAN LIFE

As noted in the Introduction, on average Z = 1.9% cosmically. Thus, 1.9% of the Galaxy has been converted by stellar processes into elements heavier than He, dispersed into the ISM by supernova and other events, and then concentrated into later "generations" of stars (e.g., Sun), and planets.

The Milky Way mass, excluding the dark halo, is of the order of $10^{11}M_O$. Thus, about $2x10^9M_O$ arose from supernova terminations of the lives of stars. If each of these had an average mass of $15M_O$ and synthesized and ejected 5% of its mass as elements heavier than He, of the order of $3x10^9$ stars had to have followed this pathway. Since the accepted age of the Milky Way is about 10^{10} yr, the average rate of supernova events must be about 1 per 3 yr. This is clearly greater than the rate (about 1 per 200 yr) observed during the past 1,000 yr for the Milky Way. It is reasonable to conclude that most of this necessary number of supernovae were events early in the life of the Galaxy. At that time, the interstellar clouds were much denser and hotter than currently. Thus, an initial larger incidence of high-mass, hence short-lived, stars would be expected.

Despite present ignorance of detail, there is no doubt that most atoms in our bodies have had origins which include the earliest and the most distant horizons of the Universe. Furthermore, once accumulated on Earth, many of these atoms have been used time and again by the same organism and even more so by different organisms since Life began. In addition, a few thousand atoms are even now being transmuted every second in each of our bodies.

We thank Dr. Adelaide M. Delluva for help in preparing this paper.

References

Allen, C. W.:1973, Astrophysical Quantities 3rd ed., Athlone, London
Aller, L. H.:1989, in Cosmic Abundances of Matter, ed. C. J. Waddington, AIP, Minneapolis, p. 224
Boothroyd, A. I., and Sackmann, I.-J.:1988, Astrophys. J. 328, 653
Brock, W. H.:1985, From Protyle to Proton - William Prout and the Nature of Matter, 1785-1985, Adam Hilger Ltd., Bristol
Chuvenkov, V. V., and Vainer, B. V.:1989, Astrophys. Space Sci. 154, 287
Dalton, J.:1810, A New System of Chemical Philosophy, Bickerstaff, Manchester

Grevesse, N., and Anders, E.:1989, Geochim. Cosmochim. Acta 53, 197

Gualtieri, D. M.:1977, Icarus 30, 234

Prout, W. (published anonymously):1816, Ann. Philos. 7, 111

Ramaty, R., and Lingenfelter, R. E.:1986, Ann. N. Y. Acad. Sci. 470, 215

Stuiver, M., Quay, P., and Ostlund, H. G.:1983, Science 219, 849

Underwood, E. J.:1981, Phil. Trans. R. Soc. Lond. 1

Weast, R. C.(ed.):1988, CRC Hand. Chem. Phys. (69th ed.) CRC Press

Weaver, T. A., Zimmerman, G. B., and Woosley, S. E.:1978, Astrophys. J. 225, 1021.

Discussion

C. MCKAY: Considering the elements essential for prokaryotic life (~17), which of these would you suppose would be absolutely required for single celled life on Mars 3.5×10^9 years ago?

R. DAVIES: As far as we know now, all of the 17 would be essential. There are small replicating systems in cells, like viruses, and mitochondria, but they need the enzymes in the cells that contain elements like iron, copper, molybdenum, etc. even though their own structures contain fewer elements (half a dozen or so). (The 17 elements needed for E. coli in rank order by mass are: O, C, H, N, P, K, Na, S, Ca, Cl, Mg, Fe, Zn, Cu, Mo, Mn, Co.)

FROM THE PHYSICAL WORLD TO THE BIOLOGICAL UNIVERSE: HISTORICAL DEVELOPMENTS UNDERLYING SETI

Steven J. Dick

U.S. Naval Observatory, Washington, D.C. 20392

ABSTRACT

More than thirty years ago the French historian of science Alexandre Koyré (1957) wrote his classic volume, *From the Closed World to the Infinite Universe*, in which he argued that a fundamental shift in world view had taken place in 17th century cosmology. Between Nicholas of Cusa in the fifteenth century and Newton and Leibniz in the seventeenth, he found that the very terms in which humans thought about their universe had changed. These changes he characterized broadly as the destruction of the closed finite cosmos and the geometrization of space. The occasion of the Third International Bioastronomy Symposium in France is an especially appropriate time to argue that the SETI endeavor represents a test for a similar fundamental shift in cosmological world view, from the physical world to the biological universe. I define the "biological universe," equivalent to what I have called before the "biophysical cosmology" (Dick, 1989), as the scientific world view which holds that life is widespread throughout the universe. In this case the biological universe does not necessarily supersede the physical universe, but a universe filled with life would certainly fundamentally alter our attitude toward the universe, and our place in it. Although Koyré mentioned life beyond the Earth as an adjunct to the revolution from the closed world to the infinite universe, only in the 1980s has the history of science begun to give full treatment to the subject. What follows is meant to be a contribution to that ongoing endeavor to understand where the extraterrestrial life debate fits in the history of science.

The modern era in the extraterrestrial life debate is normally dated from Cocconi and Morrison's paper in 1959, and though one can always find precursors, this in my view is a valid perception. Cocconi and Morrison gave definite form to SETI, Frank Drake independently first carried out the experiment, a network of interested scientists began to form and met in Green Bank in November 1961, and the most distinctive part of the modern era of the extraterrestrial life debate - the Search for Extraterrestrial Intelligence by means of radio telescopes - was off and running. In this paper, after briefly reviewing some of the long-term steps toward the biological universe, I would like to examine the immediate precursors to this modern era in the 1940s and 1950s.

LONG-TERM DEVELOPMENTS

By the 1950s there was, of course, a long tradition behind the debate over life on other worlds. Dick (1980, 1982) and Crowe (1986) have now well documented this history through the 19th century, and work on twentieth century history is now in progress (Dick, forthcoming). It is now widely accepted that the extraterrestrial life debate began and was sustained by a "cosmological connection," stretching back to the ancient atomists and Aristotle in the 4th century B.C. and revitalized by the Copernican, Cartesian and Newtonian traditions in the 17th century. Copernicus made the planets Earth-like in theory, Descartes and Newton proposed other solar systems as part of their cosmologies, and the Newtonians added the crucial factor - for them - that a universe filled with life was in agreement with an omnipotent God, the God of natural theology. Philosophical principles such as plenitude and purpose also played a role, but, I would argue, a subordinate role given meaning only in the context of these cosmologies. So in a real sense, the mental adjustment from the physical world to the biological universe was, for many natural philosophers, made by the beginning of the 18th century.

But of course proof was another matter. Though much was learned about our own planetary system in the next two centuries, most of the 19th century was spent in explorations of the philosophical or religious implications of this idea. Thomas Paine (1793) said that he who believed in Christianity and plurality of worlds had thought but little of either, and himself came down on the side of other worlds. At the opposite extreme William Whewell (1853), Master of Trinity College Cambridge rejected plurality of worlds in favor of Christianity. And as a middle ground Thomas Chalmers (1817) and others attempted to reconcile religion and extraterrestrial life.

Two great scientific developments in the latter half of the 19th century gave credence to extraterrestrial life - the rise of astronomical spectroscopy and Darwin's theory of evolution. Again and again in twentieth century discussions of life beyond the Earth, we see reference to these two 19th century achievements: astrophysics shows the elements of matter to be the same throughout the universe, and Darwin's theory of biological evolution by natural selection not only would hold for organisms throughout the universe but also may be viewed as the end product of physical evolution in the universe. It is notable that Percival Lowell was much influenced by the idea of planetary evolution, not only in his books on Mars but also in *The Evolution of Worlds* (1909). In *The Study of Stellar Evolution* (1907), George Ellery Hale pointed out that while evolution was not a new idea to astronomers in the 19th century, "it has occupied a more important position since Darwin published his great work." It did not escape Hale's notice that in 1859, the very year of the publication of the *Origin of Species*, Kirchoff began his experiments aimed at determining the chemical composition of the Sun, launching the field of stellar evolution. Despite the extreme skepticism of Lowell's claims of canals on Mars, prominent astronomers in the early 20th century such as W. W. Campbell (1920) could point to the results of spectroscopy to support the broader claim of life in the universe: "If there is a unity of materials, unity of laws governing those materials throughout the universe, why may we not speculate somewhat confidently upon life universal?" he asked. He even spoke of "other stellar systems ... with degrees of intelligence and civilization from which we could learn much, and with which we could sympathize." Such a general argument was enough to carry the day for many astronomers up to 1920.

SHORT-TERM DEVELOPMENTS

With the birth of SETI in 1959 - the centennial year of Darwin's *Origin* and Kirchoff's identification of elements in the Sun - all of these general steps to the biological universe lay in the background. From the point of view of 1959, the more immediate steps in the emergence of the biological universe stretched back less than a generation. In fact the 1950s was emerging from a 25-year period of extreme skepticism regarding life in the universe. It is significant that the general principles of the uniformity of nature and stellar evolution had not been enough for most scientists to accept life on other worlds in the first half of the twentieth century in the face of contrary theories. Just about the time Campbell wrote his article in 1920, James Jeans (1919, 1923) argued that the solar system may be unique, or at the very least "astronomy ... begins to whisper that life must necessarily be somewhat rare" in the universe. This whisper grew to a crescendo by the 1930s. Harvard Observatory Director Harlow Shapley (1923), just fresh from his triumphant use of globular clusters to show the eccentric position of the solar system in the Galaxy, held that planetary systems were unlikely and habitable planets very uncommon. And Henry Norris Russell (1926) agreed that planetary systems were infrequent and habitable planets pure speculation. From 1920 to about 1945 we see the idea of extraterrestrial life at a low point - the biological universe was in danger of extinction. The reason is to be found almost totally in a shift in theories of planetary formation, from the nebular hypothesis to the "close encounter" or "tidal" hypothesis. Developed by Chamberlin and Moulton at the University of Chicago around 1900 (Brush, 1978), this theory in the hands of Jeans gave a pessimistic view of the possibility of planetary systems, since stellar encounters would be very rare. Jeans' ideas were widely accepted in the scientific community, and his numerous popularizations of this idea spread it far and wide. As long as this idea held sway, planetary systems were freak occurrences divorced from normal stellar evolution.

At least four factors may be discerned in the modern reemergence of the biological universe. First, and arguably most importantly, a radical shift occurred once again in the estimation of the likelihood of planetary systems, from both observational and theoretical points of view. Dynamical objections to his theory led Jeans (1942) in the final years of his life to postulate a much larger primordial Sun, and therefore to conclude that stellar collisions might not be so rare after all, perhaps forming planets around one in six stars. But new developments quickly moved beyond his tidal theory; I would date the turning point for planetary systems at 1943 (Table I). In that year Russell (1943) spoke of "a radical change - indeed practically a reversal - of the view which was generally held a decade or two ago," regarding the scarcity of planetary systems. He specifically referred to the apparent discovery of planetary companions by Strand (1943) around 61 Cygni, and Reuyl and Holmberg (1943) around 70 Ophiuchi. Both used the technique of photographic astrometry to detect perturbations in the orbits of these double stars. Although their discoveries would eventually prove spurious, at the time Russell undoubtedly took them as vindication of his earlier analysis (Russell, 1935) that there were grave angular momentum problems with the close encounter hypothesis, and that some other theory must replace it. By 1944 Carl F. von Weizsäcker had come up with the beginnings of such a theory, a modified nebular hypothesis, which he elaborated (Weizsäcker, 1951) and which opened the floodgates to similar theories (Brush, 1981, 1990). The Swedish physicist Alfvén, Hoyle in Britain, Kuiper in the U.S. and others elaborated their own forms of the nebular hypothesis. Planetary systems were returned to the realm of stellar evolution. Hoyle estimated by 1950 ten million planetary systems in the Galaxy and a million habitable planets. Kuiper (1951) on the basis of binary star separation statistics, estimated a billion planetary systems in the Galaxy. The new cosmology, with its vastly expanded universe full of galaxies, also supported many planetary systems. Shapley (1958) - formerly so skeptical - detailed the arguments here and also concluded for billions of planetary systems in the Galaxy. Struve's estimate in 1961 of billions of planetary systems in the Galaxy was therefore quite common by that time.

Observational research by Otto Struve also lent support to the view of many planetary systems. Struve's work (1930, 1950) on stellar rotation showed that there was a discontinuity at the F spectral type where stellar rotation slowed. Although several braking mechanisms were possible, by the 1950s Struve (1952), his student Su-Shu Huang (1957, 1959) and others were plausibly surmising that the angular momentum might have gone into planetary systems. In Huang's words " ... planetary systems emerge as axial rotation declines. According to this view, planets are formed around the main sequence

Table I. Estimates of Frequency of Planetary Systems, 1920-1961

Author	Argument	# Planetary Systems in Galaxy	# Habitable Planets in Galaxy
Jeans (1919, 1923)	tidal theory	unique	1
Shapley (1923)	tidal theory	"unlikely"	"uncommon"
Russell (1926)	tidal theory	"infrequent"	"speculation"
Jeans (1941)	# stars	10^2	-
Jeans (1942)	> diameter of Sun	1 in 6 stars	abundant
Russell (1943)	observation of companions	very large	$> 10^3$
Page (1948)	Weizscker	$>10^9$	$> 10^6$
Hoyle (1950)	supernovae	10^7	10^6
Kuiper (1951)	binary star statistics	10^9	-
Hoyle (1955)	stellar rotation	10^{11}	-
Shapley (1958)	nebular hypothesis	10^9	10^6
Huang (1959)	stellar rotation	10^9	10^9
Hoyle (1960)	stellar rotation	10^{11}	10^9
Struve (1961)	stellar rotation	$> 10^9$	-

stars of spectral types later than F5. Thus, planets are formed just where life has the highest chance to flourish. Based on this view we can predict that nearly all single stars of the main sequence below F5 and perhaps above K5 have a fair chance of supporting life on their planets. Since they compose a few percent of all stars, life should indeed be a common phenomenon in the universe." By 1952 Struve (1952) even published a "Proposal for a Project of High-Precision Stellar Radial Velocity Work," designed to detect planets at the level of a few hundred meters/sec. It is interesting that although both Kuiper's binary star separation statistics and Struve's F5 stellar rotation discontinuity were known in the 1930s, it was not until the 1950s - after the downfall of the rare encounter hypothesis - that they were used as arguments for many planetary systems.

Whether or not one accepted these specific theoretical and observational arguments for planetary systems, with increasing knowledge of stars and stellar evolution one could still argue more generally, as Struve himself did in 1955, that the physical properties of the Sun resembled in every respect other stars of similar type, right down to axial rotation. He argued that we must infer that this similarity also extends to star formation, and accompanying planets. "Since we cannot adduce a proof one way or the other, we must rely upon what seems to be the most logical hypothesis. And this is without doubt the assumption that all, or at least most, dwarf stars of the solar type have planetary systems. The total number of planets in the Milky Way may thus be counted in the billions." Taking the solar system as an example, where one of nine planets clearly has life, one (Mars) may have had life, and one (Venus) may have life in the future, Struve concluded that the number of planets in the Milky Way with some form of life might also number in the billions. Planetary systems, as supported by renewed forms of the nebular hypothesis and theories of stellar evolution, by the apparent indirect observations of planets, by the facts of stellar rotation, and by the new cosmology, were thus the first and primary factor in the reemergence of the biological universe.

Secondly, the evidence for life in the solar system grew increasingly positive, feeding hope that this was an indication of the case for the broader universe. It is true that Rupert Wildt's postulation in 1940 of a greenhouse effect on Venus was gradually accepted and finally eliminated that planet from consideration as a biological habitat. But conditions on Mars were still believed to be acceptable, if harsh, for life, and in 1947 at the famous University of Chicago conference on planetary atmospheres Kuiper (1949) postulated plants similar to lichens on Mars, and in 1957 and 1959 Sinton gave his widely accepted spectroscopic proof of vegetation on Mars. Again and again astronomers reasoned that if life had developed on two sites in our solar system, then it was most likely common throughout the universe.

Thirdly, regarding the crucial question of the origin of life, the idea of chemical evolution was gaining widespread acceptance in the 1950s. Oparin's work, begun in the 1920s in Russia, was first published in English in 1953. In the same year Miller (1953), stimulated by Harold Urey's conclusion that the primitive Earth must have had a reducing atmosphere, published his results of their first experiments on the formation of organic compounds under conditions of a reducing atmosphere. In 1957 the first International Symposium on the Origin of Life was held in Moscow, and that field was off and running just prior to the launching of SETI. Melvin Calvin, a representative of this chemical evolution tradition, participated in discussions of life beyond the Earth at the Lunar and Planetary Exploration Colloquia at least as early as 1959 (Calvin, 1959), and was present at the 1961 Green Bank meeting, during which he was notified he had received the Nobel Prize.

Finally, we should not forget that it was a technological development - the emergence of radio astronomy - that made SETI possible for the first time over large scales. Whatever other influences they may have felt, it is certain that Drake and his SETI successors were most directly influenced by the emergence and development of this field (Drake, 1960, 1961; Morrison, Billingham and Wolfe, 1977, and many others). They may or may not have been influenced by some of the other developments we have mentioned, though the influence of Struve on Drake, at least to the extent of allowing Drake to perform the experiment at the NRAO, was obvious and direct. Bracewell also emphasized the importance of Struve in his own development and that of other SETI pioneers, and Calvin acknowledged the influence of Shapley (Swift, 1990). The importance of ideas in leading to the use of radio telescopes for SETI is obvious, but without that radio technology SETI would not be the well-developed program it is today.

SUMMARY

In summary, as Cocconi and Morrison wrote their landmark paper and Drake made his first radio search for extraterrestrial life, they had behind them - whether they knew it or not - a widespread acceptance of the idea of such life dating from the 17th century, of the principles of evolution and uniformity of nature dating from the 19th century, but only fifteen years of the reemerging tradition that planetary systems and life were likely. Belief in vegetation on Mars was at a high point due largely to the work of Kuiper and Sinton. Belief in abundant planetary systems was widespread due to the work of Russell, von Weizäscker, Struve, Hoyle and others. And belief in the ability of life to develop via chemical evolution was on the rise due to the work of Urey and Miller on synthesis of amino acids under conditions of a primitive atmosphere. Moreover, at least two prominent astronomers explicitly acknowledged at an early stage the potentially revolutionary character of these ideas. By 1958 Shapley characterized the existence of extraterrestrial life as a possible "Fourth Adjustment" that humanity would have to make in its overall view of the universe. In 1961 Struve wrote that astronomy has had three great revolutions in the past 400 years: Copernicus' removal of the Earth from the center of the solar system, Shapley's removal of the solar system from the center of the galaxy, and the revolution occurring now, embodied in the question "Are we alone in the universe?"

While the biological universe has been widely debated for more than two millennia, and widely accepted for more than two centuries, it has fallen to the last half of the twentieth century to provide observational proof for the hypothesis that there is more to the universe than matter in motion. Although difficult, and although labeled by some a pseudoscience (Tipler, 1987, 1988), that task in my view falls squarely in the tradition of the history of science which frames hypotheses and attempts to test them. In a significant local result, the Viking test for life on Mars failed. It is up now to SETI programs to test the hypothesis on the cosmological scale, and to determine whether we, or future generations, will really need to make the shift to a new scientific world view, from the physical world to the biological universe. If we do make that shift, I predict the effect on astronomy and culture will be even more profound than the move from the closed world to the infinite universe three centuries ago.

References

Brush, Stephen: 1978, Journal for the History of Astronomy, 9, 1-41, 77-104.
Brush, Stephen: 1981, *Space Science Comes of Age*, ed. Paul Hanle and Von del Chamberlain, Smithsonian Press: Washington, D.C.
Brush, Stephen: 1990, Reviews of Modern Physics, 62, 43-112.
Calvin, Melvin: 1959, Proceedings of the Lunar and Planetary Exploration Colloquium, April 25, 1959, 1, no. 6, 8-18.
Campbell, W. W.: 1920, Science, 52 (December 10, 1920), 550.
Chalmers, Thomas: 1817, *A Series of Discourses on the Christian Revelation, Viewed in Connexion with Modern Astronomy*, Edinburgh.
Cocconi, Giuseppe and Philip Morrison: 1959, Nature, 184, 844.
Crowe, Michael: 1986, *The Extraterrestrial Life Debate, 1750-1900: The Idea of a Plurality of Worlds from Kant to Lowell*, Cambridge University Press, Cambridge.
Dick, Steven J.: 1980, Journal of the History of Ideas, 1-27.
Dick, Steven J.: 1982, *Plurality of Worlds: The Origins of the Extraterrestrial Life Debate from Democritus to Kant*, Cambridge University Press, Cambridge.
Dick, Steven J.: 1989, The Planetary Report, March-April, 13-17.
Dick, Steven J.: forthcoming, *The Twentieth Century Extraterrestrial Life Debate: A Study of Science at its Limits*, Cambridge University Press, Cambridge.
Drake, Frank D.: 1960, Sky and Telescope, 19, 140-43.
Drake, Frank D.: 1961: Physics Today, 14, 40.

Hale, George Ellery: 1907, *The Study of Stellar Evolution*, University of Chicago Press, Chicago, 2.

Hoyle, Fred: 1950, *The Nature of the Universe.*, 26, 101.

Hoyle, Fred: 1955, *Frontiers of Astronomy.*, 83, 104-05.

Hoyle, Fred: 1960, *The Nature of the Universe*, 2d ed., 32, 81, 90.

Huang, Su-Shu: 1957, Publications of the Astronomical Society of the Pacific, 69, 427.

Huang, Su-Shu: 1959, Publications of the Astronomical Society of the Pacific, 71, 421.

Jaki, Stanley L.: 1978, Planets and Planetarians.: *A History of Theories of the Origin of Planetary Systems*, Scottish Academic Press, Edinburgh.

Jeans, James: 1919, *Problems of Cosmogony and Stellar Dynamics*, Cambridge University Press, Cambridge, 290.

Jeans, James: 1923, *The Nebular Hypothesis and Modern Cosmogony*, Cambridge University Press, Cambridge, 30.

Jeans, James: 1942a (written 1941), Science, 95, 589.

Jeans, James: 1942b, Nature, 149, 695.

Koyré, Alexandre: 1957, *From the Closed World to the Infinite Universe*, Johns Hopkins University Press, Baltimore.

Kuiper, Gerard P.: 1949, in Atmospheres of the Earth and Planets, University of Chicago Press, Chicago.

Kuiper, Gerard P.: 1951, ch. 8 in Astrophysics, (McGraw Hill, New York) ed. J. A. Hynek, 416-417.

Lowell, Percival: 1909, *The Evolution of Worlds*. (New York).

Miller, Stanley L.: 1953, Science, 117, 528-29.

Morrison, Philip, John Billingham and John Wolfe, *The Search for Extraterrestrial Intelligence* (NASA, 1977).

Page, Thornton: Physics Today (October, 1948), 12-24.

Paine, Thomas: 1793, *Age of Reason*.

Reuyl, Dirk and Erik Holmberg: 1943, Astrophysical Journal, 97, 41-45.

Russell, Henry Norris: 1935, *The Solar System and Its Origin*.

Russell, Henry Norris: 1943, Scientific American (July, 1943), 18-19.

Russell, Henry Norris, R. S. Dugan and J. Q. Stewart: 1926, Astronomy, vol. 1, 468.

Shapley, Harlow: 1923, Harper's Monthly Magazine, 146, 716-22.

Shapley, Harlow: 1958, *Of Stars and Men*, Beacon Press, Boston,104-114.

Sinton, William: 1957, Astrophysical Journal, 126, 231-39.

Sinton, William: 1959, Science (6 November, 1959), 1234.

Strand, Kaj: 1943, Publications of the Astronomical Society of the Pacific, 55, 29-32.

Struve, Otto: 1930, Astrophysical Journal, 72, 1.

Struve, Otto: 1950, *Stellar Evolution*, 150-51, 231-39.

Struve, Otto: 1952, Observatory, 72, 199-200.

Struve, Otto: 1955, Sky and Telescope, 14, 137-140, 146.

Struve, Otto: 1961, *The Universe*, MIT Press, Cambridge, Mass., 157-59.

Swift, David: 1990, *SETI Pioneers*, University of Arizona Press, Tucson, 54-85; 130, 141-44.

Tipler, Frank: 1987, Physics Today, December, 1987, 92.

Tipler, Frank: 1988, Physics Today, September, 1988, 14-15 and 142-44.

von Weizsäcker, C. F.: 1944, Zeitschrift fr Astrophysik, 22, 319-55.

von Weizsäcker, C. F.: 1951, Astrophysical Journal, 114, 165-86.

Whewell, William: 1853, *Of the Plurality of Worlds: An Essay*, London.

Discussion

D. LATHAM: What was the impact of Percival Lowell's enthusiastic popularization of the case for life on Mars on the attitudes of the professional community? Was there much of a backlash that made planetary research unfashionable?

S. DICK: The standard interpretation is that Lowell's work was detrimental to solar system astronomy in the 20th century. I am not so sure this is true. It is true that solar system astronomy was carried out at a low level following Lowell, but this may have been due to other factors, such as the application of spectroscopic techniques to stellar astronomy. Those same techniques, as well as radiometry, were also applied to the planets. Although Lowell's idea of canals on Mars was discredited by the time of his death in 1916, his idea of vegetation on Mars lived on. One can still see the persistence of his legacy in the Viking project. So I think one can argue just as well that Lowell had a stimulating effect on solar system astronomy in the long run.

A POSSIBLE WAY FOR EVOLUTION OF THE INTELLECT OF INTELLIGENT BEINGS

Leonid V. Ksanfomality

Space Research Institute, Academy of Science of USSR

84/32 Profsoyuznaya Str., 117810 Moscow, USSR

Summary

The development of technology shows some interesting trends and laws that have been repeated in many areas of knowledge. These trends are a temporal character of the functioning of the most effective developing technologies. A specific feature of these trends is that the decay of some old technology and its replacement by new technology just coincides (as a law) with the period of most perfect state of the old one. Together with this replacement, the progress in technology sometimes leads to the creation of new scientific and technological branches and even to some unexpected social phenomena.

There are grounds to suppose that as a result of the development of science and technological innovations, it is possible that in the not too distant future on the earth, the individuals will appear, that together with many features of the human's way of thinking, will have a special property, that we call the superintellect. This is not a service robot, the appearance of which is already determined by the development of technology. The superintellect could arise at the convergence of branches of knowledge, including electronics, optoelectronics, software and the physics of the brain. Some of these frontiers remain absolutely untouched; others seem achievable even now. The appearance on Earth of superintellect will introduce a lot of social, ethical and other problems but will also raise the intellectual power of our civilization to a tremendous extent.

SETI AS A SCIENCE

David W. Swift
University of Hawaii, Dept of Sociology
Honolulu, HI 96821 USA

Summary

Interviews with pioneers in the search of extraterrestrial intelligence clarify questions about the nature of the search.[1] Despite its 'far out' objective, it shares characteristics with other, more conventional activities which are considered to be "science".

However, SETI does face greater difficulties than many other new fields, as suggested by questions such as the following:

Does the phenomenon exist?

When can results be expected?

What useful purposes will the project serve?

Reference

1. Swift, D. W., 1990: SETI Pioneers, University of Arizona Press, Tucson, AZ.

Discussion

N. EVANS: I think that the main reason SETI may not be science is that it will be very difficult to falsify the hypothesis that extraterrestrial intelligence exists. Even if all our planned searches fail, I doubt that most of us would conclude that no ETI exists, and we would be even less likely to conclude that no extraterrestrial life (e.g., bacteria) exists. Rather than science, I think SETI is something more important: a voyage of discovery.

D. SWIFT: Are Discovery and Science mutually exclusive?

MAY THERE BE AN ULTIMATE GOAL TO THE COSMIC EVOLUTION?

Michael D. Papagiannis
Department of Astronomy, Boston University, Boston, MA 02215,USA

Summary

The 20th century has been an era of miracles, such as going from the first flight of the Wright brothers in 1903, to Sputnik I in 1957 and to walking on the moon in 1969. We can now follow with considerable detail the 15 billion year evolution of the Universe and the 3.5 billion year evolution of life on Earth. The initiation of life was made possible by some basic properties of matter. The availability, e.g., of only 1 neutron for every 7 protons, at the Big Bang prevented the conversion of all hydrogen into helium. Also the decay of Be-8 in only 2×10^{-16} sec prevented the buildup of large quantities of heavy elements, which would have led to massive planets inhospitable to life. Actually the instability of Be-8 is surprising, since all other elements that are made of alpha particles, from C-12 to Ca-40, are highly stable. Life is based primarily on H,O,C,N that are the four most common chemically-active elements in the Universe, which suggests that life may be quite common. The predisposition of the Universe for life and intelligence is called the Anthropic Principle (1).

The presence of water, which is a polar molecule and hence an excellent medium for chemical reactions, and the formation of organic compounds by natural energy sources (solar energy, etc.) made possible the origin of life in the oceans of the Earth. Through natural selection, life developed photosynthesis, which slowly changed the atmosphere of the Earth into one with free oxygen, a characteristic sign of the presence of life. This led also to an ozone layer that allowed life to conquer also the land about 400 million years ago (m.y.a.) The land became dominated by the reptiles, but about 150 m.y.a. the first mammals appeared, that not only were warm blooded, and hence more capable to deal with temperature changes, but also through long gestation they were able to develop much larger brains for their weight and hence higher intelligence. Finally in a mass extinction 67 m.y.a., the reptiles vanished and the mammals populated the whole Earth. In the last 3 million years we have the emergence of our ancestors that had a higher intelligence than the apes, showing that as matter led naturally to life, life led also to intelligence and technology.

The rapid development of technology has given us tremendous advantages over all other forms of life. This accounts for the rapid increase of the population, in spite of two world wars, and for the damage we have been inflicting on our environment (pollution, deforestation, decrease of the ozone layer, increase of the greenhouse effect, etc.) (2). These trends are obviously leading to self destruction, which may be Nature's way to guide intelligent life away from less materialistic and selfish attitudes and towards more spiritual and unselfish goals. It appears, therefore, that cosmic evolution may have an even loftier goal, called the Theistic Principle (3), and may help stellar civilizations overcome their materialistic desires and develop closer spiritual ties with the Creator of the cosmos.

References

(1) Carter, B., IAU Symposium 63, pp 291-298, 1974.
(2) Papagiannis, M.D, Q. Jl. R. Astro. Soc., 25, 309-318. 1984.
(3) Harrison, E.D., Cosmology, pp 3988-400, Cambridge University Press, 1981.

EVOLUTIONARY APPROACH TO THE SETI PROBLEM

V.V. Burdyuzha

Space Research Institute Academy of Sciences, USSR

V.I. Maron

Moscow Gubkin Oil and Gas Institute, USSR

M.D. Nussinov

125195 Moscow, Belomorskaja Str.22, Build.3, Apt. 351, USSR

Paper Presented by V. V. Burdyuzha

Summary

We examined the substance's evolution impulse which led to the origin of life of the Earth by the example of the chain of self-connected molecular objects such as "...-molecule-macro-molecule (polymer)-biopolymer (RNA and DNA in genomes of living organisms beginning from unicellular microorganisms till Homo Sapiens) (M.D. Nussinov et al., JBIS, 38, NII,494 (1985). These molecular objects as the main formshaping factor appeared in Metagalaxy successively in time. The evolution of above objects had a self-organizational character which is proved by the decrease of informational entropy at the same length of informational "text" ("programme") or by decrease of number of the informational's "text alpha-bet letter". The last one is the most important regularity of substance's self-organization in Shennon understanding of such a process. The further evolution of molecular objects shall be created by Homo Sapiens intelligent as the "genomes" of self-organizing technologies.

The above mentioned molecular branch of substance's evolution passed in both connection and interrelation with simultaneously passed there evolution of the chain of astrophysical (conditional atomic) systems for which the main formshaping factor is already the gravitational interactions.

We estimated through an electromagnetic impulse area the characteristic time of impulse existence as equal to $\tau \sim (1$ to $2) \cdot 10^{16}$s. Hence one can get the maximal characteristic linear size of the region of Metagalaxy which provides both substance and energy for supplying the self-organization of molecular systems. As it coincides roughly with the linear size of supercluster of galaxies ($R \sim 100$ Mpc) and proves to be much less then Hubble radius ($R_H \sim 2 \times 10^{28}$cm). It suppose the nonuniqueness of similar impulses in our Metagalaxy. The rough minimal estimation of the number of such impulses (that would mean also the number of extraterrestrial intelligence) gives their value about 10^5.

PEACE AMONG THE FACTIONS: COOLING TEMPERS IN THE EXTRATERRESTRIAL LIFE DEBATE

D. Brin

Heritage Research, Los Angeles, CA 90007 USA

Summary

Critics of SETI argue that its assumptions and propositions are intrinsically unfalsifiable and therefore unscientific. While this overly-narrow definition of science can easily be refuted, those of us hoping for progress in exobiology are under some obligation to admit that we actually have very little idea what we are talking about. Personal identification with a particular model or theory (always a danger in science) is especially detrimental in this field.

While exobiology comes nominally under the domain of astronomy, the Earth itself remains its central source of both fact and metaphor. The planet's distribution of biomass appears to fit neither the "oasis in a desert" model, nor that of "universal fecundity." Here it is suggested that we may derive patterns of ecological scaling by observing our homeworld. These patterns may offer insight into possible distributions on a galactic scale.

Conjectures about detectable sentient civilizations must be distinguished from those concerning exobiology in general. While the SETI proposition remains healthy it has been forced to mature under heavy criticism during the 1970s and 1980s, and now must deal with severe upper-limit constraints. At best, the universe is "sparse" in detectable ETI civilizations. A categorization of those explanations proposed so far shows no reason to believe anyone has yet found "the" answer.

THE LIKELY ORGANIZATIONAL ORDER OF
ADVANCED INTELLIGENCES

Peter Schenkel
Centro Internacional de Estudios Superiores de Comunicacion
para America Latina
Cas. 6064 CCI, Quito, Ecuador

Summary

Civilizations, highly advanced in science and technology, are bound to possess also a highly refined organizational order. Otherwise they would not have survived. The human analogy supports this proposition. Though still beset by intra-species conflict, our world is already pregnant with tendencies moving it toward a global political and economic order. Contact with a superior extraterrestrial intelligence would therefore serve mankind not only as a thrilling scientific but also as a powerful organizational and cultural stimulus.

THE ROLE OF COMPREHENSIVE PERCEPTION IN TRADITION

Mircea Pfleiderer
Institute for Astronomy, Leopold-Franzens-University, A-6020 Innsbruck, Austria

Paul Leyhausen
Auf'm Driesch 22, D-5227 Windeck 1, Germany

Paper Presented by M. Pfleiderer

1. INTRODUCTION

Our perception integrates an enormously complex situation into one "whole," which then represents that situation in our experience without concern for details. Thus it is able to condense a million-bit information to a one or few-bit "label," so to speak, which in turn may trigger off a behaviour capable of coping with the whole. It is this kind of perceptual experience that enables us to understand a complex situation (the "complex quality," Krueger, 1926) intuitively, compare it with others without any need to notice all the details consciously, and to comprehend their intricate relationships. The advantage is obvious: A point-by-point procedure would take far too much time and would be inadequate because it could not take into account the involved network of interrelations between the parts.

Often, however, perception does not leave us with an unpatterned whole but crystallizes parts of it into a configuration (gestalt), while relegating the rest to the irrelevancy of "background."

In this way, perception sorts out the overwhelming chaos of environmental stimuli impinging on the sensory system, and organizes it into manageable and recognizable portions that can be labelled, stored, retrieved, compared, and combined and recombined into more complex experiences. In fact, there is no imaginable sphere of human knowledge and experience that is not dependent first and last on the functions of perception described above. Therefore, when discussing the consequences of long-term trips into space, we must ask whether factual and technical knowledge can be preserved without being continually reinforced by actual perception and experience.

2. THE MECHANISMS OF RECOGNITION

In many cases, animals recognize biologically relevant situations and objects by a few characteristic "sign stimuli" that separately or jointly elicit an adequate response in the animal thus stimulated (Lorenz, 1943). For example, the red spot on the underpart of the herring gull's bill signals the food source to its chick, which starts pecking at it when it approaches. The red belly, spread fins, and vertical posture of the male stickleback convey a threat to its rival. The intensity or duration of the response depend on the intensity or size of the stimuli and their number (law of heterogeneous summation). The absence of one or more of the sign stimuli typical of a situation or object does not nullify the effect of the remainder. The power to translate such sign stimuli into appropriate activity stems from a special kind of perceptive mechanism called an Innate Releasing Mechanism (IRM). Every animal species possesses a number of these, although some more and some fewer, and humans are no exception. In particular, the "intuitive" understanding of postural and facial expressions is mediated by a specific IRM (Leyhausen, 1967). Innate means that these mechanisms are constitutive of the species and independent of individual experience.

There are, however, cases where a few signs are insufficient to characterize a relevant object precisely. For example, a young gosling must be able to attach itself to its mother within a few minutes of hatching. It has to know her unerringly, without having time and opportunity to compare her with other geese. The process which provides the IRM with the necessary additional information is called imprinting. Imprinting processes play a major role in the so-called socialization of many, if not all, higher animals, probably including man. However, it was so far impossible to prove this incontrovertibly because, for ethical reasons, the methods used to demonstrate imprinting in animals may not be applied when investigating human behaviour.

For decades, the question whether imprinting was just a special kind of learning or belonged in a category of its own was a bone of contention among ethologists. Recent experiments have shown that, in the chicken at least, the residues of imprinting and learning are stored in different parts of the brain (P. Bateson, in press). The kind of information acquired, the process of acquisition, and the brain mechanisms involved in imprinting are all different from those conducive to learning.

It appears that a great many fixations within the areas of socially approved codes of manners and honour, moral rules, religious beliefs, and taboos are similar to the results of imprinting. They are extremely resilient and hard to alter or eradicate. For the reason given, we cannot decide whether they owe their existence to imprinting or to processes of a different nature. For the time being, we had best speak of "quasi-imprinting".

There is no doubt that human cognition and recognition rely predominantly on the kind of perception described in the introduction and the knowledge acquired through it. Learned knowledge largely governs our decisions and actions. Yet we must not consider the less rational and unlearned modes of cognition as negligible. More often than not they determine the motivation and the direction of our learning. Lorenz called them the "innate schoolmasters" (Lorenz, 1959).

3. PERCEPTION, COGNITION, KNOWLEDGE AND DECISION-MAKING

Identifying a very complex situation or object by means of a few characteristic marks or a comprehensive "label" also has its dangers. Even an IRM may be deceived, for example by some structural feature of anatomy "mimicking" expression. One of the very few human IRMs which has been experimentally analyzed is that responding to the expression of smiling and laughter. In both, the mouth is widened and its corners are tilted upwards, making the whole face appear rounder and wider. Although our experience tells us otherwise, a round face with lips tilted up invariably conveys the impression of a friendly, genial vis-a-vis, even if this is only because the owner is built that way and in fact is not in the least expressing jollity or benevolence.

The richer possiblities of holistic and gestalt perception naturally multiply the chances of error. But they also provide redundancy of information, the opportunity to compare a great wealth of data and to check one experience against others. Thus all current as well as stored and retrieved information is tested several times over before perception turns into cognition, cognition consolidates into knowledge, and knowledge leads to decision and action. All this can happen on the basis of the intuitive understanding mentioned above, and before causal analysis, rational comprehension and balanced decision set in.

Most technical devices offer very little tolerance. Their operation is thus easily endangered by relatively small errors of the operator and his or her consequent reaction. The device allows for one response only, not for a range of choices depending on changing situations. Therefore, there is a strong tendency to couple the "right" response to a few or even a single stimulus only, thus creating a kind of artificial releasing mechanism analogous to an IRM. There is no need to comprehend the intricacies of the device in order to be able to operate it successfully, as long as it does not break down. There are two dangers: First, thorough knowledge of that particular device, of "how it works" and what is necessary to manufacture it, may become rare in a community and even be lost altogether. Second, not many individuals are trained to know any technical device thoroughly and the ability may eventually die out.

As an example of the consequences of the second case, I should like to point to traffic lights at crossroads. They were originally invented to prevent accidents and traffic jams. With them, it is no longer necessary for the individual to take in the total traffic situation; just obey the light, and no error is possible. We are trained to disregard the situation and react unthinkingly to the stimuli red, amber, or green. Multiple traffic lights seem to command stricter observance, not because they are easier to notice but because they appear to add up in a way analogous to the law of heterogeneous summation.

The present tendency of highly technical societies to reduce decisions based on understanding and rational comprehension of complex situations to yes-no choices based on mono-stimuli is a kind of retro-evolution, not to say involution, of the species.

4. LIVE EXPERIENCE AND TRADITION

In humans, the experience and knowledge of individuals is not only stored in individual brains, but can also, to a considerable extent, be made available to others through the media of speech, writing, and pictures, etc. But it must be borne in mind that no individual is capable of mediating his "all." The information channelled through the media is composed of a series of bits and must of necessity be limited. More important still is that nobody is able to express all his thoughts, experiences, methods of procedure, and results adequately and completely, whatever kind of "language" he chooses. Consequently, he cannot induce in another individual a complete knowledge of all the perceptual acts and their systemic interrelations that were necessary to create his own comprehension. He can hardly offer more than recipes for what to learn and how. Unless the receiver is capable of repeating the underlying experience for himself, he will not understand the conveyed information correctly or even at all. Since it is impossible for the receiver to share the complete, identical experience with the sender, he or she will always interpret any information received subconsciously and even consciously in terms of his or her own experience and understanding.

Nevertheless, a certain amount of fairly correct information transfer is possible. Were it not so, there could be no culture. Information transfer is particularly important between the generations. Here, however, it is even less a simple "copying over" than it is between peers. While being taught by their elders, the young continuously compare their own experience with what the older people are trying to convey. They accept some parts of it, reject others, and modify a bit here and there. In this way, tradition does not end in cultural stagnation but promotes further development. Even in a world changing as rapidly as ours, there is always a sufficient number of "constants" to prevent the overall experience of the old and the young from diverging so radically that there is no coherence. In spite of all their disputes, the generations share a common spectrum of experiences which makes communication possible and, as a rule, functional and fruitful.

To form a perceptual, conceptual, and intellectual image of the world around us, we must first get a grasp on things. This is quite literally so in the baby, whose motor activities from the first guide the senses and their development before these gradually become able to take the lead. It is the touch and feel of the real world in which we move, work, and behave that fills our sensory impressions with concrete meaning. Without this kind of incessant feedback from interaction with the outside world, our mental image of it would gradually fade and eventually deteriorate into mere fantasy.

5. CONSEQUENCES FOR INTERSTELLAR COLONIZATION

When speculating about the possibilities of colonizing celestial bodies beyond our planetary system, it is necessary to consider the frame of mind that the colonists would have upon arriving at their destination. I have often asked myself what eventually will be the mental and spiritual horizon of student generations educated within the horizon of the unvarnished concrete of our new universities. And yet, their environment is still immeasurably rich compared to that of the future space colonists. The spaceship that carried them left the earth many generations ago. Their ancestors, who originally manned it and had some genuine experience of earthly things, have long since died. Their offspring experienced these things second hand. Video and other media can never replace an environment in which you can live and move and act; they can only provide a make-believe world and not reality. However large the space transporter may be, it cannot contain all earth, human culture, and civilization. The "nth" generation, arriving on Terra X, will have lost all touch with these. They will be culturally impoverished to stone-age level. They have lived for too long in a closed, artificial "universe" where perfect technical performance is mandatory and artificial releasing mechanisms play a major role. They might even arrive at a suitable planet and be unable to recognize its potential. There is serious doubt whether the "nth" space generation would be morally and intellectually capable of creating a new Earth and a new society. On the grounds of the evidence, it would appear that man is earthbound in a sense far beyond anything we have even begun to realize.

References

Janetschek, H. 1972: Das Menschenrecht 27/1, 2
Krueger, F. 1926: Komplexqualitäten, Gestalten und Gefühle. Berlin.
Leyhausen, P. 1967: Psychol.Forschg. 31, 113-127.
Lorenz, K. 1943: Z. Tierpsychol. 5, 233-409.
Lorenz, K. 1959: Z. exp. angew. Psychologie 4, 118-165.

TECHNICAL EVOLUTION AND POSITIVE FEEDBACK

Jörg and Mircea Pfleiderer
Institute for Astronomy, Leopold-Franzens-University
A-6020 Innsbruck, Austria

Poster Paper

The technical evolution of mankind is necessarily related to the replacement or supplement of well-established, stable, "natural" regulatory processes by artificial ones that contain components of positive feedback. Such feedback is very effective in changing a situation in a relatively short time, but it must be replaced by another regulatory process after only a few time constants in order to avoid exponential explosion. In general, the interaction of many different regulatory processes gives rise to a reasonable damping of any explosive situation by a non-linear process. Technical evolution relies on inventing a positive feedback situation as well as a corresponding damping process. However, experience shows that one generally starts with the first and thinks of the latter only after the exponential situation is more clearly seen and its danger is noticeable.

Present plans for space colonization are essentially based on only the first part. For instance, one thinks of creating an atmosphere on Mars within 1000 years but not yet on how the growing of the atmospheric density could be stopped in time. The concept of galactic colonization seems to be of similar quality. Any ETI, not only ourselves, would be unable to adapt to a new environment unless it were changed within a very short time. The shorter the time and the larger the change, the more difficult it is to stop any change by damping. We propose that the present concept of galactic colonization is rather unrealistic.

CONTACT PARANOIA AND PRONOIA:
AN ANTHROPOLOGIST LOOKS AT SETI

B. Finney

Anthropology Department, University of Hawaii, Honolulu, USA

Summary

Will radio contact with an extraterrestrial civilization more advanced than ours lead to the devastation of the human spirit, or to a Golden Age for humanity? Analyses of terrestrial contact situations within our species suggest that neither the paranoid nor pronoid scenario would prevail--and that, instead, the enormous problems of translating the signal and understanding the senders would preoccupy our species for centuries if not millennia.

Editor's Note: The subject matter of this paper was given as a public lecture following the closing dinner at the Symposium. The related paper may be found in Acta Astronautica, Vol. 21, No. 2, pp. 117-121 (1990).

SPECIAL SESSION: POST DETECTION PROTOCOL

THE DECLARATION OF PRINCIPLES FOR ACTIVITIES FOLLOWING THE DETECTION OF EXTRATERRESTRIAL INTELLIGENCE

John Billingham
NASA Ames Research Center, Moffett Field, California 94035 USA

Michael Michaud
American Embassy (SCI), Paris, France

Jill Tarter
SETI Institute, Mountain View, California 94043 USA

Paper Presented by John Billingham

ABSTRACT

As the science and technology of the Search for Extraterrestrial Intelligence (SETI) has slowly emerged over the last thirty years, so there has been a gradually increasing interest in examining the consequences of the detection of a signal transmitted by another civilization. As a result of many years of discussion in the SETI Committee of the International Academy of Astronautics (IAA), a SETI "Post-Detection Protocol" was drawn up as a draft international document. After many revisions, a final document entitled "Declaration of Principles Following the Detection of Extraterrestrial Intelligence" was approved by the International Academy of Astronautics and the International Institute of Space Law.

The genesis and scope of this document are presented in the paper. Attention is drawn to the importance of unambiguous international confirmation that the signal is indeed of extraterrestrial intelligent origin, on proper dissemination of the details of the discovery, on procedures for reporting and archiving the data, and on not sending replies without adequate international consultation. This last topic will be the subject for the development of a second "Declaration of Principles", which will address questions dealing with decisions on the preparation and content of a return signal. These questions are discussed.

INTRODUCTION

Over the past decade the SETI Committee of the International Academy of Astronautics has been addressing the consequences of the detection of a signal transmitted by an extraterrestrial civilization.

Two major activities were undertaken. The first was the initiation of scientific sessions dealing with the question, as a part of the annual meetings of the International Astronautical Federation (IAF) Congress. The second was the preparation of a SETI "Declaration of Principles for Activities Following the Detection of Extraterrestrial Intelligence". This document drew heavily on the papers given in the

IAA sessions at the IAF Congresses, and also on the advice of many others in the field. It is the main topic of this paper and is presented in its entirety below.

We view these results as definitive early steps in the broad general consideration of the sequelae of a successful outcome of SETI searches. There are many further steps. One of these is the establishment of studies of the impact of the discovery on mankind, both immediate and long term. A second is a number of different and complex questions about the decision on sending replies from Earth. These topics are discussed briefly at the end of the paper.

SUMMARY OF SETI POST-DETECTION PROTOCOL PAPERS

In 1986 and 1987, the SETI Committee of the IAA and representatives of the International Institute of Space Law invited speakers to the Congress of the IAF to address many different aspects of the impact of the discovery of an extraterrestrial civilization. Sixteen papers were given, covering many of the cardinal questions which emerge. They have recently been published as a special issue of Acta Astronautica, entitled "SETI Post Detection Protocol," [1] edited by two of us (JT and MM).

It is not possible in this paper to examine the content of all the papers even in summary form. It may be useful to recognize the titles and authors to summarize the broad scope of material covered.

Part I of the special issue of Acta Astronautica deals with the receipt and verification of a signal. It has papers on Risk and Value Analysis, by one of us (JB); SETI False Alerts as "Laboratory" Tests for an International Protocol Formulation, by J. Heidmann; Signal Verification in the Real World: When Is a Signal Not a Signal?, by P. B. Boyce; and The First Two Steps After the Signal Arrives: Verify and Tell the World, also by P. B. Boyce.

Part II addresses the process of announcing the detection of an ETI signal. It includes papers on Announcing the First Signal: Policy Aspects, by J. M. Logsdon and C. M. Anderson; Signals from ETI Detected - What Next?, by Z. Paprotny; A Critical Examination of Factors that Might Encourage Secrecy, by A. Tough; and Diplomatic and Political Problems Affecting the Formulation and Implementation of an International Protocol for Activities Following the Detection of a Signal From Extraterrestrial Intelligence, by A. E. Goodman.

Part III examines the impact of an announcement of an ETI signal. It has two papers on Contact: Releasing the News, by R. Pinotti; and The Impact of Contact, by B. Finney.

Part IV explores the legal principles relating to contact and communications with ETI. The papers in this section are International Law Implications of the Detection of Extraterrestrial Intelligent Signals, by V. Kopal; XII Tables for Researchers on Extraterrestrial Intelligence, by A. A. Cocca; Discovery of ETI: Terrestrial and Extraterrestrial Legal Implications, by E. Fasan; Diplomacy and the Search for Extraterrestrial Intelligence, by A. E. Goodman; and Basic Elements of an International Terrestrial Reply Following the Detection of a Signal from Extraterrestrial Intelligence, by G.C.M. Reijnen.

Part V consists of a single article by Donald Goldsmith entitled Who Will Speak for Earth? He addresses the question of the preparation of a reply from Earth, and makes the bold proposal that we prepare the reply in advance of detecting the signal.

Part VI is the "Declaration of Principles Following the Detection of Extraterrestrial Intelligence". It is presented here for the participants of the Third International Bioastronomy Symposium.

DECLARATION OF PRINCIPLES CONCERNING ACTIVITIES FOLLOWING THE DETECTION OF EXTRATERRESTRIAL INTELLIGENCE

We, the institutions and individuals participating in the search for extraterrestrial intelligence,

Recognizing that the search for extraterrestrial intelligence is an integral part of space exploration and is being undertaken for peaceful purposes and for the common interest of all mankind,

Inspired by the profound significance for mankind of detecting evidence of extraterrestrial intelligence, even though the probability of detection may be low,

Recalling the Treaty on Principles Governing the Activities of States in the Exploration and Use of Outer Space, Including the Moon and Other Celestial Bodies, which commits States Parties to that Treaty "to inform the Secretary General of the United Nations as well as the public and the international scientific community, to the greatest extent feasible and practicable, of the nature, conduct, locations and results" of their space exploration activities (Article XI).

Recognizing that any initial detection may be incomplete or ambiguous and thus require careful examination as well as confirmation, and that it is essential to maintain the highest standards of scientific responsibility and credibility,

Agree to observe the following principles for disseminating information about the detection of extraterrestrial intelligence:

1.   Any individual, public or private research institution, or governmental agency that believes it has detected a signal from or other evidence of extraterrestrial intelligence (the discoverer) should seek to verify that the most plausible explanation for the evidence is the existence of extraterrestrial intelligence rather than some other natural phenomenon or anthropogenic phenomenon before making any public announcement. If the evidence cannot be confirmed as indicating the existence of extraterrestrial intelligence, the discoverer may disseminate the information as appropriate to the discovery of any unknown phenomenon.

2.   Prior to making a public announcement that evidence of extraterrestrial intelligence has been detected, the discoverer should promptly inform all other observers or research organizations that are parties to this declaration, so that those other parties may seek to confirm the discovery by independent observations at other sites and so that a network can be established to enable continuous monitoring of the signal or phenomenon. Parties to this declaration should not make any public announcement of this information until it is determined whether this information

is or is not credible evidence of the existence of extraterrestrial intelligence. The discoverer should inform his/her or its relevant national authorities.

3. After concluding that the discovery appears to be credible evidence of extraterrestrial intelligence, and after informing other parties to this declaration, the discoverer should inform observers throughout the world through the Central Bureau for Astronomical Telegrams of the International Astronomical Union, and should inform the Secretary General of the United Nations in accordance with Article XI of the Treaty on Principles Governing the Activities of States in the Exploration and Use of Outer Space, Including the Moon and Other Bodies. Because of their demonstrated interest in and expertise concerning the question of the existence of extraterrestrial intelligence, the discoverer should simultaneously inform the following international institutions of the discovery and should provide them with all pertinent data and recorded information concerning the evidence: the International Telecommunication Union, the Committee on Space Research of the International Council of Scientific Unions, the International Astronautical Federation, the International Academy of Astronautics, the International Institute of Space Law, Commission 51 of the International Astronomical Union and Commission J of the International Radio Science Union.

4. A confirmed detection of extraterrestrial intelligence should be disseminated promptly, openly, and widely through scientific channels and public media, observing the procedures in this declaration. The discoverer should have the privilege of making the first public announcement.

5. All data necessary for confirmation of detection should be made available to the international scientific community through publications, meetings, conferences, and other appropriate means.

6. The discovery should be confirmed and monitored and any data bearing on the evidence of extraterrestrial intelligence should be recorded and stored permanently to the greatest extent feasible and practicable, in a form that will make it available for further analysis and interpretation. These recordings should be made available to the international institutions listed above and to members of the scientific community for further objective analysis and interpretation.

7. If the evidence of detection is in the form of electromagnetic signals, the parties to this declaration should seek international agreement to protect the appropriate frequencies by exercising the extraordinary procedures established within the World Administrative Radio Council of the International Telecommunication Union.

8. No response to a signal or other evidence of extraterrestrial intelligence should be sent until appropriate international consultations have taken place. The procedures for such consultations will be the subject of a separate agreement, declaration or arrangement.

9. The SETI Committee of the International Academy of Astronautics, in coordination with Commission 51 of the International Astronomical Union, will conduct a continuing review of procedures for the detection of extraterrestrial intelligence and the subsequent handling of the data. Should credible evidence of extraterrestrial intelligence be discovered, an international committee of scientists and other experts should be established to serve as a focal point for continuing analysis of all observational evidence collected in the aftermath of the discovery, and

also to provide advice on the release of information to the public. This committee should be constituted from representatives of each of the international institutions listed above and such other members as the committee may deem necessary. To facilitate the convocation of such a committee at some unknown time in the future, the SETI Committee of the International Academy of Astronautics should initiate and maintain a current list of willing representatives from each of the international institutions listed above, as well as other individuals with relevant skills, and should make that list continuously available through the Secretariat of the International Academy of Astronautics. The International Academy of Astronautics will act as the Depositary for this declaration and will annually provide a current list of parties to all the parties to this declaration.

DISCUSSION

It should be noted that the Declaration was endorsed in April of 1989 by the Board of Trustees of the IAA and the Board of Directors of the International Institute of Space Law. It is currently being considered for endorsement by the International Astronomical Union, COSPAR, the Union Radio Scientifique Internationale, and the International Astronautical Federation. It is intended that the document then be made available to all individuals and organizations conducting SETI, with a request that they consider becoming signatories. It is further intended, as indicated in Principle 9, that an international committee of experts be convened to examine the evidence collected in the aftermath of a discovery. Lists of candidates from the appropriate international institutions are being assembled by the IAA SETI Committee. It has been proposed by Donald Tarter (USA) that the Committee be established before the discovery. A similar proposal was made by Ney in 1985 [2]. Tarter's proposal [3] is for a "Contact Verification Committee". His survey of the key individuals currently active in SETI indicates a majority opinion in favor of his proposal. The IAA SETI Committee is examining the possibility of activating such a Committee in the near future, in coordination with Commission 51 of the International Astronomical Union.

It should be noted that there are some similarities between the establishment of SETI Declarations of Principles and the Planetary Protection Agreements reached in COSPAR under the aegis of their SubCommission F3 [4]. For many years these agreements have been adhered to in the design and operation of planetary spacecraft in order to minimize the dangers of "forward contamination" of other planetary environments from Earth. COSPAR has been most effective in promulgating these agreements without the need for formal treaties between nations. Perhaps the SETI Declaration will be equally effective. In any event, should nations wish to use the SETI material, as prepared and endorsed by the international space organizations, as a basis for formal international agreements or treaties, that could be a logical step. Eventually the document might be considered by the United Nations Committee on the Peaceful Uses of Outer Space in reference to its relating to the Treaty on Principles Governing the Activities of States in the Exploration and Uses of Outer Space [5].

THE SECOND DECLARATION OF PRINCIPLES

Principle 8 of the Declaration of Principles states that "No response to a signal or other evidence of extraterrestrial intelligence should be sent until appropriate international consultations have taken place. The procedures for such consultations will be the subject of a separate agreement, declaration or arrangement." The question of Earth's response to extraterrestrial intelligence is being examined by the IAA SETI Committee at the present time, and the initial proposals will be presented in a paper at the 1990 IAF Congress.

The detection of extraterrestrial intelligence could take many forms. Whatever the nature of the signal or other evidence, detection would raise some basic questions that require informed international discussion. Should the Earth send a message to the alien civilization? Who decides? Are there reasons why we should not reply? If we decide to reply, what should we say? In what form or language? Who decides?

The paper proposes approaches to dealing with these questions, centered on the international discussion of a set of principles that later might be incorporated into an agreement or other document. That document would lay out procedures to be followed for the formulation and transmission of such a message, and deciding on the sending of a message. It will also examine the principles on which the substance of a message should be based.

The authors recognize that, in the absence of compelling evidence of an extraterrestrial intelligence, it may be difficult to sustain high-level interest in this issue. Nonetheless, they believe that, given the significance of a possible future detection, it is timely and appropriate to begin the international dialogue now.

References

1. Tarter, J.C., and Michaud, M.A., "SETI Post Detection Protocol;" Special issue of Acta Astronautica, 21, #2 Pergamon Press (1990).
2. Ney, P., "An Extraterrestrial Contact Treaty?," J.Brit. Interplanetary Soc. 38, 521-522 (1985).
3. Tarter, D.E., "SETI and the Media: Views From Inside and Out," presented at the 40th IAF Conference, Malaga, Spain (October 1989).
4. DeVincenzi, D.L., and Stabekis, P.D., "Revised Planetary Protection Policy for Solar System Exploration," Advances in Space Research, 4, 12, 291-295 (1984).
5. Treaty on Principles Governing the Activities of States in the Exploration and Use of Outer Space Including the Moon and Other Celestial Bodies (The Space Treaty), UN-Doc A Res/2222(XXI) (27 January 1967).
6. Michaud, A.G., Tarter, J.C., and Billingham, J., "The Response from Earth Following the Detection of Signals from an Extraterrestrial Civilization." To be presented at the International Astronautical Congress in Dresden, DDR (October 1990).

Discussion

M. HARRIS: I note that some fundamentalist religious groups believe that the existence of extraterrestrials violates their beliefs. The panel suggested that Dr. Billingham may wish to contact organizations such as Committee for the Scientific Investigation of Claims of the Paranormal (C.S.I.C.O.P.) which have had experience in handling attacks from these groups.

J. BILLINGHAM: Thank you for the suggestion.

D. WERTHIMER: Does the protocol address what we should do if the first contact is not by radio? (Addressed to J. Billingham.)

J. BILLINGHAM: It is really addressed primarily to radio detection. The vast majority of searches are going to be in the microwave region of the spectrum.

P. BOYCE: John Billingham - you did not mention that the protocol will be very valuable in helping to educate the media about the need to make all possible checks and to verify the extraterrestrial origin of the signal. The lay press does not, by and large, understand the need for caution, skepticism and verification.

What now are your plans for approval of the protocol and what role will the members of IAU Commissioin 51 have in this process?

J. BILLINGHAM: I have to agree about the important role that the protocol will play in spreading the word about the search and about the implications of a detection, in addition to specific post-detection activities.

The protocol is now being considered for endorsement by COSPAR, the I.A.F., URSI and the I.A.U., and specifically Commission 51 of I.A.U.

G. MARX: As chairman of IAU Commission 51, I have informed the president of IAA that the Organizing Committee of Commission 51 is ready to join the Protocol. If the Business Meeting of Commission 51 confirms this intention, I'll ask IAU that at the General Assembly 1991 IAU should formally accept the Protocol.

J. BILLINGHAM: Thank you.

F. RAULIN: A question to John Billingham. Your protocol does not seem to take into account the content - if understandable - of the message. I think it should. Imagine that this message - although it is very unlikely - indicates a very near invasion of our planet by ETI.

J. BILLINGHAM: It is clearly stated in the protocol that all the information should be made available to everyone. So if the signal includes a message, that will be disseminated too.

D. BRIN: If I might interject for John, here, there are several lines in the protocol which do try to address the question just raised. For instance, the step in which information is shared only with "other signatories to this agreement" achieves two ends. First, it encourages all groups actually to sign. And second, it does allow the professionals a time interval to discuss possible ramifications before a general announcement is made. There is also a provision that any general announcement should only be made by the original discoverer. Besides, providing for simple courtesy, these steps also allow for a pause, should one prove necessary. Despite all of this, however, it remains best to emphasize openness. Concealment is more likely to do harm than candor.

J. BILLINGHAM: Thank you.

P. SCHENKEL: Because of the likelihood that zealots of all brands would try to whip world public opinion into a frenzy (which might impair efforts to deal with contact with ETI on a sober and rational basis), it was proposed that the international committee of scientists should be formed beforehand and not after the reception of a verified extraterrestrial signal. I refer to point 9 of the Protocol.

More efforts should be undertaken by the scientific community to popularize the scientific nature of SETI via mass media and the whole range of the educational system.

To the panel (W. Calvin et al): Because of its great political, strategic and military implications, the decision regarding the question if, what, who, and how mankind should answer an extraterrestrial signal or message would necessarily be taken on the highest political level. For this the committee of scientists should prepare adequate recommendations, guidelines, etc.

Contact might also occur via an extraterrestrial spacecraft attracted by the electromagnetic wave front expanding from Earth. Guidelines and procedures should be worked out for this scenario. Otherwise we might get caught unprepared, though chances for such a scenario might be small.

NO ANSWER.

V. STRELNITSKI: Does the hesitation about possibilities and necessity to reply to ETI signals, expressed in your communication, mean that you consider the previous attempts to declare our existence (Arecibo, Voyager, etc.) to be a mistake? Will you try to prohibit any attempts to send signals from the Earth (to stimulate our potential correspondents for example) in the future?

J. BILLINGHAM: It is my opinion that the previous attempts have only a remote possibility of being detected. The protocol, in Principle 8, does strongly recommend that no one should reply to a signal until extensive international consultations have taken place.

R. TERRILE: Because of the possibility of a transient signal being detected, it is important that other observers be notified immediately. When more than two people know about a potential detection, I doubt that it can be kept a secret from the press. A campaign to make the press aware of this possibility is essential. Another "Cold Fusion" fiasco is to be avoided at all cost.

J. BILLINGHAM: Thank you!

THE CONSEQUENCES OF CONTACT: VIEWS OF THE SCIENTIFIC COMMUNITY AND THE SCIENCE MEDIA

Donald E. Tarter and Walter Gillis Peacock
The University of Alabama in Huntsville
Huntsville, AL 35899 USA

Paper Presented by D. E. Tarter

ABSTRACT

Information is drawn from a 1989 survey of the search for extraterrestrial intelligence (SETI) community and the international science media as to the anticipated consequences of a successful outcome to the search for extraterrestrial intelligence. The evidence indicates that both groups expect intense public interest and excitement about the announcement of a discovery. Public fear and shock associated with an announced discovery are anticipated to be of low to moderate intensity, but the possibility of angry or even violent reactions among some extremist groups is believed possible by both the media and the science community. Rumor, confusion, and disbelief are expected to be significant problems related to an announced discovery. The study indicates that any agency established by the SETI community to verify an announced discovery should explore what its role should be in relation to public attempts to exploit and perhaps distort the facts of the discovery. This study indicates that such a verification agency would be desirable to ensure the orderly and accurate dissemination of information and to avert undesirable and perhaps dangerous consequences of an extraterrestrial intelligence discovery.

INTRODUCTION

The search for extraterrestrial intelligence (SETI) has been described in the science media as a scientific enterprise which, if successful, could produce one of the most momentous discoveries in the history of science—a discovery that will have a profound social impact on humanity. Members of the SETI community have been for sometime contemplating and planning for a possible success in their endeavors. Previous research has indicated that a SETI discovery could present a potential logistical nightmare (Franknoi 1988). Under such conditions, scientists might be tempted to prematurely release information that is poorly interpreted and misleading. Members of the media covering the announcement could strongly and, perhaps, adversely affect the public perception of the event unless steps are taken to assure a mechanism for an orderly release of accurate information. In order to do this, the SETI community has developed and finalized a protocol agreement among scientists as to appropriate actions in case of suspected contact.

In 1989, an international survey of 221 science media people and 65 individuals defined as being active in the SETI community was conducted (Tarter, 1989). Each group was sent a lengthy questionnaire dealing with SETI news and information policy and the perceived impact of SETI success. Responses were obtained from 53, or 24%, of those in the media; and from 36, or 55%, of those in the SETI community. The protocol agreement was favorably received by the majority of the media and the SETI community. However, the majority of each community felt that the protocol agreement alone would be inadequate to prevent possible premature or misleading announcements or alleviate the confusion surrounding an announcement. The majority of both the SETI community and the media favored the establishment of a standing committee to function for the purpose of contact verification and information dissemination.

Procedures have been finalized for the SETI Protocol Agreement, but as of yet no formal action has been taken on the establishment of a contact verification committee. The purpose of the proposed contact verification committee would be threefold. The committee would initially serve as a source of information about SETI programs for the media and the public. In the eventuality of an announced discovery of extraterrestrial intelligence, the committee would serve as a confirmation agency for the validity of the claim. The committee would give or withhold its endorsement for any alleged discovery. In the event of an endorsed or verified discovery, the committee would serve as a source of information and organization for the massive numbers of inquiries that would likely follow.

If such a committee were established, it would be beneficial to have some relatively sound information about what type of situations it might face. Contained within the 1989 survey were questions designed to elicit this type of information. The following section describes the basic findings from the research.

ASSESSING PUBLIC RESPONSE

An attempt was made to measure the opinion of the media and the SETI community and the types of public response anticipated in the eventuality of a SETI discovery. A question was constructed that asked both groups to rank, on a ten-point scale of intensity, nine likely public reactions to a discovery. The mean rankings of the media and the SETI community are presented in Table 1 and Figure 1.

These measures reveal a striking similarity between the media and members of the SETI community as to their anticipated public reactions. As demonstrated by the means presented in Table 1, the media tends to rank public reaction slightly more extreme than do members of the SETI community. The major differences lie in those areas labeled fear, disbelief, and shock, with the media again attributing a more extreme public reaction in regards to these than the SETI community. The only differences that are statistically significant are those concerning public fear and disbelief. It should be noted, however, that even though the media does expect significantly greater fear than the SETI community, the mean responses are 3.9 and 4.9 on a scale of 10, indicating that both groups believe that the fear response would be of low intensity.

Interestingly, the greatest difference in potential public response anticipated by these groups comes in the area of disbelief. Media representatives are significantly more likely to feel that an announcement would be doubted than are members of the SETI community. Designing convincing arguments for the reality of their discovery might well be a task that needs study by any future contact

Table 1. Mean rankings [1] of public response to discovery of extraterrestrial intelligence.

Public Response	Sample Means	Combined Community Means	SETI Media Means	Science Means	t-value
Interest	8.759	9.029	8.585		1.374
Excitement	7.885	7.647	8.038		-0.812
Rumor	7.317	6.935	7.549		-1.128
Confusion	6.583	6.121	6.882		-1.324
Disbelief	6.360	5.559	6.885		-2.616[2]
Hope	5.494	5.314	5.480		0.067
Joy	5.357	5.324	5.380		-0.120
Shock	4.919	4.353	5.288		-1.717
Fear	4.523	3.882	4.942		-2.118*

1. Respondents were asked to rate their response on a 10-point scale, with 1 indicating little and 10 indicating much; to an official announcement.

2. Denotes significance difference ($P(t) \leq 0.05$).

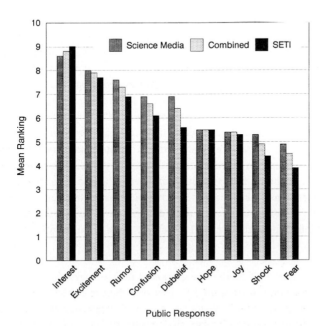

Figure 1: Mean rankings of public response to discovery of extraterrestrial intelligence.

verification committee. This would be particularly the case in the eventuality that the signal was ambiguous or carried low information content.

In scanning Table 1, it can be seen that the most intense public reactions anticipated are in the areas labeled interest, excitement, rumor, confusion, and disbelief. As indicated by previous research (Franknoi 1988; Tarter 1989), immense public interest and excitement would likely attend any public announcement of the discovery of extraterrestrial intelligence (ETI). Most importantly, both groups seem to agree that rumor, confusion, and disbelief would be rather pronounced outcomes of an ETI discovery.

These findings lend emphasis to argument that a standing organization that could serve a rumor control and clarification function would be very helpful to both the scientific community and the media. It may also be critical in allaying public disbelief and the potential for rumor and confusion. Such a function becomes even more important when we examine the potential for exploitation and angry reactions that might erupt under some types of ETI detection.

Table 2.   Percent answering Yes to questions concerning country-specific group reactions to the discovery of extraterrestrial intelligence.

Group Reactions	Combined Sample	SETI Community	Science Media	t-value
Exploit discovery	86.2%	80.8%	89.7%	-1.019
Angry or violent	59.7%	64.0%	57.1%	0.546

Table 2 indicates that both the SETI community and the media are in strong agreement that there are groups in their country that might try to exploit the discovery of ETI for the promotion of their own purposes. Specifically, 81% of the SETI community and 90% of the media respondents feel that exploitation is likely. When asked to mention specific groups that might choose to exploit the findings, the range of responses went from UFO cults and religious fanatics through various commercial enterprises to NASA and the Planetary Society. In fairness, it must be noted that only one respondent mentioned the Planetary Society and no definition of expoitation was given. The respondent's definition of exploitation could therefore have been benign or malignant. Such a wide variety of groups that could exploit a discovery suggests that any verification agency should think carefully about its role in commenting on the legitimate use of information about an ETI discovery. Little or no attention has been given to this issue.

As also indicated in Table 2, when asked if there were groups in their country that might have angry or even violent reactions to the announcement of the ETI discovery, 64% of the SETI community and 57% of the media felt that such groups exist. Speculation as to what type groups might react with anger or violence focused almost entirely on those with fundamentalist religious persuasions. A response such as this from both the media and the SETI community suggests that additional attention be given to further identifying and understanding these potentially violence-prone groups.

CONCLUSIONS

The evidence provided by the 1989 International SETI Survey indicates the expectation of intense public interest and excitement about the announcement of any discovery of ETI. Public fear and shock associated with such an announcement are believed to be of low to moderate intensity; however, among some groups with extreme belief systems, the possibility for angry or violent reactions is deemed realistic. Rumor, confusion, and disbelief are considered to be significant problems associated with a discovery announcement.

A wide variety of exploitation attempts on any announced discovery are anticipated. Not only fringe groups and cults would seek exploitation, but commercial groups, governmental groups, and space-support groups might also use such findings for their own proposes. A standing verification agency should give attention as to what role it should play in monitoring and commenting on the exploitation of the discovery of ETI.

These findings suggest that an official agency established to deal with contact announcement would have a wide variety of issues to explore in order to assure the orderly and accurate dissemination of information about an ETI discovery. Its proper functioning could avert many undesirable and perhaps dangerous consequences of an ETI discovery.

Acknowledgments

This study was made possible by internal grant from the University of Alabama in Huntsville. The authors would like to acknowledge the assistance of Eva L. Buchanan, Kathryn Smith, and Walter S. Sullins in preparation of this paper.

References

A.G. Franknoi, 1988, "What If We Succeed?: A Media Viewpoint," in G. Marx (Ed.), Bioastronomy: The Next Steps, Kluwer Academic Publishers, pp. 417-424.

D.E. Tarter, 1989, "SETI and the Media: Views From Inside and Out," paper presented at the 40th Congress of the International Astronautical Federation, Paris, France.

Discussion

J. TARTER: From a sociologist's point of view - were there any surprises in the results?

D. TARTER: I guess the major surprise was the extent of agreement among the media and the SETI community as to the importance, the type of difficulties likely to be faced at contact, and the probable public response to contact. The results seem to verify that a great deal of rumor and confusion expected by the SETI community is also expected by the media, lending strength to the argument that steps need to be taken to design a response system to minimize confusion.

It is interesting that fear is adjudged to be significantly greater by the media. Though both judged the public fear response to be in the low moderate range, the media's response, which may be somewhat more sensitive to public opinion, may indicate something valuable for the SETI community to consider.

G. MARX: To my personal understanding the Protocol implicitly excludes sending a message without previous understanding of the international (scientific) community.

Past experience shows that there will be a time gap between the first suspicional observation and the final interpretation. (See e.g. the case of the cold fusion.) This transient period will be an era of pride, secrecy, rumor misunderstanding, rivalry. According to my opinion the Protocol has rather a psychological role, to prepare the scientific community ethically, the media scientifically, and to decrease the confusion and misbehavior at the time of the impact of a "signal." That means that the Protocol should be made widely public.

SELECTION CRITERIA IN BIOASTRONOMY: EXCERPTS FROM A PANEL DISCUSSION

Panel Members: Ivan Almar (Chairman), William H. Calvin, Vladimir V. Rubtsov,
Woodruff T. Sullivan, III, Jill C. Tarter, and Dan J. Werthimer

Each panel member presented his/her views for about 5 minutes and the subsequent discussion was open to all participants. The Chairman has summarized the comments of each panel member.

INTRODUCTION - I. Almar

Objective criteria are needed to distinguish between natural and artificial phenomena of the Universe. The late V.F. Shvartsman emphasized in his talk at the Tallin conference (1) the following arguments: "I am convinced that among the several ten thousand radio sources in the catalogues of radio astronomy as well as among the several ten million optical sources on star maps there are plenty of artificial origin. These sources are recorded even today but they are misinterpreted since the recognition of ETI is not only a scientific, but also a global cultural problem." Whether or not Shvartsman is correct, searches for extraterrestrial intelligence definitely require guidelines to devise a successful strategy.

Among the few serious attempts to deal with the problem is the cautious approach of I.S. Shklovskii, who formulated "the principle of the presumption of natural origin" during the Byurakan conference (2). Shklovskii declares that the artificial origin of an object or a phenomenon should only be announced when all possible explanations by natural causes have been exhausted. While the principle is scientifically logical, it did not turn out to be constructive because one can never determine the moment when all natural explanations have been exhausted.

There are several well-known cases when new astronomical observations generated convincing explanations which were later proved to be only illusions. Consider a recent instructive example: During a 7-hour observation period in 18 January 1989, astronomers using the 4 m telescope at Cerro Tololo, discovered an extremely rapid pulsar at the location of supernova 1987A (3). In spite of the fact that the new pulsar was seen unexpectedly early, spun unexpectedly rapidly and was unexpectedly bright compared to theoretical predictions, - moreover, despite repeated attempts, no subsequent pulses were detected - a number of detailed "scientific explanations" were published in the literature. In January 1990, however, following a case when similar rapid pulses had been found as a consequence of a simple electric interference, the Cerro Tololo team disclosed that also in the case of the new pulsar at supernova 1987A a camera used to transmit images from the telescope to the control room was the culprit (4). It is clear that in this case an artificial signal was successfully interpreted as a "natural phenomenon"! It is worthwhile to mention that in a paper presented at the previous Bioastronomy conference, Hilton and Almar warned that ETI signals might be detected in connection with the outburst of this near-by supernova (5). Clearly the artificial signals - in this case of course of terrestrial origin - arrived, but were misinterpreted quickly by theoreticians.

Artificial objects and signals are more and more universal. This is commonplace if the growing number of artificial satellites and space probes on the sky is taken into consideration. Nowadays with the passing of only three decades of active space research it is easy to calculate all objects launched by our civilization and to select their voices from the big orchestra of natural radio sources, but sooner or later the situation will change and misinterpretations will occur. Therefore we need absolute criteria to distinguish between objects and signals produced by our own space activity from those originating from ETI activity.

Finally, there is a third problem which belongs to the field of interest of the biological sciences: how to distinguish between natural phenomena produced by inorganic evolution and by living creatures respectively. Up till now there is no definition of life accepted by the whole scientific community, therefore we have to wait probably until other life forms on other planets will be detected.

References

1. Shvartsman, V.F., "Problema Poiska Zhisni vo Vselennoi," p. 1986 (in Russian).
2. Shklovskii, I.S., "Problema CETI," p. 133, Mir, Moscow, 1975 (in Russian).
3. Nature, Vol. 338, p. 234, 1989.
4. Nature, Vol. 344, p. 679, 1990.
5. Hilton, W.F. and Almar, I., "Bioastronomy - the next steps," p. 377, Kluwer, Dordrecht, 1988.

Comments by Panel Members

J. Tarter: CRITERIA FOR ARTIFICIALITY

Although we speak of SETI - the search for extraterrestrial intelligence - we must remember that the search we conduct is at this moment the search for extraterrestrial technology. In the next century, space-based instruments may permit us to image terrestrial-type planets around distant stars and permit us to conduct chemical assays of their atmospheres. At that time we may be able to debate the relevance of putative chemical evidence for life and exobiology, but for the moment we are restricted to a search for evidence of distant technologies.

Because of our nature, the complex history of SETI, and the current confusion of SETI with UFOs and pseudo-science on the part of some of the public, I think we should demand that any signal or phenomenon we accept as being of ETI origin must be repeatable. No matter how intriguing and how well documented, I believe that we should not be willing to declare a short-lived signal to be credible evidence of ETI unless it is repeated and verified. Therefore our criteria will require signals or phenomena to persist or repeat themselves on a reasonable time scale.

I suggest that anthropocentric prejudices will exclude manifestations of extraterrestrial technologies whose periods of repetition exceed a human lifetime. Most of the observational data we collect today are not archived, or at least not without a great deal of filtering and compression. It is unlikely that at some future time one could examine the microwave equivalent of a Harvard Plate File to find earlier evidence of a detected signal. We will be forced to rely upon an individual's memory of having previously seen the same thing. Hence I expect that we will necessarily exclude ourselves from recognizing as artificial any signals or phenomena that do not repeat with timescales less than a human lifetime. In the end I believe that Arthur Clarke has the last word when he says, "Any sufficiently advanced technology will be undistinguishable from magic."

D. Werthimer: ON DISTINGUISHING EXTRATERRESTRIAL RADIO SIGNALS FROM MAN-MADE RADIO SIGNALS

As all SETI observers quickly learn, distinguishing an extraterrestrial signal from radio frequency interference (RFI) can be quite difficult. As an example, a simple algorithm we use in the SERENDIP data analysis is to reject radio signals which appear in two or more places on the sky. However, this algorithm does not eliminate short duration RFI (e.g., when the RFI transmitter turns off before the telescope moves). We now utilize five algorithms to identify and reject various types of RFI. Even so, some interference conspires to mislead us and appears as interesting candidate signals requiring further investigation.

Because it is so easy to be confused by interference, I concur with Jill Tarter that a signal that is seen only once or twice and never again, will not be very convincing evidence for another civilization.

W. Calvin: WHY AN INTENTIONAL ETI SIGNAL MIGHT MASQUERADE AS A FAMILIAR RADIO ASTRONOMY OBJECT

Let's look at our "Is it a real signal?" problem backwards: What would we do to catch the eye (or ear) of another civilization if we decided to avoid the assumption that they would be searching for intelligent signals? Realizing that they too would probably use automation to preselect "interesting" signals from amidst the noise, we'd probably take a lesson from categorical perception in the cognitive sciences or from interspecies communication in ethology.

Novelty is relative to what you already know: it's the frontier, the familiar as it grades into the exotic - and then into "noise". Anything too different from the usual is often not seen; it is ignored as noise, at least until we finally develop the cognitive strategies that cause us to take note of it. I'm sure that we humans are just big shadows to most species of animals; we really aren't seen in much detail. Interactions between most species of animals are like two ships passing in the night: they just get out of each other's way. To grazing animals, we're probably only seen as "other". So long as we stand upright and don't move too quickly, they mostly ignore us. But if we get down on hands and knees so as to appear to be four-legged like their usual predators, grazing animals will take notice and move away.

To get their attention, we should pretend to be something that they routinely scan for: a suitable masquerade may be required. The advice from the zoological, ethological, and cognitive science communities would probably be, "Look for the radio equivalent of an eye spot." These features of fish, amphibians, etc., are not real eyes, just pigmented skin or scales that look, from a distance, like large, all-seeing eyes.

If we were to ask an advertising agency to advise us about ways to attract attention to a message in a cluttered background, we'd discover another trick, a version of cognitive dissonance. A few years ago, somebody discovered how to make billboard ads far more memorable than they usually are:

What they do (as you can see for yourself walking through any American airport) is to make a sentence that is intentionally nonsense. Usually it is a sentence fragment, missing the verb or its object. Other times, you cannot figure out an antecedent. And so you puzzle over what you missed during the next several minutes, long after the billboard has passed out of sight. You eventually give up because there is, in fact, no right answer - it really is intentional nonsense. The hidden message was, "Write this brand name on your (mental) blackboard a dozen times, so you'll remember it!" Thanks to all this rehearsal time as you puzzle about the pseudosentence, you will recall the brand name on your next trip to the grocery store. Such ads do catch our attention, and they get past our usual barriers to visual clutter that is glimpsed for only a passing moment.

To carry the eyespot principle further: we might simulate the familiar radio characteristics of common astronomical objects but add a bizarre twist for a radio astronomer on the other end, something like that cognitive dissonance of the deceptive billboard.

I suggest that we'd conclude, if we were to design an outgoing beacon, that we really ought to mimic a pulsar in signal appearance, sending signals that a simple time-shifting analysis like autocorrelation or expectation density would easily pick up as a non-zero peak. A one per second rhythm, for example. Don't take pulsar too literally here, maybe a narrow bandwidth example, such as maser lines, would be better.

The second trick is to label a signal as unequivocally odd compared to the standard of the cover signal, e.g., a very unusual pulsar. I would suggest that an unusual feature could be incorporated with the typical pattern of a "pseudo whatnot." The unusual "whatnot" pattern would therefore catch the attention of the system designers and operators. A rhythmic beat that suddenly becomes ten times faster would certainly qualify. In this example a secondary peak would periodically show up in an autocorrelation or expectation density, then disappear.

The third trick, after getting their attention, is to convince the distant analysts that no physical system is likely to produce such signal: You want them to throw up their hands (or whatever) and shout, "Impossible!" For example, let the fast bursts have a prime number of pulses, and the next burst have the next-higher prime number. So it goes 2, 3, 5, 7, 11, 13, 17, 19, 23 and so forth. If you later want to send pictures in a 79 x 111 raster, just repeat 79 as you go past it, and repeat 111 before going on to 113. Then pause and send your 79 x 111 raster with the tutorial information in it, a la the Pioneer spacecraft message. This might only occupy one percent of your transmission time, with a long informative message transmitted once a month.

Flipping this back around again to the topic of this panel, we can see that what I'd suggest for our own automated search criteria is really a form of science-as-usual: concentrating on finding naturally interesting objects in the sky but looking for truly odd features in a few of them. It's like the preoccupation of the 18th century natural historian with the oddities, those freaks of nature. And like the 19th century's Charles Darwin, who saw variability as the engine that drove the evolution of new species. I'm not sure that variability in radio sources is the leading edge of radio astronomy anymore, but it is where I'd expect to find the intentional signals of extraterrestrial civilizations, as unnatural variants on what first appears natural. And even if you don't get a signal, maybe you'll still get a nice catalog of narrow-band pulsars or some such, to write home about.

W. Sullivan: CAN WE DISTINGUISH ETI'S "MUSIC" FROM NATURE'S?

Philip Morrison has described the SETI enterprise as the "archeology of the future" (and has urged us to dig!), but our present discussion is more akin to the archeology of the earliest human sites, where there are often great debates about whether a find represents artifact or simply an odd-shaped stone.

In my brief remarks I wish to focus on the investigation of a strong, continuous signal that everyone accepts as emanating from a fixed sidereal position, not RFI, and not a hoax. I also presume that the signal does not contain a string of millions of well-defined bits, repeated over and over. In such a case, even without any specific decoding, I believe there would be general agreement that we had found ETI.

My overall view is that we tend to be too optimistic when thinking about the problems of distinguishing a signal of ETI origin from one naturally produced. One cannot state too strongly that there exists no well-defined phenomenological boundary between natural and artifactual signals. For example, we usually define artifactual signals as having a high degree of coherence (in frequency, time, polarization, etc.), but the physical universe also supplies an abundance of natural signals of this sort. I can imagine a variety of cases that could remain perplexingly ambiguous perhaps for years or decades, driving us crazy! Here are some examples:

(1)   Suppose that the contingencies of history had led to the discovery of extraterrestrial radio waves and then shortly later pulsars, before we had any detailed understanding of stellar evolution and structure (including the notion of small, extremely dense stars). ETI would then have been an extremely likely explanation for these pulses, amongst scientists as well as the public. In fact, recall that the Cambridge radio astronomers in 1967 did for a time seriously consider the pulsars to be of ETI origin - they called the first pulsars LGM 1 and LGM 2 ("Little Green Men"). Now shift to our current situation with SETI. Simply because our present astrophysics knows of no natural way to produce one-hertz-wide signals, does this mean that if we receive them, we should automatically treat them as artifactual? Exactly how many theorists thinking for how long need to be baffled before we give up and call it ETI?

(2)   Suppose that we receive a pattern of narrowband signals that are separated in frequency by quantities that are multiples of a certain basic frequency-interval. The multiples turn out to be 1, 2, 3, 5, 7, 11, 13... Or perhaps 1, 2, 4, 8, 16, 32... Or 1, 1, 2, 3, 5, 8, 13... Is this ETI? It's the kind of thing we usually talk about expecting, but it's sobering to recall that Nature is also full of such patterns, and that indeed one important aspect of science is searching for patterns and explaining them in a "natural" way. For example, late nineteenth-century spectroscopists grappled with trying to understand exactly such neat relationships exhibited by different materials. One material emitted narrowband signals at frequencies proportional to

$$1/2\,2 \;-\; 1/(n+2)^2 \qquad \text{for } n = 1, 2, 3....$$

("Balmer's Law" of 1885 was in fact inspired by observations of the spectrum of Sirius.)

(3) Going somewhat beyond radio signals: How perfect a circle would we have to detect in order to be certain it was artifactual? Such a circle might be detected as part of a signal's raster pattern or observed on a planetary surface or actually resolved as an image on the sky. This sort of criterion is extremely problematic, for Nature makes some pretty good circles: Saturn's rings are circular to a part in 10^4 and stars to 1 in 10^5. Note that Kepler was convinced that the moon was inhabited and that one of his arguments was based on the latest high-tech results from Italy: Galileo's drawings of the moon as seen through a telescope, revealing a crater (Albategnius) so perfectly circular that it had to be an artificial embankment.

Science is a social activity searching for consensus about how the natural world behaves. We follow in the footsteps of Pythagoras and Kepler as we search for the "Music of the Spheres", by which I mean "harmonic" relations that are both mathematical and aesthetically pleasing. But this is equally true for SETI or "normal" science. Any candidate ETI signal will be exhaustively studied, and only accepted as truly ETI after scientists reach a consensus that it could not possibly be of natural origin - extraordinary conclusions require extraordinary evidence and cogitation. But what will be the criteria? How will the consensus be reached? Indeed, it may be profoundly difficult to distinguish ETI's music from Nature's.

V. Rubtsov: ON THE QUESTION OF PALEOVISITS

I'd like to discuss here the question of paleovisits, that is hypothetical ancient visits of extraterrestrials to the Solar System in general and to the Earth in particular. It is in fact of vital importance to our SETI field of research. After the works of Hart, Tipler, Singer, it seems we should choose between two possibilities: either the Solar System has been visited in the past, or the Earth possesses one of the first (i.e., one of the oldest) civilizations in the Galaxy. Both possibilities are rather interesting.

I will call "paleovisitology" that direction of investigations which is aimed at studying the paleovisit question. It is a common opinion that searches for ancient extraterrestrial artifacts (ETAs) should be the main task of paleovisitology. Yet there is in fact no real search; there have been only a few attempts to apply rather vague ideas of extraterrestrial artifacts to certain enigmatic (or pseudo-enigmatic) finds. I will try to sketch here some elements of a more rigorous approach to this task.

To find an extraterrestrial artificial object, we need, apart from criteria of its artificiality, some criteria of its "alienness" which would enable us to select against the background of similar terrestrial artificial phenomena. Such criteria may be implied if, for example, the object under examination was found in the "pre-human" strata of the "geological chronicle" of the Earth. Let us remember that there are two alternatives to the assumption of the object's extraterrestrial artificial origin: either it could be natural, or artificial, but nonetheless terrestrial.

The task can be especially difficult if one cannot establish the age of the object. We find ourselves before a pitfall: an object, whose technological level is only slightly beyond the well known capabilities of terrestrial civilization, might be secretly made in an advanced laboratory here on Earth.

An object manufactured by technologies whose level far exceeds the terrestrial ones, could very likely be mistaken for something natural or might even be unrecognizable as something worthy of note.

When we have only some spoilt fragments of an ETA, we must restore their original state and then reconstruct the ETA. Whether or not this task can be fulfilled depends on the extent of preservation of these fragments, their representativeness as regards the original ETA, the researcher's ability to build up a good theoretical model of the latter, etc. Certainly, the result of such restoration will be quite ambiguous.

One can imagine a situation, where it is impossible to restore the whole device from its remnants. To assert that the object is an ETA, one must prove: (1) a geological (or at least prehistorical) age of the find; (2) an artificial origin; and (3) a super-high (but intelligible to us) technological level.

The result of this kind of investigation will seldom be considered as fully reliable. Generally we will have several alternative hypotheses on the nature of an object under examination. There will appear, on the basis of these hypotheses, a set of various research programs, whose development may lead us to the final answer. An interdisciplinary study of the object, including physical, chemical, technological, structural and other types of analysis would be combined to form a system model (or a number of models) of the ETA.

In the not-so-distant future we may accidentally discover possible ETAs. In that event it would be necessary to set up an efficient system for gathering and evaluating reliable information. This system would enable us to accumulate a data bank and to develop systematic theoretical and experimental research in this field. We should plan for this work now. If we do not, future discoveries of possible ETAs will remain nothing more than funny curiosities.

General Discussion

M. HARRIS: The justification for SETI is really that it stimulates us to look for new strategies and new natural phenomena. Could SETI research be judged on the likelihood that it will yield new or unanticipated results?

J. TARTER: NASA's Microwave Observing Project will look at the Universe through a new set of narrow band filters that have never before been used. It may lead to the discovery of a new "class A" phenomenon (M. Harwit's classification), but it is not trivial to estimate the "probability of a discovery." Therefore we still have to sell the program on the basis of SETI itself.

D. BLAIR: Putting yourself in the minds of another civilization trying to make contact with us, one must recognize the incredible space in frequency that has to be searched. How do you make that easier? A traditional idea is to find an interstellar communication channel, i.e. one or several frequencies which may be definable on the basis of fundamental and absolute phenomena. The one we have selected to search is pi times the hydrogen frequency. There are clearly many other numbers to be considered.

P. ZARKA: I have several questions and remarks. First to Dr. Tarter. I would like to know why the time of repetitivity is limited to the timespan of a human life?

J. TARTER: We don't have complete historical records in radio astronomy. Interesting events are often thrown out of the data because they ruin what the observer was originally looking for.

P. ZARKA: Question to Dr. Werthimer. Do you intend to identify each feature in the dynamic spectra and then to construct an algorithm to eliminate any that could be RFI?

D. WERTHIMER: We currently have algorithms to get rid of RFI, but not to classify and find what it is coming from; e.g., if the telescope moves and we see a feature on two different places on the sky, then we eliminate the possibility of an interesting signal. If we see a very broadband thing, with lots of features, then we throw that out because it is too complicated to figure out.

P. ZARKA: I am wondering if one might define a ratio of information over power? In the case of pulsars or H-lines this ratio is very small.

W. SULLIVAN: Sure, you can define various systems but a priori I don't think you could get consensus on it. I think that we should not expect "one million bit" information to begin with, but let us say only two frequencies going back and forth with a one second periodicity. Can that type of signal be explained naturally or not?

P. ZARKA: Question to Dr. Rubtsov. Are your considerations consistent with the criteria of repetitivity, of consensus and of Shklovskii's "principle of presumption of natural origin"? Is there any ancient object which can stand a thorough analysis?

V. RUBTSOV: Why not? We are dealing with objects or phenomena, not only signals; the main criteria of selection between natural and artificial origin must be, however, the same. Nevertheless in some sense it is more complicated to select extraterrestrial artifacts than artificial signals since any object found on Earth will be among natural and artificial bodies.

P. SCHENCKEL: There are other possibilities that have to be taken into consideration, not only radio signals. I think the electromagnetic radiation which is continuously beamed out from the Earth, reaching about 60 light years at present, might be intercepted by a cruising extraterrestrial ship. What happens if suddenly a message would be received asking permission for landing? Mankind ought to be prepared for such a case.

I. ALMAR: I would like to remark that the problem of selection between natural and artificial phenomena is even broader if we take into account the possible existence of astroengineering activity of supercivilizations (Dyson spheres etc.). M. Harwit's book on Cosmic Discovery claims that there is only a limited number of really different astronomical phenomena; I would like to mention that if considerable astroengineering activity is present then the number of phenomena would be larger or even infinite.

D. SCHWARTZMAN: The question of intentionality is very important and this relates to the meaning of negative results. There should be a distinction made between the search for intentional signals of an advanced civilization and the search for leakage of radiation from an emerging civilization. Unfortunately we lack the sensitivity to do a systematic search for emerging civilizations in the near future but what about unintentional radiation of an advanced civilization? Those kind of signals may be the most likely to be detected in the near future.

J. TARTER: The bottom line is that you can't conceive what you can't conceive. If you are talking about manifestations of an advanced technology you are going to approach that question from the framework of what <u>we</u> know and what <u>we</u> can do. You ask yourself what must an extraterrestrial advanced civilization do? It presumably needs power, has to do something for power generation, and may or may not have a need and capability for interstellar travel, etc. SETI people have looked at this over the last 30 years, investigating what might be the ramifications of an advanced technology. You have a problem interpreting the negative results, because any result is very model-dependent (e.g., how much fusion power, how much fission power). I don't see how you get away from it once you go beyond what we do now. If one uses the Earth as an analogue, one can make significant statements like: If I fail to detect the following phenomenon then I will have shown there is no Earth analogue within a certain distance.

S. VON HOERNER: A few years ago in a Chinese restaurant I ordered my fortune cookie and it said: "The clever thing is to be prepared for the unexpected."

I. ALMAR: Thank you for the interesting contributions.

SUMMARY OF SCIENTIFIC CONTENT OF THE SYMPOSIUM

Frank D. Drake
University Of California, Santa Cruz, CA

In my summary I shall review in succession the six sessions along which the program was articulated.

SESSION 1: COSMIC EVOLUTION

This part was characterized by the high quality of the long-term programs undertaken to detect other planetary systems. These show real promise for solving the key problem of the effective existence of planets, including those favorable for the existence of life as we know it on Earth. The work of Campbell and Latham, among others, was provocative and stimulating.

The infrared and optical observations by Backman, Terrile and others, revealed striking results on dust disks around stars, disks from which planets may form or have already formed.

Also, important information has been gathered on the conditions on early Mars and Earth, which will be of great value in evaluating the processes that led to the development of life.

SESSION II: ORGANIC AND PREBIOTIC EVOLUTION

Val Cenis produced a remarkable watershed of facts and hypotheses in this domain. We are no longer confined to Oparin's warm pond for the production of the precursors of life. From the review of Irvine and Brown, we recognize a great richness in interstellar and prebiotic chemistry, a multitude of chemical pathways in the interstellar medium, with still further help from processes going on at the surface of interstellar grains.

Polycyclic aromatic hydrocarbons and other complex molecules in the interstellar medium are molecules of special interest. Is HCN polymerization a viable pathway to complex biochemistry? Are we HCN-based life after all?   Matthews' increasing evidence for the idea that comets played an important role in prebiotic chemistry is to be noted.

Maurette presented a tantalizing suggestion that prebiotic chemistry took place in micrometeorites as they entered the atmosphere. Even clays may play a prime role in the origin of life.

SESSION III: PRIMITIVE EVOLUTION

Before Val Cenis we were aware of only the tip of the iceberg of primitive evolution. Complex but key phenomena occurred when comets and asteroids impacted the young Earth. Chyba, et al, suggest that organics may have been delivered, but how much? This is a new and exciting field.

The papers by Doyle, Schwartzman and Whitmore described complicated atmospheric phenomena that may well extend and stabilize the continuously habitable zone around stars and the conditions on planets favorable for life. New understanding of the physical and chemical conditions of Mars and Titan was provided by McKay, Cabrol, Grin, and Coustenis. From Schwartzman we learned that life may have accelerated the chemical weathering sink for carbon dioxide, thus producing a climatically favorable atmosphere.

As we study the chemistry of the Galaxy, it appears that we have more chemical "friends" than we ever imagined, and long ago, they became us.

SESSION IV: SEARCHING FOR EVIDENCE

Remarkable progress have been made in two ways in the search for extraterrestrial intelligence:

(1)   The raw observing power available for detecting extraterrestrial signals has increased tremendously; the radio sensitivities up by a thousand times, and the channels available from 1 to 10^7, and soon to 10^8. This improvement has continued for these decades to follow a precise exponential law (Klein).

(2)   Recognizing that a real-time, prompt response to candidate signals is very important, we have developed a set of thoroughly planned observational programs:

- The NASA SETI microwave observing program, highly developed and based on a sophisticated software, will soon begin;
- META I and II are running in Massachusetts and Argentina;
- SERENDIP is engaged in piggy-back searches;
- A search program is underway at the University of Western Australia and CSIRO.

One of the new (but unconfirmed) candidates for an ETI signal is from the Parkes radiotelescope (Blair). Added attention is being given to aids such as special times and frequencies of observation (Boyce, Sullivan).

SESSION V: TANTALIZING (AND FUN!) SPECULATIONS

Possibilities for advanced evolution were considered by various authors:

The search for Dyson spheres using the IRAS catalogue, by Jugaku; gravitational lenses as ultimate transmitter and receiver for space signals, by Eshleman; the search for light or gamma-ray bursts

bursts from interstellar spacecraft, by Harris and Arkhipov; trash as the ubiquitous signature of intelligent activity, by Holmes.

Speculations on the possible variety of cognitive processes were produced following on a number of them: Darwinian thought, by Calvin; communication with Alex the parrot, by Pepperberg; the necessity for transfer of skills in complex comprehension and in space colonization, by the Pfleiderers.

SESSION VI: BIOASTRONOMY AND PHILOSOPHY

Bioastronomy appears, in the end, to be the birthplace of sciences. To understand life in the universe, and especially ourselves, one is led to study all the sciences, including:

- Nucleosynthesis,
- Aerodynamics,
- Newtonian and relativistic mechanics,
- Radio technology, electromagnetic theory,
- Chemistry of all kinds,
- Molecular biology,..
- ...and many more subjects, even philosophy and ethics (Papagiannis).

The papers by Davies and Ksanformality challenge our cosmological world views from the role of nuclear synthesis in the origin of life to speculations about machine copies of the human intellect. Dick examines the history behind the concept of a biological universe. Brin concludes that bioastronomy reflects a human search for novelty; it is just human beings behaving normally. Bioastronomy can and does serve as a foundation and stimulus for progress in all of science and science education.

LIST OF PARTICIPANTS

THIRD BIOASTRONOMY SYMPOSIUM ATTENDEES

1.	D. Latham	40.	S. Dick	79.	D. Deamer
2.	E. Martin-Ge	41.	A. Natta	80.	L. Laming
3.	S. Gulkis	42.	J. Pfleiderer	81.	C. McKay
4.	Unidentified	43.	J. Goudou	82.	D. Whitmire
5.	N. Jennerjohn	44.	S. Aiello	83.	J. Lepine
6.	J. Billingham	45.	G. Daigne	84.	Mrs. Delsemme
7.	P. Lagrange	46.	N. Evans	85.	R. Terrile
8.	J. Dreher	47.	Unidentified	86.	L. Doyle
9.	J. Tarter	48.	S. Aiello	87.	S. F. Likhachev
10.	D. Brin	49.	J. C. Ribes	88.	V. R. Eshleman
11.	D. Backman	50.	Unidentified	89.	H. Quintana
12.	C. Griffith	51.	T. Wilson	90.	B. Campbell
13.	C. Chyba	52.	G. Rudnitskij	91.	V. Stretlnitskij
14.	F. Drake	53.	D. Toublanc	92.	C. Maccone
15.	C. Michaux	54.	S. Isobe	93.	S. Slysh
16.	E. J. Hysom	55.	D. Werthimer	94.	P. Schenkel
17.	T. McDonough	56.	J. Heidmann	95.	P. Carricaburu
18.	D. Holmes	57.	Y. Chliamovitch	96.	W. Calvin
19.	J. Jugaku	58.	M. Mayor	97.	E. Ortiz
20.	M. Klein	59.	G. Marx	98.	P. Kafka
21.	V. V. Rubtsov	60.	M. Papagiannis	99.	D. Benest
22.	W. Davis	61.	E. Olsen	100.	P. Allen
23.	M. C. Maurel	62.	P. Boyce	101.	H. Yokoo
24.	S. Von Hoerner	63.	M. A. Heidmann	102.	M. J. Harris
25.	A. D. Fokker	64.	F. Biraud	103.	D. Scott
26.	W. Sullivan III	65.	J. A. Villalobos	104.	A. C. Levasseur-Regourd
27.	D. Blair	66.	I. Almar	105.	N. Hallet
28.	G. Beaudet	67.	Unidentified	106.	A. Leger
29.	A. Lecacheux	68.	B. Irvine	107.	D. Souilhac
30.	F. Raulin	69.	N. Cabrol		
31.	F. Cerceau	70.	Unidentified		
32.	M. Pfleiderer	71.	L. Marochnik		
33.	H. Davies	72.	I. Coulson		
34.	N. Vogt	73.	C. Matthews		
35.	D. Tarter	74.	V. Burdyuzha		
36.	S. Tine	75.	R. Brown		
37.	P. Zarka	76.	E. Grin		
38.	R. Davies	77.	A. Delsemme		
39.	D. Schwartzman	78.	A. Duquennoy		

Next page: The participants of the Val Cenis symposium enjoy the sunshine in front of the Lanslevillard church. (photo by J. Simier)

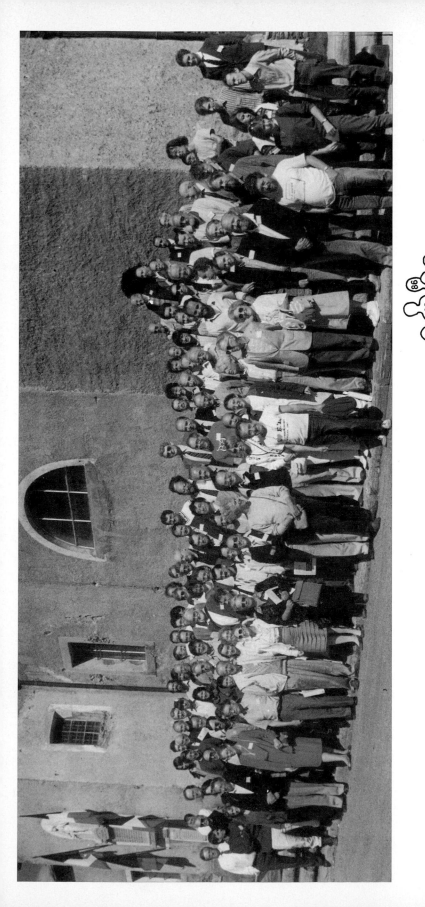

LIST OF PARTICIPANTS

Aiello, Santi	Univ. di Firenze, Italy
Allen, Peter	Walton-on-Thames, Surrey, UK
Almar, Ivan J.	Konkoly Observatory, Budapest, Hungary
Arkhipov, A.V.	Academy of Sci. Ukrainian SSR, Kharkov, USSR
Backman, Dana E.	NASA-Ames Res. Center, Moffett Field, CA USA
Beaudet, Gilles	Univ. de Montreal, Canada
Benest, Daniel	Observatoire de Nice, France
Billingham, John	NASA-Ames Res. Center, Moffett Field, CA USA
Biraud, François	Observatoire de Paris-Meudon, France
Blair, David G.	Univ. of Western Australia, Nedlands, Australia
Boyce, Peter	American Astronomical Soc., Washington, DC USA
Brin, David	Los Angeles, CA USA
Brown, Ron	Monash University, Australia
Burdyuzha, Vladimir	Lebedev Physical Insitute, Moscow, USSR
Cabrol, Nathalie	Observatoire de Paris-Meudon, France
Calvin, William	Univ. of Washington, Seattle, WA USA
Campbell, Bruce	Univ. of Victoria, BC Canada
Carricaburu, Pierre	Saint-Mandé, France
Chiamovitch, Pierre	Hermance, Switzerland
Chyba, Christopher F.	Cornell University, Ithaca, NY USA
Coulson, Iain M.	Joint Astronomy Centre, Hilo, HI USA
Coustenis, Athena	Observatoire de Paris-Meudon, France

Cudaback, David	Univ. of California, Berkeley, CA USA
Daigne, Gérard	Observatoire de Bordeaux, Floirac, France
Davies, Helen C.	Univ. of Pennsylvania, Philadelphia, PA USA
Davies, Robert E.	Univ. of Pennsylvania, Philadelphia, PA USA
Davis, Wanda L.	SETI Institute, Mt. View, CA USA
Deamer, David W.	Univ. of California, Davis, CA USA
Delsemme, Armand H.	Univ. of Toledo, Ohio, USA
Denis, Laurent	Station de Radioastronomie de Nancay, France
Devincenzi, Donald L.	NASA-Ames Res. Center, Moffett Field, CA USA
Dick, Steven J.	U.S. Naval Observatory, Washington, DC USA
Doyle, Laurance R.	SETI Institute, Mt. View, CA USA
Drake, Frank D.	Univ. of California, Santa Cruz, CA USA
Dreher, John W.	NASA-Ames Res. Center Moffett Field, CA USA
Duquennoy, Antoine	Observatoire de Genève, Switzerland
Eshleman, Von R.	Stanford University, Palo Alto, CA USA
Evans, Neal J.	Univ. of Texas, Austin, TX USA
Finney, Ben	Univ. of Hawaii, Honolulu, HI USA
Fokker, A.D.	Al Bilthoven, The Netherlands
Goudou, J.	Parc du Soleil et du Cosmos, France
Griffith, Caitlin	SUNY, Stony Brook, NY USA
Gulkis, Samuel	Jet Propulsion Laboratory, Pasadena, CA USA
Hallet, Nichole	Observatoire de Paris-Meudon, France
Harris, Michael J.	Naval Research Lab., Washington, DC USA
Heidmann, Jean	Observatoire de Paris-Meudon, France

Holmes, Diane L.	Univ. College, London, England
Hoerner, Von S.	Esslingen, RFA
Huart, Pierre	District de Haute-Maurienne, Lanslebourg, France
Hysom, Edmund E.	Cambridge, UK
Irvine, William M.	FCRAO, Univ. of Massachusetts, Amherst, MA USA
Isobe, Syuzo	National Astr. Obser., Tokyo, Japan
Jugaku, Jun	Tokai Univ., Kanagawa, Japan
Kafka, Peter	Max-Planck-Inst. für Astrophysik, Garching, RFA
Klein, Michael J.	Jet Propulsion Laboratory, Pasadena, CA USA
Ksanfomality, Leonid V.	Academy of Sciences, Moscow, USSR
Laming, Lionel	Paris, France
Larson, Harold P.	Univ. of Arizona, Tucson, AZ USA
Latham, David W.	Center For Astrophysics, Cambridge, MA USA
Lecacheux, Alain	Observatoire de Paris-Meudon, France
Léger, Alain	Univ. Paris, Paris, France
Lepine, Jacques	Inst. Astron. e Geofis., Univ. São Paulo, Brazil
Leschine, Susan	FCRAO, Univ. of Massachusetts, Amherst, MA USA
Levasseur-Regourd, Chantal	Univ. Paris, Verrières-le-Buisson, France
Likhachev, S. F.	Soyuznauka State Ctr, State Com. For People Educ. of USSR, Moscow, USSR
Maccone, Claudio C.	Torino, Italy
Marochnik, Leonid	Space Res. Inst., USSR Acad. Sciences, Moscow, USSR
Martin, Eduardo L.	Inst. d'Astrophysique de Paris, Paris, France
Marx, Gyorgy	Eötvös Univ., Budapest, Hungary

Mathews, Clifford	Univ. of Illinois at Chicago, Chicago, IL USA
Maurel, Marie-Christine	Inst. Jacques Monod, Paris, France
Maurette, Michel	CSNSM, Batiment, Orsay, France
Mayor, Michel	Observatoire de Genève, Switzerland
McDonough, Thomas	California Inst. of Technology, Pasadena, CA USA
McKay, Christopher P.	NASA-Ames Res. Center, Moffett Field, CA USA
Michaud, Michael A.G.	American Embassy (SCI), Paris, France
Michaux, Claude	Jet Propulsion Laboratory, Pasadena, CA USA
Montebugnoli, Stelio	Inst. di Radioastronomia, Bologna, Italy
Olsen, Edward T.	Jet Propulsion Laboratory, Pasadena, CA USA
Oritz, Emilio	Paris, France
Prusan, Albert	Le Canet, France
Papagiannis, Michael D.	Boston University, Boston, MA USA
Pfleiderer, Jorg	Inst. für Astronomie, Innsbruck, Austria
Pfleiderer, Mircea	Inst. für Astronomie, Innsbruck, Austria
Quintana, Hernan	Univ. Catolica, Santiago, Chile
Raulin, François	Univ. Paris val de Marne, Creteil, France
Ribes, Jean-Claude	Observatoire de Lyon, France
Roeder, Robert C.	Southwestern Univ., Georgetown, TX USA
Rubtsov, Vladimir V.	Ukrainian Extramural Polytechnical Institute, Kharkov, USSR
Rudnitskij, G.M.	Moscow State University, USSR
Schenckel, Peter	Centro Internacional de Estudios, Quito, Ecuador
Schneider, Jean	Observatoire de Paris-Meudon, France
Schwartzman, David W.	Howard Univ., Washington, DC USA

Scott, D.R.D.	Queens College, Cambridge, UK
Slysh, V.I.	USSR Academy of Sciences, Moscow, USSR
Souilhac, Dominique	Ecole Nationale Supérieure d'Electronique et de Mécanique, Nevers, France
Strelnitskij, Vladimir	Astronomical Council, Acad. of Sci., Moscow, USSR
Suchkov, A.A.	Rostov State Univ., USSR
Sullivan, Woodruff T.	Univ. of Washington, Seattle, WA USA
Swift, David W.	Univ. of Hawaii, Honolulu, HI USA
Tarter, Donald E.	Univ. of Alabama, Huntsville, AL USA
Tarter, Jill C.	NASA-Ames Res. Center, Moffett Field, CA USA
Terrile, Richard J.	Jet Propulsion Laboratory, Pasadena, CA USA
Tine, Stefano	Univ. di Firenze, Italy
Toublanc, Dominique	Observatoire de Bordeaux, Floirac, France
Vallee, J.P.	Herzberg Inst. of Astrophysics, Edinburgh, UK
Villalobos, Jose A.	Ciudad Rodrigo Facio, San Jose, Costa Rica
Vogt, Nikolaus	Pontificia Univ. Catolica, Santiago, Chile
Werthimer, Dan J.	Univ. of California, Berkeley, CA USA
Whitmire, Daniel P.	Univ. of Southwestern Louisiana, Lafayette, LA USA
Wilson, Thomas L.	Max-Planck-Inst. für Radioastronomie, Bonn, Germany
Yokoo, Hiromitsu	Kyorin University, Mitaka, Japan
Zakirov, U.N.	USSR Academy of Sciences, Kazan, USSR
Zarka, Philippe	Observatoire de Paris-Meudon, France
Zinnecker, Hans	Inst. für Astro. und Astroph., Wurzburg, Germany

Bioastronomers view *Mont SETI*. Located a mere 20 km North-Northeast from Val Cenis, the 3153-m high mountain was sighted by those who hiked to the *Refuge des Evettes*. (photo by P. Huart)

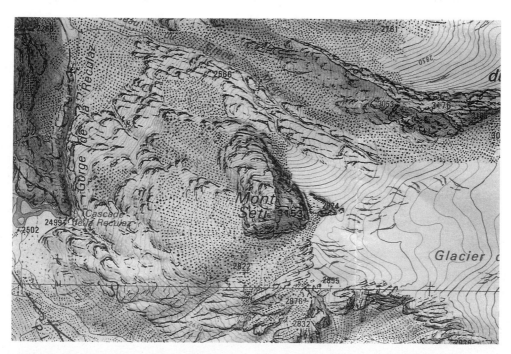

The name *Mont SETI* is probably 2000 years old and of Gaulish origin. But it could also originate from the Latin *saeta, seta:* a bristle ... to bristle: to raise the head and show defiance. SETI, symbol of a challenge? (Map reproduced with the kind permission of the Institut Geographique National @, Paris, 1988)